MATHEMATICS
The Basic Skills

Also published by Stanley Thornes (Publishers) Ltd:

E. Smith: EXAMPLES IN MATHEMATICS FOR GCSE FOUNDATION TIER

MATHEMATICS
The Basic Skills

WITH ANSWERS

S. Llewellyn
A. Greer

FIFTH EDITION

Stanley Thornes (Publishers) Ltd

First published in 1983 by:
Stanley Thornes (Publishers) Ltd
Second edition 1983
Third edition 1987
Fourth edition 1991
Fifth edition 1996

Reprinted in 2001 by:
Nelson Thornes Ltd
Delta Place
27 Bath Road
CHELTENHAM
GL53 7TH
United Kingdom

02 03 04 05 / 15 14 13 12 11 10 9

A catalogue record of this book is available from the British Library

ISBN 0-7487-2509-1

Page make-up by Tech-Set Ltd

Printed and bound in Great Britain

Contents

Introduction

Mathematics is used in:

<u>construction</u> — building houses, towers, monuments, etc.;

<u>surveying</u> — making maps, finding the heights of mountains, etc.;

<u>commerce</u> — trading in shops and business;

<u>navigation</u> — finding the way at sea and in the air;

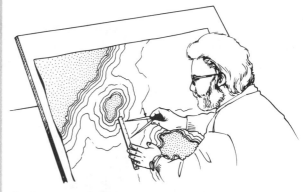

<u>the home and the community</u> — home decorating, shopping, income tax, rates, VAT, mortgages, insurance, etc.

We can only make guesses *why* mathematics evolved. Some think that people wanted to keep a record of their property in case it was stolen!

As mathematics developed so mathematical *tools* developed too.

The abacus is an early counting aid which is still in use in some parts of the world such as China and Malaysia. Some of them are made in beautiful hardwoods and sold as souvenirs to tourists.

Clay tablets were used to *record* numbers using a stone or stick to scratch the surface. When the clay was soft it was easy to make marks and then the clay was baked hard. The marks used were not the same as the symbols used today. It took many years for people to find the best way of writing numbers.

As people began to trade, different *money* systems grew up. Goods were *weighed* and *measured*.

Tally sticks with notches on them were used to *compare* records.

The modern tools are pen, paper, calculators, computers and many other scientific and engineering instruments.

1

Number

THE NUMBER SYSTEM

We use the symbols, 0, 1, 2, 3, 4, 5, 6, 7, 8 and 9 (arabic numbers), and the *position* of the figures gives the value of the number:

HUNDREDS	TENS	UNITS
100		1
100	10	1
100	10	1
3	2	4

324 is three hundred and twenty-four
324 is 3 hundreds 2 tens and 4 units

One hundred is ten times ten. Ten is ten times one.

The word *unit* means one.

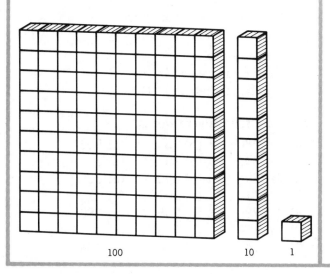

| 100 | 10 | 1 |

The Romans found it much more difficult to work with roman numbers because they did not use *place value* in the same way that we do. (Place value is the idea of a number having a different value when it is in a different place.)

In 614 the 4 means 4 *units*
In 146 the 4 means 4 *tens*
In 461 the 4 means 4 *hundreds*

In these three numbers the 4 stands for a different value when it is in a different place.

It is important to remember that the *smallest* numbers, the *units*, are always on the *right hand side* of the whole number.

Arabic numbers in

Words	Figures
one	1
two	2
three	3
four	4
five	5
six	6
seven	7
eight	8
nine	9
ten	10
eleven	11
twelve	12
thirteen	13
fourteen	14
fifteen	15
sixteen	16

seventeen	17
eighteen	18
nineteen	19
twenty	20
twenty-one	21
twenty-two	22
twenty-three	23
twenty-four	24
twenty-five	25
twenty-six	26
twenty-seven, etc.	27

These examples show how important 0 is in our number system:

205 is 2 hundreds 0 tens and 5 units

250 is 2 hundreds 5 tens and 0 units

In 205 the zero keeps the place for the missing *tens*

In 250 the zero keeps the place for the missing *units*

Counting in tens:

10	ten
20	twenty
30	thirty
40	forty
50	fifty
60	sixty
70	seventy
80	eighty
90	ninety
100	one hundred

In words:

205 is two hundred and five

250 is two hundred and fifty

The Romans found it quite hard to do arithmetic with roman numbers. They did not have a symbol for 0.

Roman numbers	Values
I	1
II	2
III	3
IIII (later IV)	4
V	5
VI	6
VII	7
VIII	8
VIIII (later IX)	9
X	10
L	50
C	100
D	500
M	1000

Roman numbers are written by putting the letters side by side:

MMCCCV	2305
MDCLVI	1656
CCCLXIII	363

Dates (on cinema films and TV programmes) and numbers on clocks are often written using roman numbers.

——————— **EXERCISE 1** ———————

Write the following numbers (i) in figures, (ii) in roman numbers

1) (a) thirty-five
   (b) three hundred and five
   (c) three hundred and fifty

2) (a) seventy-nine
   (b) seven hundred and nine
   (c) seven hundred and ninety

3) (a) sixty-five
   (b) six hundred and five
   (c) six hundred and fifty.

Write the following in words

4) (a) 980	(b) 98	(c) 908
5) (a) 800	(b) 80	(c) 8
6) (a) 89	(b) 890	(c) 809
7) (a) 110	(b) 101	(c) 11
8) (a) 78	(b) 708	(c) 780
9) (a) 130	(b) 103	(c) 13
10) (a) 14	(b) 140	(c) 104.

Write the following in figures

11) (a) twenty-six
    (b) two hundred and six
    (c) two hundred and sixty

12) (a) fifty-five
    (b) five hundred and fifty
    (c) five hundred and five

13) (a) twelve
    (b) one hundred and twenty
    (c) one hundred and two

14) (a) twenty-one
    (b) two hundred and ten
    (c) two hundred and one

15) (a) seven hundred
    (b) seventy
    (c) seven.

16) Copy the table below and fill in the missing numbers and words.

Number in words	Figures		
	H	T	U
twenty-four			
nineteen			
seventy-six			
eight			
one hundred and sixty-four			
seven hundred and eight			
three			
		6	1
	2	5	4
		5	8

Fill in the missing numbers

17) 542 is ___ hundreds ___ tens ___ units

18) 902 is ___ hundreds ___ tens ___ units.

19) What does the number 5 mean in
    (a) 580   (b) 56   (c) 35   (d) 159?

20) What does the number 3 mean in
    (a) 113   (b) 300   (c) 35 (d) 13?

21) What does the number 9 mean in
    (a) 98   (b) 897   (c) 689   (d) 902?

22) For which of these numbers does the number 7 mean 7 units?
    (a) 709   (b) 79   (c) 867   (d) 17

23) For which of these numbers does the number 3 mean 3 tens?
    (a) 173   (b) 832   (c) 113   (d) 31

24) For which of these numbers does the number 8 mean 8 hundreds?
    (a) 678   (b) 890   (c) 834   (d) 508

25) Put these numbers in order of size with the smallest first
    (a) 123   (b) 321   (c) 132

26) Put these numbers in order of size with the smallest first
    (a) 305   (b) 35   (c) 350   (d) 530
    (e) 503   (f) 53

27) Put these numbers in order of size with the smallest first
    (a) 103   (b) 301   (c) 130   (d) 31
    (e) 13   (f) 310

28) Put these numbers in order of size with the largest first
    (a) 51   (b) 15   (c) 510   (d) 150
    (e) 105   (f) 501

29) Put these numbers in order of size with the largest first
    (a) 123   (b) 132   (c) 321   (d) 312
    (e) 231   (f) 213.

---

999 is the largest whole number using hundreds, tens and units. For larger numbers we use *thousands*:

1000 is *one thousand*
10 000 is *ten thousand*
100 000 is *one hundred thousand*

For large numbers a gap is left between each group of three numbers counting from the right. Care must be taken when there are noughts in the middle of the number:

5008 is five thousand and eight
16 012 is sixteen thousand and twelve
505 040 is five hundred and five thousand and forty

EXERCISE 2

Write the following in figures

1) two thousand, four hundred and fifty-one
2) five thousand, three hundred and eleven
3) six thousand, one hundred and twelve
4) one thousand and fourteen
5) six thousand and nine
6) five thousand three hundred
7) fourteen thousand
8) twenty-four thousand
9) seventy thousand
10) sixteen thousand, two hundred and eight
11) seventy-five thousand, one hundred and forty-two
12) forty thousand and sixty-four
13) three hundred thousand
14) one hundred and ninety-five thousand, five hundred and eighty
15) one hundred and fourteen thousand, six hundred and thirteen
16) three hundred and fifty-one thousand, six hundred and forty-five
17) seven hundred and twelve thousand, two hundred and one.

Write the following in words

18) 5763
19) 6721
20) 9884
21) 16 000
22) 65 000
23) 809 000
24) 72 050
25) 7009
26) 8002
27) 7080
28) 8090
29) 6050
30) 8900
31) 7800
32) 65 332
33) 14 500
34) 163 404
35) 900 009
36) 452 986
37) 304 021.

38) Copy the table below and fill in the missing numbers

Number in words	Hundred thousands	Ten thousands	Thousands	Hundreds	Tens	Units
five thousand						
sixty-two thousand						
three hundred thousand						
seventy-four thousand and nine						
six hundred thousand two hundred						
seventy thousand and fifty						
ninety-nine thousand						
six thousand						

Numbers larger than 999 999 are counted in *millions*:

1 000 000 is *one million*
10 000 000 is *ten million*
100 000 000 is *one hundred million*

6 789 435 is six million, seven hundred and eighty-nine thousand, four hundred and thirty-five

16 987 853 is sixteen million, nine hundred and eighty-seven thousand, eight hundred and fifty-three

235 634 798 is two hundred and thirty-five million, six hundred and thirty-four thousand, seven hundred and ninety-eight

Again care must be taken when there are noughts in the middle of the number:

5 082 406 is five million, eighty-two thousand, four hundred and six

14 405 600 is fourteen million, four hundred and five thousand, six hundred

300 012 015 is three hundred million, twelve thousand and fifteen

After 999 999 999 numbers are counted in billions:

1 000 000 000 is *one billion*. (This is the American billion. The English billion is 1000 times greater but is seldom used.)

──────── EXERCISE 3 ────────

Write the following in figures

1) eight million
2) nineteen million
3) six hundred million
4) three hundred and twenty-four million, five hundred and sixty-seven
5) three million, seven hundred thousand and thirty
6) four million and forty
7) sixty-four million, three hundred thousand, six hundred and sixty-six
8) nine hundred and ninety-nine million
9) four hundred and sixty-four million, three hundred and thirty thousand, two hundred and six.

Write the following in words

10) 7 000 000
11) 57 000 000
12) 189 000 000
13) 7 012 506
14) 5 869 014
15) 167 189 112
16) 900 003 040.
17) Copy the table below and fill in the missing numbers.

Number in words	Billions	Hundred millions	Ten millions	Millions	Hundred thousands	Ten thousands	Thousands	Hundreds	Tens	Units
one billion, six hundred and ten million										
eight million, thirty thousand and fifty										
seven hundred and six million										

──────── **THE CALCULATOR** ────────

There are very many types of calculator on the market but one like that shown in the diagram is suitable for most purposes.

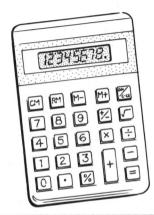

The keyboard of this calculator has ten number keys marked 0, 1, 2, 3, 4, 5, 6, 7, 8 and 9, and there are six function keys marked +, −, ×, ÷, √ and %. In addition there are four memory keys M+ (memory plus), M− (memory minus), MR (memory recall) and MC (memory clear). There is also an = (equals) key and two keys marked CE (correct error) and C (clear the calculator).

ROUNDING NUMBERS

When numbers greater than 10 have to be added, subtracted, multiplied or divided a calculator is often used. A calculator in good condition does not make mistakes but human beings do. We need a method for making sure that the answer produced by the calculator is sensible. The rounding of numbers allows us to do this.

It is usual to round off

a number between 10 and 99 to the nearest number of tens

a number between 100 and 999 to the nearest number of hundreds

a number between 1000 and 9999 to the nearest number of thousands

EXAMPLES

38 would be rounded to four tens, i.e. 40
732 would be rounded to seven hundreds, i.e. 700
8648 would be rounded to nine thousands, i.e. 9000

When a number lies half-way between tens, hundreds, thousands, etc., the number is always rounded up. Thus 6500 would be rounded up to 7000 and 35 would be rounded up to 40.

EXERCISE 4

Round off the following numbers to the nearest number of tens

1) 62   2) 7   3) 75   4) 83   5) 66

Round off the following numbers to the nearest number of hundreds

6) 796   7) 804   8) 76   9) 450   10) 895
   800   800   100   500   900

Round off the following numbers to the nearest number of thousands

11) 3632   12) 891   13) 7289
    4000   1000   7000
14) 7500   15) 6865
    8000   7000

FACTORS

One of the most basic ideas about numbers is how they break down into "factors":

6 is a *factor* of 36 because it divides into 36 without leaving a remainder.

You can easily see this either by using a dot pattern:

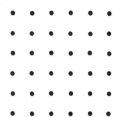

or by looking at tins stacked in layers of six tins:

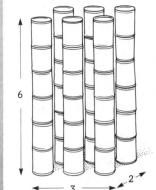

$6 \times 6 = 36$

× is the "*times*" sign (it means "lots of")

Chess boards have $8 \times 8 = 64$ squares:

= is the *"equals"* sign (it means *"is the same as"*)

8 is a factor of 64 because it goes into 64 without leaving a remainder.

However, 7 *is not* a factor of 15 because there will be a remainder of 1 if we divide 7 into 15:

$$\begin{array}{r} 2 \text{ rem } 1 \\ 7\overline{)15} \end{array}$$

A number is a factor of another number if it goes into that number without leaving a remainder.

Here are some more examples:

5 is a factor of 20
3 is a factor of 12
7 is not a factor of 17
2 is not a factor of 11
6 is not a factor of 19.

─────── **EXERCISE 5** ───────

1) Is 6 a factor of
   (a) 12   (b) 18   (c) 21   (d) 27
   (e) 36   (f) 42   (g) 54   (h) 56?

2) Is 5 a factor of
   (a) 3   (b) 10   (c) 15   (d) 17
   (e) 20   (f) 22   (g) 27   (h) 30?

3) Is 8 a factor of
   (a) 16   (b) 20   (c) 27   (d) 36
   (e) 32   (f) 40   (g) 56   (h) 58?

4) Is 9 a factor of
   (a) 18   (b) 20   (c) 27   (d) 36
   (e) 44   (f) 45   (g) 54   (h) 56?

5) Is 10 a factor of
   (a) 15   (b) 20   (c) 25   (d) 30
   (e) 45   (f) 50   (g) 99   (h) 100?

6) Which of the following numbers are factors of 15?
   (a) 2   (b) 3   (c) 5   (d) 10

7) Which of the following numbers are factors of 18?
   (a) 2   (b) 3   (c) 4   (d) 5
   (e) 6   (f) 7   (g) 8   (h) 9
   (i) 10   (j) 12

8) Which of the following numbers are factors of 50?
   (a) 2   (b) 3   (c) 5   (d) 10
   (e) 25

9) Which of the following numbers are factors of 56?
   (a) 7   (b) 8   (c) 9   (d) 10

───────────────────────────

Reminder: *A number is a factor of another number if it goes into that number without leaving a remainder.*

3 *is* a factor of 12
5 *is not* a factor of 11

The factors of 60 are:

1, 2, 3, 4, 5, 6, 10, 12, 15, 20, 30 and 60

Notice that 1 and 60 are counted as factors. They "pair off" from each end if the factors are arranged in order of size:

$1 \times 60 = 60$
$2 \times 30 = 60$
$3 \times 20 = 60$
$4 \times 15 = 60$, etc.

The factors of 12 are:

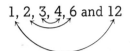

1, 2, 3, 4, 6 and 12

You can check you have all the factors by pairing them off from each end.

Prime numbers have only two factors, 1 and the number itself. (1 is not prime because it has only 1 factor.)

The factors of 8 are 1, 2, 4 and 8 (four factors) so 8 is not prime.
The factors of 7 are 1 and 7 (just two factors) so 7 is prime.

──────────── **EXERCISE 6** ────────────

1) Write down all the factors of these numbers. Arrange them in order of size and check that you have all the factors by pairing them off from each end.
   (a) 10   (b) 14   (c) 15   (d) 16
   (e) 20   (f) 25   (g) 36   (h) 42

2) Write down the factors of these numbers and state which of them are prime numbers.
   (a) 1   (b) 3   (c) 5   (d) 6
   (e) 11   (f) 17   (g) 19   (h) 22
   (i) 24   (j) 26   (k) 31   (l) 32

For more difficult numbers it is helpful to know about the rules which will be described next.

This work is important because it helps you to become "at ease" with numbers (and often makes the working of problems easier too).

You almost certainly know this rule already:

If a number ends in 0, 2, 4, 6 or 8 then it is an even number and 2 will divide into it without leaving a remainder.

EXAMPLE

The numbers

   4, 16, 12, 18, 22 and 28

are all even numbers.

If 2 will divide into a number without leaving a remainder we say that the number *"is divisible by 2"*.

──────────── **EXERCISE 7** ────────────

1) Which of the following numbers are divisible by 2?
   (a) 6   (b) 8   (c) 9   (d) 11
   (e) 26   (f) 35   (g) 41   (h) 72
   (i) 190   (j) 789   (k) 1764   (l) 3156
   (m) 6583   (n) 2018   (o) 4102   (p) 6155

The rule for 3 is not so well known. It sounds quite difficult but is really quite easy as the examples will show. First we explain the meaning of the words "digit sum".

The *digits* of a number are the single figures in that number, for example the digits of 612 are 6, 1 and 2.

The *digit sum* of a number is found by adding the digits together so the digit sum of 612 is 6 + 1 + 2 which is 9. (+ is the "plus" sign; it means "add".)

Sometimes the digit sum is not a single figure so we keep adding the digits together until we have a single figure.

EXAMPLE

Find the digit sum of 1768.

Add the digits 1 + 7 + 6 + 8 to get 22.

Add the digits again 2 + 2 to get 4.

The digit sum of 1768 is 4.

EXAMPLE

Find the digit sum of 667.

Add the digits $6 + 6 + 7$ to get 19.

Add the digits again $1 + 9$ to get 10.

Add the digits again $1 + 0$ to get 1.

The digit sum of 667 is 1.

―――――――――― EXERCISE 8 ――――――――――

Write down the digit sums of these numbers

1) (a) 611   (b) 502   (c) 304   (d) 321
2) (a) 506   (b) 410   (c) 175   (d) 56
3) (a) 28   (b) 912   (c) 108   (d) 820
4) (a) 1653 (b) 1998 (c) 2346 (d) 7653.

3 will go into a number without leaving a remainder if it goes exactly into the digit sum of the number.

EXAMPLE

Does 3 go into 411 without leaving a remainder?

The digit sum of 411 is $4 + 1 + 1$ which is 6.

3 goes into 6 exactly so it also goes into 411 without leaving a remainder.

EXAMPLE

Does 3 go into 5734 without leaving a remainder?

Adding the digits of 5734:

$5 + 7 + 3 + 4 = 19$
$1 + 9 = 10$
$1 + 0 = 1$

So the digit sum of 5734 is 1.

3 does not go into 1 exactly so 3 will not go into 5734 without leaving a remainder.

If 3 goes into a number without leaving a remainder we say that the number is "divisible" by 3.

―――――――――― EXERCISE 9 ――――――――――

By using the digit sum rule discover which of these numbers are divisible by 3

1) (a) 314   (b) 405   (c) 512   (d) 891
2) (a) 689   (b) 321   (c) 613   (d) 144
3) (a) 172   (b) 540   (c) 153   (d) 181
4) (a) 108   (b) 72   (c) 85   (d) 192
5) (a) 1886 (b) 2665 (c) 5184
6) (a) 7101 (b) 1200 (c) 5103
7) (a) 5004 (b) 1823 (c) 6102.

Numbers that are divisible by 3 are called *"multiples"* of 3.

―――――――――― EXERCISE 10 ――――――――――

Which of the following numbers are multiples of 3 and are also even numbers?

1) (a) 651   (b) 288   (c) 512
2) (a) 504   (b) 36   (c) 1002
3) (a) 109   (b) 901   (c) 108
4) (a) 864   (b) 5121 (c) 4122

The divisibility rule for 5:

5 will go into a number without leaving a remainder if the number ends in 0 or 5 and will not do so otherwise.

EXAMPLE

15, 25, 35, 40, 65 and 100 are divisible by 5 (they all end in 0 or 5).

16, 27, 141 and 269 are *not* divisible by 5.

―――――――――― EXERCISE 11 ――――――――――

Which of the following numbers are divisible by 5?

1) (a) 10   (b) 30   (c) 45   (d) 31
   (e) 66
2) (a) 125 (b) 150 (c) 175 (d) 109
3) (a) 105 (b) 501 (c) 615 (d) 803

4) (a) 10 000   (b) 65 655   (c) 555 551
5) (a) 561 100   (b) 13 455   (c) 5 000 000

Numbers which are not even are called *"odd numbers"*. Odd numbers end in 1, 3, 5, 7 or 9.

──────── **EXERCISE 12** ────────

1) Copy these numbers and underline the odd numbers
   (a) 33   (b) 61   (c) 40   (d) 53
   (e) 62   (f) 77   (g) 95   (h) 88
   (i) 99   (j) 612   (k) 303   (l) 450
   (m) 1567 (n) 88 993.

2) Using the divisibility rules or otherwise find which of the following numbers are divisible by (i) 2 and 5, (ii) 2, 3 and 5?
   (a) 15   (b) 30   (c) 45   (d) 36
   (e) 42   (f) 40   (g) 56   (h) 60
   (i) 48   (j) 90   (k) 100   (l) 70
   (m) 54   (n) 75   (o) 120

3) All of the following numbers are (i) divisible by 2 or (ii) divisible by 3 or (iii) divisible by 5 or (iv) not divisible by 2, 3 or 5.
   Draw up a chart with four columns headed (i), (ii), (iii) and (iv) and put the numbers in the correct columns.
   (a) 8   (b) 9   (c) 4   (d) 7
   (e) 25   (f) 16   (g) 35   (h) 33
   (i) 34   (j) 104   (k) 19   (l) 206
   (m) 215   (n) 321

Numbers which end in 0 are divisible by 10.

EXAMPLE

60, 100, 500 and 70 000 are all divisible by 10.

71, 609 and 3598 are *not* divisible by 10.

──────── **EXERCISE 13** ────────

1) Which of the following numbers are divisible by 10?
   (a) 98   (b) 755   (c) 400   (d) 1000
   (e) 35   (f) 670   (g) 1001 (h) 80

2) Which of the following numbers *are not* divisible by 10?
   (a) 72   (b) 80   (c) 85   (d) 900
   (e) 1857

6 will go exactly into a number only if both 2 and 3 also go exactly into the number.

──────── **EXERCISE 14** ────────

Using the divisibility rules find which of these numbers are divisible by 6.

1) (a) 14   (b) 24   (c) 30   (d) 36
   (e) 42

2) (a) 45   (b) 54   (c) 60   (d) 66
   (e) 72

3) (a) 285   (b) 336   (c) 405   (d) 512

──────── **TYPES OF NUMBERS** ────────

Positive numbers have a value greater than zero.

They either have no sign at all in front of them or a + sign.

Some examples of *positive whole numbers* are:

   3   8   +7   +11   +190   +345

Negative numbers have a value less than zero. They have a — sign in front of them. Negative numbers are used on thermometers to show centigrade temperatures below freezing point.

5 °C

0 °C   Freezing point

−5 °C

12

Negative numbers are also used on graphs:

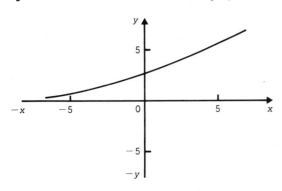

Some examples of *negative whole numbers* are:

$$-6 \quad -11 \quad -45 \quad -100$$

Positive and negative numbers do not have to be whole numbers. *Parts of a whole number are shown by using either "fractions" or "decimals".*

Some examples of *positive fractions* are:

$$\tfrac{1}{2} \quad \tfrac{3}{4} \quad \tfrac{7}{8}$$

We talk about $\tfrac{1}{4}$ hour.

 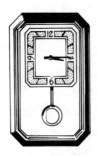

Some examples of *positive decimals* are:

$$0.10 \quad 0.5 \quad 0.75 \quad 0.9$$

Decimal is short for "decimal fraction". Decimals are just another way of writing fractions.

10 p written as part of a pound is £0.10 and a very small weight might be measured as 0.75 grams. Decimals are used in everyday life especially when using money, weighing or measuring.

Some examples of *negative fractions* are:

$$-\tfrac{1}{2} \quad -\tfrac{3}{4} \quad -\tfrac{5}{8}$$

and some *negative decimals* are:

$$-0.6 \quad -0.03 \quad -0.028$$

─────── **EXERCISE 15** ───────

Describe the following in words (for example negative whole number, positive fraction, etc.)
1) -6
2) $\tfrac{1}{2}$
3) 0.10
4) $+7$
5) $-\tfrac{3}{4}$
6) -0.03
7) 51
8) -0.1
9) $\tfrac{7}{8}$

When writing a whole number and a fraction or a whole number and a decimal then the *whole number is always written before the fraction or decimal point.*

In these numbers the whole number has been underlined:

$$\underline{3}\tfrac{1}{2} \quad \underline{14}\tfrac{1}{2} \quad \underline{3}.5 \quad \underline{45}.1 \quad \underline{450}.01$$

─────── **EXERCISE 16** ───────

Copy these numbers and underline the whole numbers.
1) $6\tfrac{1}{2}$
2) $13\tfrac{1}{2}$
3) $509\tfrac{1}{2}$
4) 4.1
5) 45.2
6) 390.12

DIRECTED NUMBERS

Positive and negative numbers (those with +, − or no sign in front of them) are called *directed numbers*.

EXAMPLE

The temperature is −2°C.
It rises by 5°C.
What is it now?

Count 5 degrees *upwards* on the thermometer because it is rising.

The temperature now is 3°C.

EXAMPLE

The temperature was 4°C.
It falls by 5°C.
Write down the new temperature.

Count 5 degrees *downwards* because it is falling. The temperature is now −1°C.

EXERCISE 17

1) The temperature is −2°C. It rises by 3°C. What is it now?

2) The temperature is 4°C. It falls by 6°C. What is it now?

3) The temperature is −4°C. It rises by 4°C. Give the new temperature.

4) The temperature is −3°C. It falls by 2°C. What is it now?

5) The temperature is −9°C. It rises by 7°C. What is the new temperature?

6) The temperature is −7°C. It rises by 10°C and then falls by 2°C. What is it now?

7) (Take care!) After a fall of 5°C the temperature is 3°C. What was it before?

8) After a rise of 4°C the temperature is 2°C. What was it before?

9) After a rise of 3°C the temperature is −2°C. What was it before?

10) By how much must the temperature rise for it to become 5°C if it was −3°C before?

EXERCISE 18

1) Write two hundred and forty in figures.

2) Write 605 in words.

3) In 6008 what does the figure 6 mean?

4) Does 7 mean 7 units in 7512?

5) Write forty thousand and eighteen in figures.

6) Write 50 020 in words.

7) Is 8 a factor of 72?

8) Is 6521 an even number?

9) Is 301 an odd number?

10) Is 5 a factor of 35?

11) Is 381 divisible by 3? (Use the digit sum rule.)

12) Is 49 a multiple of 7?

13) Does 5 go into 1000 without leaving a remainder?

14) Write down all the factors of
    (a) 9   (b) 21   (c) 23   (d) 29
    and state whether or not they are prime numbers (have only two factors).

15) Is −16 a positive whole number?

16) Is +8$\frac{1}{2}$ negative?

17) Is 0.5 a positive decimal?

18) Is 60 a multiple of 5?

19) Does 7 go into 42 without leaving a remainder?

20) Is 199 an odd number?

21) Write down all the even numbers between 11 and 21.

22) Write down all the multiples of three between 5 and 25.

23) Does 8 go into 96 without leaving a remainder?

24) Is 56 a multiple of 7?

25) Is -0.6 a negative decimal?

26) Is $+800$ a positive whole number?

27) Underline the whole numbers in
(a) $2\frac{1}{2}$ (b) 12.5.

28) Are 6 and 5 both factors of 30?

29) Are 3 and 8 both factors of 24?

30) Write 60 050 in words.

31) Fill in the missing numbers
312 is ___ hundreds ___ tens ___ units.

32) Fill in the missing words
3002 is 3 _____ 2 _____ .

33) Write down the digit sum of
(a) 512   (b) 604.

34) Which of these numbers are divisible by both 2 and 5?
(a) 560   (b) 650   (c) 905.

35) The temperature is $5\,°C$. It falls by $9\,°C$. What is the new temperature?

36) After a fall of $6\,°C$ the temperature is $-8\,°C$. What was it before?

37) By how much must the temperature rise to become $4\,°C$ if it was $-7\,°C$ before?

38) The temperature changes from $-2\,°C$ to $-6\,°C$. Has it risen or fallen and by how much?

39) The reading on a scale changes from -2 to $+5$. By how many units has it risen?

40) How many degrees are there between $-9\,°C$ and $-5\,°C$?

2

The Four Rules for Number

The arithmetic *signs* are:

+   plus
−   minus
×   times
÷   divided by

It is very important to have the ideas associated with these signs clearly understood.

ADDITION

Addition is concerned with putting things together.

For example putting discs together on a table:

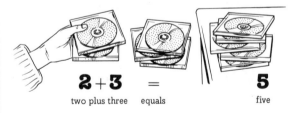

2 + 3   **=**   **5**

two plus three   equals   five

or putting distances together on a map:

Birmingham   18   Coventry   24   Leicester

Distances in miles

It is $18 + 24 = 42$ miles from Birmingham to Leicester via Coventry.

When adding it is important to make sure that the *units* are underneath each other by lining up the figures on the *right hand sides* of the whole numbers. (Otherwise your answers will be wrong.)

EXAMPLE

Add $785 + 43 + 1889$.

```
  785       The units figures 5, 3 and 9 have
   43       been lined up underneath each other.
 1889 +
 ─────
 2717       Start adding at the right hand
 1 2 1      column.
```

You can either begin at the *top* of the units column and add *downwards* (5 and 3 is 8, 8 and 9 is 17 which is 1 *ten* and 7 *units* so 1 is carried forward into the tens column and 7 is put down in the units column, etc.), or begin at the *bottom* and add *upwards* (9 and 3 is 12, etc.).

When you get to the tens column *do not forget* to add the 1 that has been carried. It can be put underneath the line to remind you or it can be put above the line by the 8.

The answer is called the *sum*. The sum of 785, 43 and 1889 is 2717.

EXERCISE 1

These addition sums have already been lined up correctly. Copy the numbers (writing the figures clearly) and add the numbers together.

16

1)   12
  124
———

2)   78
 131
———

3)   72
 291
———

4) 125
  35
———

5)   419
3 558
———

6)   798
7 903
———

7) 1 891
  121
———

8) 980
8 976
———

9) 9 087
  26
———

10) 10 877
15 099
———

11) 23
198
———

12) 6 789
1 009
———

13) 45 672
1 097
8 003
———

14) 88 007
982
54
———

15) 91 900
7
2 065
———

16) 31 111
202
55
———

17) 111 902
23 034
3 167
———

18) 900 876
54
1 098
———

19)   89
564
17 885
———

20) 9 008
1 076
888
———

21) 345 001
308 577
456 098
———

22) 981 176
897 101
22 439
———

23) 892
8 345
1 540
———

24) 448
 21
———

25) 253 261
303 511
109 907
———

26) 1 019
76
8 004
———

27) 976 309
707 521
———

28) 17 156
81 043
———

29) 109 824
340 198
98 008
33 557
———

30) 119 987
219 998
976 695
808 009
———

31) 8 999
45 101
11 933
67 003
———

32) 389 197
202 070
54 562
717 122
———

33) 188 709
335 007
4
25
———

34) 1 008
50 901
6 121
300
———

35) 112 091
56 083
7 071
21 890
———

36) 7 933
55
7
908 002
———

Line these figures up correctly and add the numbers

37) 145 + 2035
38) 4381 + 117
39) 156 + 142
40) 92 + 305
41) 1089 + 745
42) 2704 + 8009
43) 3865 + 198 + 1008
44) 270 + 303 + 57
45) 505 + 72 + 4881

46) $109 + 7 + 36$
47) $11 + 8 + 940$
48) $3 + 8 + 137$
49) $5 + 7 + 21 + 52$
50) $11 + 24 + 56 + 78$
51) $613 + 5008 + 8 + 27$
52) $5008 + 4978 + 303 + 6921$
53) $117\,908 + 897\,113$
54) $56\,707 + 3008 + 45\,098$
55) $24\,687 + 56 + 3008$
56) $1\,233\,405 + 2\,876\,881$
57) $245\,640\,707 + 33\,998\,707$
58) $11\,987 + 1988 + 27 + 6$
59) $16 + 5 + 303 + 23\,998 + 7665$
60) $91\,187 + 34\,998 + 98\,666.$

ADDITION USING A CALCULATOR

Before attempting to add using a calculator, we should always make a rough estimate of the answer to make sure that the answer produced is sensible.

It is suggested that all the exercises involving addition, subtraction, multiplication and division be first worked out using paper and pen or pencil. The answers should then be checked by using a calculator.

EXAMPLE

Add 74, 82 and 476.

   Rough estimate,

   $70 + 80 + 500 = 650$

The calculator is used as follows. (Take care to press the keys in the proper order and check the display as you proceed.)

Input	Display
74	74.
+	74.
82	82.
+	156.
476	476.
=	632.

Hence,

   $74 + 82 + 476 = 632$

The rough estimate shows that the answer is sensible, but is it accurate?

Any answer produced by a calculator should be carefully checked. One way of checking the answer in the example is to input the numbers into the calculator in reverse order, i.e. $476 + 82 + 74$. If the same answer is produced we can be confident that this answer is correct.

--- EXERCISE 2 ---

Using a calculator work out the answers to question numbers 37 to 60 inclusive of Exercise 1.

--- SUBTRACTION ---

Subtraction means taking away.

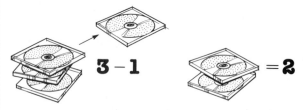

We read $3 - 1 = 2$ as "three minus one equals two" or "the *difference* between 3 and 1 is 2". If I buy some goods costing £2.80 and pay with a £5 note then I *take £2.80 away from* £5.00 to find the change I should have:

$$
\begin{array}{r}
£ \\
5.00 \\
2.80 - \\
\hline
2.20 \\
\hline
\end{array}
$$

There should be £2.20 change.

When subtracting it is again important to make sure that the *units* are underneath each other by lining up the figures on the *right hand sides of the whole numbers.*

18

EXAMPLE

Subtract 56 from 845.

```
845
 56 —
789
```

The *units* figures (5 and 6) have been lined up underneath each other. Start at the *right hand* end.

We say "6 from 5 won't go, so we *borrow from the next column*. We borrow a ten. Adding this to the 5 we get 15; 6 from 15 is 9 so we put the figure 9 down underneath the 6 in the units column".

There are two methods in general use for "paying back" the ten that has been borrowed.

In the first method the ten is "paid back at the bottom":

```
845
 56 —
  1 1
789
```

We say "5 and 1 is 6, 6 from 4 won't go so we borrow again from the next column; 6 from 14 is 8; put 8 down and pay back again at the bottom; 0 and 1 is 1; 1 from 8 is 7; put 7 down".

In the second method the figures are crossed off at the top:

```
 7 3
8̸4̸5
 56 —
789
```

We say "cross the 4 off and put 3 in its place; 5 from 3 won't go so we borrow again from the next column; 5 from 13 is 8; put 8 down, cross off the 8 and put 7; 0 from 7 is 7; put 7 down".

You should use the method you know.

Difficulty sometimes arises in the second method when there is a 0 in the next column.

EXAMPLE

Subtract 61 from 5000.

```
 4 9 9
 5̸0̸0̸0
   61 —
 4939
```

We say "1 from 0 won't go so borrow from the next column, but this is a nought so we go across the columns until we come to the 5; cross off the 5 and put 4 and put 1 by the 0 in the next column. Now we can borrow from this since it has become a 10, so cross off the 10 and put 9 with a 1 by the 0 in the tens column. Again we can borrow from it so cross off the 10 and put 9 with a 1 by the units column. Now we can take 1 from 10 which is 9; put 9 down and go to the tens column; 6 from 9 is 3; put 3 down; nothing from 9 is 9; put 9 down; nothing from 4 is 4; put 4 down".

This sounds very complicated but becomes automatic after some practice. Notice the effect is to cross off 500 and put 499 in its place with a 1 by the 0 in the units column.

──────── EXERCISE 3 ────────

Subtract the following numbers.

1)
```
47
15 —
─
```

2)
```
29
16 —
─
```

3)
```
56
23 —
─
```

4)
```
69
36 —
─
```

5)
```
78
45 —
─
```

6)
```
82
39 —
─
```

7)
```
56
38 —
─
```

8) 94
   28 −
   ⎯

9) 63
   37 −
   ⎯

10) 82
    36 −
    ⎯

11) 156
     42 −
    ⎯

12) 459
    308 −
    ⎯

13) 672
    439 −
    ⎯

14) 567
    396 −
    ⎯

15) 165
     97 −
    ⎯

16) 785 − 314
17) 987 − 238
18) 7000 − 82
19) 5000 − 57
20) 3000 − 61
21) 8000 − 89
22) 5000 − 63
23) 7652 − 6187
24) 5109 − 1089
25) 6003 − 4552
26) 5090 − 4113
27) 6080 − 123
28) 9971 − 325
29) 3050 − 441
30) 5581 − 303
31) 4561 − 224
32) 6670 − 119
33) 1000 − 507

34) 8713 − 511
35) 9981 − 119
36) 2000 − 402
37) 3000 − 517
38) 90 000 − 87 111
39) 14 090 − 13 020
40) 54 131 − 506
41) 11 103 − 7009
42) 38 561 − 29 140
43) 95 613 − 20 517
44) 11 012 − 9091
45) 92 945 − 51 014
46) 11 011 − 11 009
47) 30 000 − 5008
48) 55 000 − 4098

A subtraction can be checked by adding the bottom two numbers. Their sum should equal the top number.

EXAMPLE

    4281   Check:
    2113 −   2113
    2168   2168 +
                4281

──────────── **EXERCISE 4** ────────────

Subtract these numbers and check each answer.
 1) 356 − 121
 2) 502 − 319
 3) 559 − 381
 4) 691 − 312
 5) 5567 − 1009
 6) 3398 − 2897
 7) 567 − 94
 8) 192 − 17
 9) 8132 − 6109
10) 5101 − 321
11) 6000 − 3081
12) 4000 − 292

SUBTRACTION USING A CALCULATOR

EXAMPLE

Use a calculator to subtract 964 from 3942.

Rough estimate $= 4000 - 1000 = 3000$

Input	Display	
3942	3942.	
$-$	3942.	
964	964.	
$=$	2978.	This is the answer
$+$	2978.	
964	964.	
$=$	3942.	This is the check

Hence

$3942 - 964 = 2978$

If the answer is correct then

$2978 + 964 = 3942$

This check has been included in the above program.

──────── EXERCISE 5 ────────

Using a calculator work out the answers to the questions in Exercise 3.

COMBINED ADDITION AND SUBTRACTION

To find the value of

$24 + 5 - 7 + 6 - 18$

first pick out all the *positive* numbers (the ones with no sign or a + sign in front of them) and add these together:

$$\begin{array}{r} 24 \\ 5 \\ \underline{6+} \\ 35 \\ {\scriptstyle 1} \end{array}$$

and then pick out all the *negative* numbers (the ones with a $-$ sign in front of them)

and add these together:

$$\begin{array}{r} 7 \\ \underline{18+} \\ 25 \\ {\scriptstyle 1} \end{array}$$

and take the sum of the negative numbers away from the sum of the positive numbers:

$$\begin{array}{r} 35 \\ \underline{25-} \\ 10 \end{array}$$

so that:

$24 + 5 - 7 + 6 - 18 = 10$

──────── EXERCISE 6 ────────

1) $6 + 5 - 2 - 1$
2) $8 + 4 - 3 - 2$
3) $8 - 4 + 3 - 1$
4) $17 - 5 + 11$
5) $8 - 6 + 2$
6) $18 + 11 - 17$
7) $21 + 18 - 20 - 10$
8) $17 - 5 + 4 - 3$
9) $25 - 6 + 8 - 11$
10) $13 + 8 - 9 - 4 - 2$
11) $6 + 3 - 1 + 4 - 7$
12) $17 + 1 + 8 - 2 - 6 - 5$
13) $10 + 3 + 6 - 2 - 9$
14) $5 - 6 + 4 - 2 + 3$
15) $28 + 42 - 56 - 3$

USING A CALCULATOR FOR COMBINED ADDITION AND SUBTRACTION

When using a calculator we can input positive and negative numbers as they occur. There is no need to add the positive and negative numbers separately.

EXAMPLE

Find the value of $33 - 37 - 42 + 85 + 38 - 20$.

$$\text{Rough estimate} = 30 - 40 - 40 + 90 + 40$$
$$- 20$$
$$= 160 - 100$$
$$= 60$$

Input	Display
33	33.
—	33.
37	37.
—	−4.
42	42.
+	−46
85	85.
+	39.
38	38.
—	77.
20	20.
=	57.

To check the answer, it is 57, input the numbers in reverse order, that is

$$-20 + 38 + 85 - 42 - 37 + 33$$

If we obtain the same answer we can be sure that:

$$33 - 37 - 42 + 85 + 38 - 20 = 57$$

EXERCISE 7

Using a calculator, work out the answers to each of the questions in Exercise 6.

MULTIPLICATION

Multiplication is a quick way of adding equal numbers.

$$
\begin{array}{r}
6 \\
6 \\
6 \\
6\,+ \\
\hline
24 \\
\end{array}
$$

$4 \times 6 = 24$ (four times six equals twenty-four)

24 is called the *product* of 4 and 6.

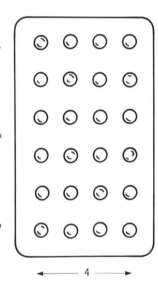

You do not have to count the number of pills in this packet in order to tell there are 24. From the picture you can see that $6 \times 4 = 4 \times 6 = 24$.

There are six bricks on each layer of this tower and there are four layers altogether. There are 24 bricks. On the bottom layer there are two rows of three bricks or $2 \times 3 = 6$ bricks in each layer. It would be possible to write 24 as the product of three factors:

$$24 = 2 \times 3 \times 4$$

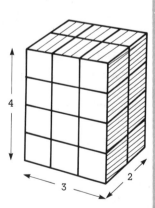

The order in which the factors are written down does not matter:

$24 = 3 \times 2 \times 4$
or $\quad 4 \times 3 \times 2$
or $\quad 4 \times 2 \times 3$

(Check for yourself that this is so.)

The important things to remember about multiplication are:

1) it is repeated addition,
2) the order in which the numbers are multiplied does not matter (this is called the *commutative law*).

TABLES

The multiplication table

1	2	3	4	5	6	7	8	9	10
2	4	6	8	10	12	14	16	18	20
3	6	9	12	15	18	21	24	27	30
4	8	12	16	20	24	28	32	36	40
5	10	15	20	25	30	35	40	45	50
6	12	18	24	30	36	42	48	54	60
7	14	21	28	35	42	49	56	63	70
8	16	24	32	40	48	56	64	72	80
9	18	27	36	45	54	63	72	81	90
10	20	30	40	50	60	70	80	90	100

To find 6×8 look along the *sixth* row and down the *eighth* column.

There you find the product of 6 and 8 which is 48

$6 \times 8 = 48$

This table is useful for revising your tables.

You should try to make sure that you know up to 10×10 without looking at the table.

1) 7×8
2) 5×6
3) 3×7
4) 9×7
5) 8×4
6) 6×8
7) 7×6
8) 9×5
9) 3×8
10) 4×9
11) 5×8
12) 6×5
13) 8×2
14) 9×9
15) 7×7
16) Multiply these numbers by (i) 2 (ii) 3 (iii) 4 (iv) 6
    (a) 2   (b) 3   (c) 4   (d) 5.
17) Multiply these numbers by (i) 5 (ii) 6 (iii) 9
    (a) 6   (b) 7   (c) 8   (d) 9.

SHORT MULTIPLICATION

EXAMPLE

Multiply 67 by 5. This means "how many is 5 lots of 67?"

```
H T U
  6 7
    5 ×
  3 3 5
    3
```

H stands for hundreds, T for tens and U for units. We say 5 times 7 which is 35 (from tables). We put the *units* figure down (which is 5) and carry the *tens* figure (which is 3) and usually put it down under the line to remind ourselves to add it in next time. We then multiply 5 by 6 to get 30 and add in the carried 3 to make 33 and put this down with one 3 in the tens column and the other in the hundreds column.

23

This is the pattern of working:

 You should always make sure that your figures are lined up very carefully.

1) 35
   2×
   ―

2) 23
   3×
   ―

3) 48
   3×
   ―

4) 52
   5×
   ―

5) 64
   7×
   ―

6) 81
   9×
   ―

7) 37
   8×
   ―

8) 83
   5×
   ―

9) 54
   6×
   ―

10) 56
    7×
    ―

EXAMPLE

Multiply 214 by 8.

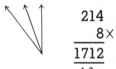
```
   214
     8×
  ────
  1712
   1 3
```

We say "8 times 4 is 32 (from tables); put 2 down and carry 3; then 8 times 1 is 8 and the carried 3 makes 11; put 1 down and carry 1; then 8 times 2 is 16 and the carried 1 makes 17 so put 17 down".

1) 567 × 8
2) 405 × 3
3) 319 × 7
4) 562 × 6
5) 680 × 5
6) 452 × 3
7) 198 × 4
8) 926 × 9
9) 307 × 7
10) 456 × 6
11) 135 × 5
12) 758 × 7
13) 672 × 8
14) 903 × 6
15) 877 × 8
16) 241 × 9
17) 526 × 8
18) 753 × 4
19) 697 × 6
20) 584 × 8

To multiply a whole number by 10 add a nought (this has the effect of moving the figures along one place to the left in the columns).

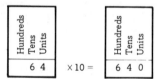

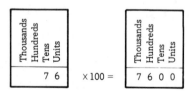

EXAMPLES

$5 \times 10 = 50$
$24 \times 10 = 240$
$356 \times 10 = 3560$

To multiply by 100 add two noughts

$6 \times 100 = 600$
$34 \times 100 = 3400$
$456 \times 100 = 45\,600$

Multiply these numbers by (i) 10 (ii) 100.

1) 8
2) 65
3) 17
4) 331
5) 151
6) 4000
7) 4990
8) 51 199

To multiply a whole number by a number such as 50, 60, 70, etc., multiply by the single figure and then add a nought.

EXAMPLE

32×40.

$32 \times 4 = 128$
$32 \times 40 = 1280$

EXERCISE 12

Multiply the following numbers by (i) 20 (ii) 30 (iii) 50 (iv) 80.

1) 34
2) 72
3) 151
4) 789
5) 45
6) 60
7) 301
8) 6553
9) 56
10) 31
11) 681
12) 2008

— LONG MULTIPLICATION —

EXAMPLE

Multiply 173 by 35.

The pattern is

```
      173
       35 ×
      865
     5190
   0 6055
```

(This is the pattern the eye follows.)

We say 5 times 173 (as for short multiplication) and put 865.

Next we multiply by 30: put a nought and multiply by 3: 3 times 173 is 519. Add the two rows to get 6055.

Some people multiply by 30 first and then 5:

```
      173
       35 ×
     5190
      865
     6055
```

and some people leave a blank instead of putting a nought:

```
      173
       35 ×
      865
      519
     6055
```

You should use whichever method you are used to.

EXERCISE 13

1) 473
   24 ×
   ——

2) 568
   31 ×
   ——

3) 481
   28 ×
   ——

4) 512
   16 ×
   ——

25

5) 721
   78×
	7) 582
	   15×
		9) 605
		   33×

6) 789
   14×
	8) 285
	   35×
		10) 293
		   75×

11) Multiply 105 by 26
12) Multiply 314 by 23
13) Multiply 634 by 36
14) Multiply 718 by 25
15) Multiply 346 by 45
16) Multiply 316 by 82

17) 628×56
18) 378×75
19) 306×35
20) 513×56

EXAMPLE

A harder problem: multiply 143 by 276.

$$\begin{array}{r} 143 \\ 276\times \\ \hline 858 \\ 10\,010 \\ 28\,600 \\ \hline 39\,468 \end{array}$$

First multiply by 6. Next by 70 (put a nought and multiply by 7). Then by 200 (put two noughts and multiply by 2). Add the three rows to get 39 468.

Multiplying by a number like 520 or 406 results in a row of noughts which can be left out if desired.

——— EXERCISE 14 ———

1) 162
  284×

2) 314
  406×

3) 3010
   123×

4) Multiply 1050 by 301
5) Multiply 512 by 251
6) 304×520
7) 709×319
8) 1020×708
9) 987×410
10) 126×218

MULTIPLICATION USING A CALCULATOR

EXAMPLE

Using a calculator find the value of 244 × 493.

Rough estimate = 200 × 500 = 100 000

Input	Display
244	244.
×	244.
493	493.
=	120292.

The rough estimate shows that the answer is sensible. To check the accuracy of the answer multiply in reverse order, i.e. 493 × 244.

EXAMPLE

Find the product of 27, 18 and 53.

Rough estimate = 30 × 20 × 50 = 30 000

Input	Display
27	27.
×	27.
18	18.
×	486.
53	53.
=	25758.

Hence

Product = 27 × 18 × 53 = 25 758

The rough estimate shows that the product is sensible. To check its accuracy multiply in reverse order, i.e. 53 × 18 × 27.

——— EXERCISE 15 ———

Use a calculator to find values for each of the following

1) Multiply 127 by 39.
2) Find the product of 436, 519 and 17.
3) Find the value of 367 × 294.

4) What is the product of 387 and 587?
5) Find the value of $879 \times 429 \times 35$.
6) Work out the value of $745 \times 39 \times 233$.
7) Multiply 73 469 by 63.
8) Work out 29 804 times 508.

DIVISION

Division is concerned with sharing into equal parts.

EXAMPLE

Share £6 equally between three people.

$6 \div 3 = 2$ dot pattern

∴ ∴ ∴

(six divided by three equals two)

Each person will get £2.

EXAMPLE

Cut a piece of string, 15 cm long, into three equal pieces. How long will each piece be?

$15 \div 3 = 5$ dot pattern

∴ ∴ ∴

∴ ∴ ∴

∴ ∴ ∴

∴ ∴ ∴

∴ ∴ ∴

Each piece will be 5 cm long:

| 5 cm | 5 cm | 5 cm |

There are many ways of writing division sums:

15 divided by 3 may be written $15 \div 3$ or $3 \overline{)15}$ or $\frac{15}{3}$:

$24 \div 8$ is the same as $8 \overline{)24}$ or $\frac{24}{8}$.

$36 \div 3$ is the same as $3 \overline{)36}$ or $\frac{36}{3}$

$81 \div 9$ is the same as $9 \overline{)81}$ or $\frac{81}{9}$

$110 \div 11$ is the same as $11 \overline{)110}$ or $\frac{110}{11}$

$56 \div 8$ is the same as $8 \overline{)56}$ or $\frac{56}{8}$

SHORT DIVISION

EXAMPLE

Divide 432 by 8.

This says "how many 8s are in 432?"
The numbers have special names:

8 is called the *divisor*
432 is called the *dividend*
the answer is called the *quotient*.

$\begin{array}{r} 54 \\ 8 \overline{)432} \end{array}$ 8s into 4 won't go; 8s into 43 go 5 (from tables). Put 5 above the 3; 5 times 8 is 40; take 40 away from 43 leaving 3; put 3 by the 2 to make 32; 8s into 32 go 4; put 4 above the 2, so that:

$432 \div 8 = 54$.

EXAMPLE

Divide 4205 by 3.

$\begin{array}{r} 1401 \text{ rem 2} \\ 3 \overline{)4205} \end{array}$ In this example there is a remainder. When 3 is divided into 4205 there will be 2 left over.

For determining remainders using a calculator see p. 61.

EXERCISE 16

1) $2 \overline{)7128}$
2) $2 \overline{)316}$
3) $4 \overline{)512}$
4) $6 \overline{)368}$
5) $5 \overline{)653}$
6) $7 \overline{)343}$
7) $8 \overline{)344}$
8) $5 \overline{)509}$
9) $5 \overline{)625}$
10) $7 \overline{)245}$
11) $9 \overline{)819}$
12) $11 \overline{)671}$
13) $203 \div 7$
14) $200 \div 8$
15) $281 \div 3$
16) $451 \div 4$
17) $3418 \div 8$
18) $8190 \div 6$
19) $4187 \div 6$
20) $5008 \div 7$
21) $9087 \div 8$
22) $6151 \div 3$
23) $5091 \div 9$
24) $3050 \div 5$

25) $\dfrac{506}{5}$ 29) $\dfrac{2345}{5}$

26) $\dfrac{4517}{4}$ 30) $\dfrac{3456}{8}$

27) $\dfrac{2880}{9}$ 31) $\dfrac{2319}{9}$

28) $\dfrac{7343}{7}$ 32) $\dfrac{1451}{6}$

——— LONG DIVISION ———

EXAMPLE

Divide 1032 by 24.

This says "how many 24s in 1032?"

```
    43
24)1032
   96
   72
```

We say "24s into 1 won't go; try 24s into 10; 24s into 10 won't go; try 24s into 103; 24s into 103 go 4" (you work this out by working out the 24 times table:

$$1 \times 24 = 24$$
$$2 \times 24 = 48$$
$$3 \times 24 = 72$$
$$4 \times 24 = 96$$
$$5 \times 24 = 120$$

you can see that 5 times 24 would be too big.)

Put the 4 above the 3 and take 96 away from 103. This leaves 7. In short division the 7 would go up by the 2 but in long division to make the working easier we bring down the 2 and put it by the 7 to make 72. The next step is to say "how many 24s in 72?" By looking at the 24 times table you can see that the answer is 3; put 3 next to the 4 above the line so that:

$$1032 \div 24 = 43.$$

Long division can always be checked by multiplying back:

```
   43
   24×
  172
  860
 1032
    1
```

Also a rough check can be made:

24 is roughly 25
1032 is roughly 1000

there are 40 times 25 in 1000 so the answer is not *obviously* wrong.

EXAMPLE

Divide 350 000 by 50.

```
      7 000
50)350 000
   350
```

$$350\,000 \div 50 = 7000.$$

50 will not divide into 3 or 35 so we try 50 into 350. This goes 7 exactly leaving no remainder. The remaining three zeros are written above the line.

EXAMPLE

Divide 92 579 by 44.

```
    2 104 rem 3
44)92 579
   88
    45
    44
   179
   176
     3
```

44s into 92 go 2 remainder 4. Bring down the 5. 44s into 45 go 1 remainder 1. Bring down the 7. 44s into 17 won't go. Put a 0. Bring down the 9. 44s into 179 go 4 remainder 3.

——— EXERCISE 17 ———

1) 21)567   8) 61)1159
2) 32)512   9) Divide 75 000 by 50
3) 54)1242   10) Divide 46 800 by 18
4) 37)505   11) Divide 16 500 by 22
5) 56)616   12) Divide 3253 by 25
6) 71)3410   13) Divide 90 343 by 43
7) 25)825   14) Divide 16 695 by 45

DIVISION USING A CALCULATOR

EXAMPLE

Divide 567 by 21

Input	Display	
567	567.	
÷	567.	
21	21.	
×	27.	This is the answer
21	21.	
=	567.	This is the check

Hence $567 \div 21 = 27$

EXERCISE 18

Use a calculator to work out the answers to questions 1–3, 5, 7–11 and 13–14 of Exercise 17.

— SEQUENCE OF OPERATIONS —

In working out mathematical statements of the following types:

$2 \times (3 + 2)$
$4(7 + 11)$
$16 - (5 - 1)$
$4 + 2 \times 3$
$6 + 12 \div 3$
$3 \times 5 - 2 \times 4$

a certain sequence must be observed.

Brackets must be worked out first.
Multiplication and division must be done before addition and subtraction.

The order of working out can be remembered by the letters BODMAS which gives the initial letters of **B**rackets, then the **O**rder is **D**ivision and **M**ultiplication followed by **A**ddition and **S**ubtraction.

EXAMPLES

$2 \times (3 + 2) = 2 \times 5 = 10$
   (3 is added to 2 first)

$4(7 + 11)$ means $4 \times (7 + 11)$
$4 \times (7 + 11) = 4 \times 18 = 72$

We get the same answer (and sometimes make the working easier) if we *expand the bracket.*

$4(7 + 11) = 4 \times 7 + 4 \times 11$
   (do the multiplications first)
$\qquad\quad = 28 + 44 = 72$

$16 - (5 - 1) = 16 - 4 = 12$
   (1 is taken from 5 first)

$4 + 2 \times 3 = 4 + 6 = 10$
   (2 times 3 is done first)

$6 + 12 \div 3 = 6 + 4 = 10$
   (12 ÷ 3 is done first)

$3 \times 5 - 2 \times 4 = 15 - 8 = 7$
   (3 times 5 is done first then
   2 times 4 is done next)

Scientific calculators and computers work out the correct order automatically.
The answer to $4 + 2 \times 3 =$
on a scientific calculator would be 10.
Note that $4(7 + 11)$ must be entered as $4 \times (7 + 11)$. You cannot leave out the '×'.
Simple calculators will not work out $4 + 2 \times 3$ in the correct order so you have to do this yourself.

EXERCISE 19

1) $4 \times (5 + 3)$
2) $5 \times (6 + 2)$
3) $9 \times (1 + 3)$
4) $7 \times (3 + 4)$
5) $6 \times (5 + 2)$
6) $9 \times (3 + 4)$
7) $8 \times (6 + 3)$
8) $5 \times (5 + 4)$
9) $3(2 + 7)$
10) $4(3 + 5)$
11) $11(3 + 5)$

12) $8(9+3)$
13) $15-(3-1)$
14) $17-(4-3)$
15) $24-(6-2)$
16) $35-(16-12)$
17) $22-(15-3)$
18) $33-(10-8)$
19) $45-(17-12)$
20) $20-(9-6)$
21) $15-(8-4)$
22) $13-(9-6)$
23) $20-(15-5)$
24) $14-(6-4)$
25) $5+2\times6$
26) $7+4\times3$
27) $9+3\times2$
28) $8+7\times4$
29) $3+8\times3$
30) $5+8\times9$
31) $4+7\times7$
32) $3+6\times7$
33) $2+3\times7$
34) $8+7\times5$
35) $7+9\times6$
36) $9+7\times9$
37) $5+20\div4$
38) $3+24\div8$
39) $5+36\div6$
40) $9+54\div9$
41) $5+32\div4$
42) $8+45\div5$
43) $6+15\div3$
44) $5+63\div7$
45) $8+72\div8$
46) $5+60\div12$
47) $4+16\div2$
48) $8+40\div8$
49) $7\times9+6\times3$
50) $11-(5-3)$
51) $5+81\div9-6$
52) $6\times4+64\div8$
53) $6+2\times3-5$
54) $(7+5)\div2+1$
55) $(3+4)\times5$
56) $(8+2)\div(2+3)$
57) $(7+8)-(3-2)$
58) $6\times7+2\times5$

59) $(24+36)\div12$
60) $(19+14)-(27-16)$
61) $22\div11+44\div4$
62) $6\times(7+2)-9$
63) $5+4\times3+1$
64) $2+4\times8-2$
65) $3\times5-2+1$
66) $27\div3+27\div9$
67) $25-(30-10)$
68) $4+2\times7+6$

USING THE MEMORY KEYS
——— ON A CALCULATOR ———

Quite complicated arithmetic operations can be done on a simple calculator by using the memory keys.

EXAMPLE

Work out $47-3\times(20-5)+36\div9$

Input	Display
47	47.
M+	47.
20	20.
−	20.
5	5.
×	15.
3	3.
M−	45.
36	36.
÷	36.
9	9.
M+	4.
MR	6.

So $47-3\times(20-5)+36\div9=6$

Now try Exercise 19 using a calculator.

CHECKING THE ACCURACY OF A CALCULATION

Always make sure your answers are sensible! You can often see if there is a mistake by looking at the *size* of the answer. (If you invest £100 for 2 years at $12\frac{1}{2}$% interest you would not expect to end up with twenty-five million pounds!)

If you round your numbers to the nearest ten (or to the nearest hundred or thousand if they are large) then you can make a quick estimate to see if your answer is likely to be right.

EXAMPLE

Make a rough check to see if this bill is correct:

	£	
Fittings	199	this is roughly 200
Pipes	389	this is roughly 400
Plaster	201	this is roughly 200
Tiles	590	this is roughly 600 +
TOTAL ..	1779	1400

The bill is incorrect.

EXERCISE 20

Make a rough check and say if these sums are correct.

```
1)    5 099          4)  20 000
      3 999               5 989
      1 001 +             1 303 +
     90 099              19 392

2)   35              5)     689
     15                     372
     45 +                   110 +
     95                  11 171

3)  6001             6)  20
     300                 19
     999 +               31 +
    7300                 80
```

```
7)  5002             8)  4 001
    3001                 2 002
    1001 +               2 999 +
      94               90 002
```

Long multiplication can be checked by using digit sums.

Reminder: to find the *digit sum* of a number such as 698 add the digits together:

$$6 + 9 + 8 = 23$$

and add again to get a single figure:

$$2 + 3 = 5$$

So the digit sum of 698 is 5.

EXAMPLE

Multiply 503 by 65 and check the answer.

```
     503
      65 ×
    2 515
   30 180
   32 695
```

If you multiply the digit sums of 503 and 65 and reduce this to a single figure then this should equal the digit sum of the answer.

Check: the digit sum of 503 is 8. The digit sum of 65 is 2.

$$8 \times 2 = 16$$

Reducing this to a single figure:

$$1 + 6 = 7$$

so the digit sum of the answer *should* be 7.

Now we check:

$$3 + 2 + 6 + 9 + 5 = 25$$

Reducing this to a single figure:

$$2 + 5 = 7$$

The long multiplication is likely to be correct.

Multiply the following and check each answer using the digit sum check.

1) 651×17
2) 282×13
3) 312×15
4) 519×23
5) 204×17
6) 328×34
7) 561×12
8) 708×29
9) 526×34
10) 386×17
11) 6128×43
12) 5298×75

1) Add 5, 7, 3 and 4.
2) Add 2, 4, 9, 6 and 5.
3) Add 35 and 47.
4) Add 46, 17 and 39.
5) Subtract 13 from 46.
6) Subtract 72 from 100.
7) Subtract 151 from 160.
8) Multiply 15 by 4.
9) Multiply 72 by 6.
10) Divide 45 by 9.
11) Divide 72 by 8.
12) Multiply 45 by 12.
13) Divide 372 by 12.
14) Increase 300 by 100.
15) Take 19 away from 50.
16) What is 6 times 7?
17) Does $6 \times 7 = 7 \times 6$?
18) Add $8 + 8 + 8 + 8$.
19) How many 12s in 48?
20) Increase 1400 by 20.

1) Add these sets of figures

6078	3263	613
234	4609	534
19	3579	234
3009 +	4007 +	609 +
———	———	———
———	———	———

2) Subtract the following
   (a) $400 - 56$ (b) $356 - 79$
   (c) $6005 - 227$ (d) $4651 - 303$
   (e) $12\,301 - 11\,809$ (f) $11\,601 - 9876$

3) Multiply
   (a) 654×7 (b) 128×9
   (c) 516×8 (d) 803×3
   (e) 889×6

4) Divide
   (a) 343 by 7   (b) 819 by 9
   (c) 216 by 6   (d) 850 by 5

5) Multiply these numbers and check your answers (i) with a rough check and (ii) by using digit sums
   (a) 512×5 (b) 708×25
   (c) 314×97 (d) 105×73
   (e) 4132×91

6) Work out these answers by using long division
   (a) $330 \div 15$ (b) $187 \div 17$
   (c) $234 \div 13$ (d) $2400 \div 25$
   (e) $782 \div 34$

7) $7 + 5 \times 3$
8) $15 - 2 \times 4$
9) $3 \times 4 - 2 \times 3$
10) $20 \div 4 + 5$
11) $5 + 20 \div 5$
12) $56 - 13 + 12 - 15$
13) $13 + 11 - 14 + 25 - 9$
14) $10 + 13 - 9 - 5 - 2 + 3$

The questions in this exercise are all of a practical nature. They all depend on the four rules for their solution.

1) The following diagram shows a shaft for a lathe. Find its overall length.

2) Five resistors in an electrical system are connected in series. Their values are 898, 763, 1175, 72 and 196 ohms. Their total resistance is found by adding these five values — what is it?

3) Two holes are drilled in a steel plate as shown in the diagram below. Find the dimension marked A.

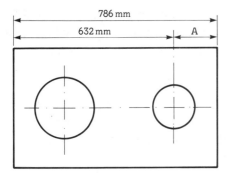

4) A small tank holds 798 litres of oil when full. If 39 litres are run off, how much oil remains in the tank?

5) A housewife went shopping and bought 500 grams of tomatoes, 250 grams of butter, 3000 grams of potatoes and 1200 grams of bacon. What was the total weight of her purchases?

6) A metal casting has a weight of 138 kilograms. A piece having a weight of 9 kilograms is removed. What is the weight of the casting remaining?

7) The diagram shows a shaft intended for a diesel engine. Calculate the dimension marked X.

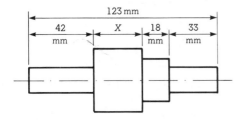

8) A screw has a weight of 12 grams. What is the total weight of 1250 such screws?

9) A bale of cloth contains 189 metres. Lengths of 15 metres, 36 metres and 29 metres are cut from the bale. How much cloth remains on it?

10) 19 hole centres are to be marked off 34 millimetres apart. What is the distance between the first and last hole centres?

11) A plank of wood 918 millimetres long is to be divided into 51 equal parts. What is the length of each part?

12) A shop buys 27 similar dresses for £351. How much does each dress cost?

13) The recipe for a small cake uses 85 grams of butter, 114 grams of sugar, 142 grams of plain flour and 14 grams of baking powder. What is the total weight of these ingredients?

14) 98 steel forgings have a total weight of 8134 kilograms. How much does each forging weigh?

15) A furniture shop on one day sells a dining room suite for £532, two easy chairs for £183 each, a settee for £348 and a divan bed for £93. How much are the total sales for the day?

16) A steel bar is 936 millimetres long. How many lengths, each 32 millimetres long, can be cut from the bar and what length remains?

17) A woman in a clothing factory cuts three patterns in 15 minutes. How many patterns can she cut in an 8 hour shift allowing 20 minutes for starting and 10 minutes for finishing the shift and 1 hour for her lunch break?

18) In the first 2 hours of a shift an operator makes 32 soldered joints per hour. In the next 3 hours the operator makes 29 joints per hour. In the final 2 hours 26 joints are made per hour. How many joints are made altogether in the 7 hour shift?

19) A farmer digs 23 210 kilograms of potatoes which are then placed in bags, each bag containing 55 kilograms. How many bags of potatoes does the farmer obtain?

20) 15 hole centres are to be marked off 39 millimetres apart. If 18 millimetres are to be allowed between the centres of the end holes and the edges of the plate, find the total length of plate required.

21) A woman is on a diet and is allowed 850 calories per day. Below is given a list of foods, their weights and the number of calories they contain.

Food	Weight (grams)	Calories
Cabbage, boiled	113	8
Carrots, boiled	113	8
Cauliflower, boiled	113	10
Chicken, roast	113	167
Coffee with milk	cup	25
Cucumber	57	6
Egg, poached	57	86
Grapefruit	113	25
Ham, lean	113	136
Lettuce	57	7
Orange	one	44
Plaice, steamed	113	105
Tea with milk	cup	20
Tomatoes	113	16
Toast with butter	slice	136

For breakfast she has: 1 grapefruit, a poached egg, a slice of toast with butter and 1 cup of tea with milk. For lunch she has: 113 g of boiled carrots, 113 g of cauliflower, 113 g of steamed plaice and 1 cup of coffee with milk.
   (a) Calculate the number of calories in the breakfast.
   (b) Calculate the number of calories in the lunch.
   (c) Find the number of calories she can consume for dinner.

22) A typist can type 80 words per minute. She has to type a manuscript consisting of 48 000 words. How long will it take her to type the manuscript?

23) A typist finds that she can get 14 words on one line and 28 lines per page of typing. If she types out a manuscript consisting of 39 200 words, how many pages will she type?

24) A car travels 9 kilometres on 1 litre of petrol. How many litres will be needed for a journey of 324 kilometres?

25) A wall is 342 bricks long. If it has 23 courses of bricks, how many bricks have been used in building the wall?

In mathematics there is not always one correct way of doing things. (Some ways may be more efficient than others!) The following exercise gives you the opportunity of explaining patterns in your own words.

--- **EXERCISE 25** ---

You may use a calculator for this exercise.

1) What numbers can you make just using 2, 3 and 6 and the signs $+$, $-$, $\times$ and $\div$? (One example would be $2 + 6 - 3 = 5$.)

2) $3 \times 5 = 15$ and $15 \div 3 = 5$. Give some other examples to show the same pattern. Explain why this always works by drawing some sketches, perhaps using dot patterns.

3) $8 \times 7 = 56$ and $7 \times 8 = 56$. Using a calculator give some more examples to show this symmetric property of multiplication. Demonstrate it by drawing up part of a multiplication square.

4) $1 \times 9 = 9$ and $\quad\quad 9 = 9$
$2 \times 9 = 18$ and $1 + 8 = 9$
$3 \times 9 = 27$ and $2 + 7 = 9$
Work through the nine times table showing this pattern, explaining what happens when the answer has more than two digits.

5) Investigate the pattern made by the last digit of multiples of 4 and make comments.

6) Explain using sketches why if $5 + 6 = 11$ then $11 - 5$ must equal 6. Demonstrate that this works using other numbers.

7) In your own words explain about digit sums and how they can be useful.

8) $3 \times 3 = 9$ ends in a 9
$3 \times 3 \times 3 = 27$ ends in a 7
$3 \times 3 \times 3 \times 3 = 81$ ends in a 1

Continue until a pattern emerges and comment.

9) Can you write the numbers up to 20 just using 2, $+$, $-$, $\times$, $\div$ and brackets?

For example
$$1 = 2 \div 2$$
$$2 = 2 + 2 - 2$$
$$3 = 2 \div 2 + 2$$
$$4 = 2 \times 2$$

Try to use 2 not more than six times in each sum.

EXERCISE 26

These questions are for practice in the correct use of $+$, $-$, $\times$ and $\div$. Show all working emphasising which signs you have used.

1) A journey from Huddersfield to Leeds via Wakefield took 75 minutes. It took 32 minutes to go from Huddersfield to Wakefield. How long did it take to go from Wakefield to Leeds?

2) A builder estimates his costs as £360 parts and £400 labour. VAT is £133. What will his total estimate be (a) without VAT (b) with VAT?

3) A 21 m long rope is cut into three parts. The first part is 16 m and the second is 2 m. How long is the third part?

4) A typist has a speed of 60 words per minute. She types 1200 words, has a break of 5 minutes and then types another 360 words. How long would this take altogether?

5) It takes 2 minutes to make a machine part of type A and 3 minutes to make a machine part of type B. How long would it take to make 9 parts of type A and 4 parts of type B?

6) A man writes 800 words in 2 hours. How many words would he write at the same speed in (a) 1 hour (b) 4 hours?

7) There are 17 sheets of cardboard in a stack which is 51 mm high. What is the thickness of (a) one sheet of cardboard (b) 29 sheets of cardboard?

8) A page has room for 50 lines of type. 18 lines are taken up by a picture and 2 by underlining. The rest of the lines have type on them. How many lines of type are there?

3

The Four Rules for Fractions

We talk about $\frac{1}{2}$ lb of butter or $\frac{1}{4}$ hour. These are examples of fractional amounts.

A fraction is a fragment or a small piece.

SKETCHING FRACTIONS

The number on the bottom (the denominator) tells you into how many equal pieces to divide the whole:

$\frac{1}{6}$ means divide the whole into six equal pieces.

The number on the top (the numerator) tells you how many parts to take:

$\frac{5}{6}$ means take five parts.

EXERCISE 1

1) Here are some drawings of fractions of a disc:

$\frac{1}{2}$ $\frac{1}{5}$

$\frac{3}{4}$ $\frac{3}{5}$

This is what three of those discs would look like from the top:

Which one has been left out?

2) Here is a sketch of a disc with $\frac{1}{4}$ shaded (i) side view (ii) from the top:

(i)   (ii)

In a similar way sketch these portions of a disc

(a) $\frac{3}{8}$ (b) $\frac{1}{5}$ (c) $\frac{1}{12}$ (d) $\frac{4}{5}$

(e) $\frac{5}{12}$ (f) $\frac{5}{8}$

3) Draw three sketches of a circle and shade
(a) $\frac{1}{3}$ (b) $\frac{2}{3}$ (c) $\frac{5}{6}$

36

CHANGING TO EQUIVALENT ─── FRACTIONS ───

In these diagrams $\frac{1}{2}$ a circle and $\frac{3}{6}$ of a circle have been shaded:

You can see that the amount of shaded area is the same in both diagrams. So:

$$\frac{1}{2} = \frac{3}{6}$$

To change $\frac{1}{2}$ into $\frac{3}{6}$ we have multiplied top and bottom by 3.

If you drew more sketches of fractions you would find that *if you multiply top and bottom of a fraction by the same (non-zero) number you do not alter its value.*

EXAMPLES

(a) $\dfrac{1}{2} = \dfrac{5}{10}$ (b) $\dfrac{3}{4} = \dfrac{6}{8}$

─── **EXERCISE 2** ───

Draw suitable sketches to show that

1) $\dfrac{4}{5} = \dfrac{8}{10}$ 5) $\dfrac{3}{4} = \dfrac{9}{12}$

2) $\dfrac{1}{3} = \dfrac{3}{9}$ 6) $\dfrac{1}{3} = \dfrac{4}{12}$

3) $\dfrac{2}{3} = \dfrac{4}{6}$ 7) $\dfrac{3}{5} = \dfrac{6}{10}$

4) $\dfrac{1}{5} = \dfrac{2}{10}$ 8) $\dfrac{1}{3} = \dfrac{2}{6}$

9) $\dfrac{2}{3} = \dfrac{8}{12}$

10) Say what fraction of each circle has been shaded

(a)   (b)   (c)   (d)

There is a definite pattern when changing fractions:

$\dfrac{3}{7}$ $\overset{5\times3=15}{\underset{\text{7s into 35 go 5}}{\xrightarrow{\hspace{1cm}5\hspace{1cm}}}}$ $\dfrac{15}{35}$ $\dfrac{3}{7} = \dfrac{15}{35}$

$\dfrac{4}{9}$ $\overset{4\times4=16}{\underset{\text{9s into 36 go 4}}{\xrightarrow{\hspace{1cm}4\hspace{1cm}}}}$ $\dfrac{16}{36}$ $\dfrac{4}{9} = \dfrac{16}{36}$

$\dfrac{2}{3}$ $\overset{4\times2=8}{\underset{\text{3s into 12 go 4}}{\xrightarrow{\hspace{1cm}4\hspace{1cm}}}}$ $\dfrac{8}{12}$ $\dfrac{2}{3} = \dfrac{8}{12}$

─── **EXERCISE 3** ───

Copy these fractions and fill in the missing numbers.

1) (a) $\dfrac{1}{4} = \dfrac{}{20}$ (b) $\dfrac{2}{3} = \dfrac{}{6}$

    (c) $\dfrac{5}{6} = \dfrac{}{12}$ (d) $\dfrac{4}{5} = \dfrac{}{20}$

2) (a) $\dfrac{3}{4} = \dfrac{}{16}$ (b) $\dfrac{1}{2} = \dfrac{}{10}$

    (c) $\dfrac{3}{8} = \dfrac{}{24}$ (d) $\dfrac{3}{5} = \dfrac{}{10}$

3) (a) $\dfrac{5}{6} = \dfrac{}{18}$ (b) $\dfrac{1}{4} = \dfrac{}{24}$

    (c) $\dfrac{6}{7} = \dfrac{}{42}$ (d) $\dfrac{9}{11} = \dfrac{}{88}$

4) (a) $\dfrac{4}{9} = \dfrac{}{27}$ (b) $\dfrac{6}{11} = \dfrac{}{22}$

    (c) $\dfrac{5}{8} = \dfrac{}{56}$ (d) $\dfrac{3}{4} = \dfrac{}{20}$

5) (a) $\dfrac{5}{6} = \dfrac{}{36}$ (b) $\dfrac{6}{7} = \dfrac{}{21}$

    (c) $\dfrac{3}{5} = \dfrac{}{25}$ (d) $\dfrac{7}{8} = \dfrac{}{32}$

6) (a) $\dfrac{3}{8} = \dfrac{}{40}$ (b) $\dfrac{2}{3} = \dfrac{}{18}$

    (c) $\dfrac{5}{9} = \dfrac{}{45}$ (d) $\dfrac{3}{5} = \dfrac{}{30}$

7) (a) $\dfrac{5}{11} = \dfrac{}{33}$ (b) $\dfrac{3}{8} = \dfrac{}{64}$

    (c) $\dfrac{1}{2} = \dfrac{}{40}$ (d) $\dfrac{2}{3} = \dfrac{}{27}$

8) (a) $\dfrac{4}{5} = \dfrac{}{25}$ (b) $\dfrac{7}{8} = \dfrac{}{56}$

    (c) $\dfrac{3}{7} = \dfrac{}{49}$ (d) $\dfrac{5}{12} = \dfrac{}{60}$

REDUCING A FRACTION TO ITS
—— LOWEST TERMS ——

By drawing you can see

(a) $\dfrac{4}{6} = \dfrac{2}{3}$ (b) $\dfrac{3}{12} = \dfrac{1}{4}$

We can divide the top and the bottom of a fraction by the same (non-zero) number without altering the value of the fraction.

(Drawing does not provide a proof of the rule but it shows you, perhaps, why it works.)

Fractions like $\frac{2}{3}$ and $\frac{1}{3}$ are said to be in their *lowest terms* (there is no number which will divide into top and bottom).

EXAMPLE

Reduce (a) $\frac{4}{6}$ (b) $\frac{3}{12}$ to their lowest terms.

(a) $\dfrac{4}{6} = \dfrac{2}{3}$ (dividing top and bottom by 2)

(b) $\dfrac{3}{12} = \dfrac{1}{4}$ (dividing top and bottom by 3)

Sometimes top and bottom can be divided more than once:

EXAMPLE

Reduce $\frac{42}{56}$ to its lowest terms.

$$\dfrac{42}{56} = \dfrac{6}{8} \quad \text{(dividing top and bottom by 7)}$$

$$= \dfrac{3}{4} \quad \text{(dividing top and bottom by 2)}$$

—————— **EXERCISE 4** ——————

1) By drawing show that

   (a) $\dfrac{6}{9} = \dfrac{2}{3}$ (b) $\dfrac{6}{8} = \dfrac{3}{4}$

   (c) $\dfrac{4}{12} = \dfrac{1}{3}$.

Reduce the following to their lowest terms

2) $\dfrac{3}{9}$ 4) $\dfrac{15}{25}$

3) $\dfrac{5}{20}$ 5) $\dfrac{5}{15}$

6) $\dfrac{6}{12}$ 17) $\dfrac{14}{21}$

7) $\dfrac{10}{30}$ 18) $\dfrac{18}{24}$

8) $\dfrac{36}{42}$ 19) $\dfrac{12}{20}$

9) $\dfrac{42}{49}$ 20) $\dfrac{6}{10}$

10) $\dfrac{12}{16}$ 21) $\dfrac{5}{30}$

11) $\dfrac{56}{64}$ 22) $\dfrac{7}{49}$

12) $\dfrac{21}{35}$ 23) $\dfrac{11}{33}$

13) $\dfrac{10}{40}$ 24) $\dfrac{15}{30}$

14) $\dfrac{16}{20}$ 25) $\dfrac{16}{24}$

15) $\dfrac{12}{24}$ 26) $\dfrac{9}{18}$

16) $\dfrac{15}{20}$

HIGHEST COMMON FACTOR (HCF)

The highest common factor of two numbers is the largest number that will go into them both exactly. Cancelling by the HCF reduces a fraction to its lowest terms most quickly but it is not necessary to do this. Fractions can be cancelled by stages.

You can find the HCF by writing down the factors of each number and finding the highest one common to both.

EXAMPLE

Find the HCF of 6 and 15

  6 has factors 1, 2, 3, 6
  15 has factors 1, 3, 5, 15

  3 is the highest number occurring in both so the HCF is 3.

──────────── **EXERCISE 5** ────────────

Find the HCF of

1) 4 and 12   6) 8 and 12
2) 5 and 10   7) 9 and 12
3) 6 and 9   8) 15 and 25
4) 8 and 10   9) 24 and 36
5) 7 and 14   10) 25 and 30

─── TYPES OF FRACTIONS ───

$\frac{8}{5}$ is known as a *top-heavy* fraction. The number on top is larger than the number underneath. (Sometimes top-heavy fractions are known as "improper fractions".)

$1\frac{3}{5}$ is known as a *mixed fraction*. It is a whole number and a fraction:

$$\frac{8}{5} = \frac{5}{5} + \frac{3}{5} = 1\frac{3}{5}$$

$$\frac{5}{5} = 1$$

We have a rule for making top-heavy fractions into mixed fractions:

Divide by the number underneath. The answer goes as the whole number. The remainder goes on top of the fraction and the number underneath stays the same.

EXAMPLE

Turn $\frac{17}{5}$ into a mixed fraction.

5s into 17 go 3
remainder 2:

$$\frac{17}{5} = 3\frac{2}{5}$$

--- **EXERCISE 6** ---

Turn the following top-heavy fractions into mixed fractions

1) $\frac{9}{4}$

2) $\frac{11}{3}$

3) $\frac{15}{7}$

4) $\frac{17}{8}$

5) $\frac{20}{7}$

6) $\frac{51}{8}$

7) $\frac{5}{2}$

8) $\frac{7}{3}$

9) $\frac{11}{4}$

10) $\frac{17}{6}$

11) $\frac{32}{5}$

12) $\frac{25}{4}$

13) $\frac{19}{9}$

14) $\frac{35}{6}$

15) $\frac{27}{4}$

16) $\frac{43}{8}$

17) $\frac{29}{11}$

18) $\frac{57}{8}$

19) $\frac{13}{4}$

20) $\frac{18}{5}$

21) $\frac{82}{9}$

22) $\frac{57}{5}$

23) $\frac{37}{4}$

24) $\frac{53}{12}$

25) $\frac{15}{8}$

26) $\frac{11}{5}$

27) $\frac{43}{5}$

28) $\frac{23}{7}$

29) $\frac{21}{4}$

30) $\frac{17}{4}$

31) $\frac{26}{5}$

32) $\frac{35}{4}$

33) $\frac{13}{2}$

34) $\frac{24}{5}$

35) $\frac{3}{2}$

36) $\frac{33}{8}$

To make mixed fractions top-heavy, multiply the whole number by the number underneath the fraction and add the number on top of the fraction. This goes on top. The number underneath stays the same.

EXAMPLE

$$2\frac{3}{4} = \frac{11}{4} \quad \text{(2 times 4 plus 3 is 11)}$$

pictorially:

--- **EXERCISE 7** ---

Make the following mixed fractions top-heavy

1) $3\frac{1}{2}$

2) $5\frac{3}{4}$

3) $6\frac{1}{3}$

4) $4\frac{2}{3}$

5) $7\frac{1}{2}$

6) $2\frac{1}{2}$

7) $3\dfrac{2}{3}$

8) $4\dfrac{2}{5}$

9) $5\dfrac{1}{3}$

10) $8\dfrac{5}{6}$

11) $9\dfrac{3}{5}$

12) $6\dfrac{7}{8}$

13) $6\dfrac{2}{7}$

14) $3\dfrac{1}{5}$

15) $9\dfrac{5}{6}$

16) $8\dfrac{4}{7}$

17) $2\dfrac{1}{7}$

18) $3\dfrac{5}{9}$

19) $3\dfrac{2}{9}$

20) $11\dfrac{3}{5}$

21) $2\dfrac{5}{7}$

22) $3\dfrac{7}{8}$

23) $5\dfrac{3}{5}$

24) $8\dfrac{9}{10}$

25) $4\dfrac{3}{7}$

26) $9\dfrac{5}{7}$

27) $3\dfrac{1}{3}$

28) $8\dfrac{6}{7}$

29) $9\dfrac{3}{8}$

30) $15\dfrac{1}{2}$

31) $2\dfrac{3}{5}$

32) $9\dfrac{1}{4}$

33) $7\dfrac{4}{5}$

34) $5\dfrac{4}{5}$

35) $6\dfrac{2}{3}$

LOWEST COMMON MULTIPLE
(LCM)

The LCM of a set of numbers is the smallest number into which all of the numbers will divide exactly.

EXAMPLES

(a) The LCM of 2, 3 and 4 is 12.

(b) The LCM of 5 and 15 is 15.

(c) The LCM of 4, 5 and 6 is 60.

The LCM can be found by writing multiples of the largest number, checking each time if this will do:

(a) 4, 8, 12; 3 and 4 go into 12.

(b) 15; 5 and 15 go into 15.

(c) 6, 12, 18, 24, 30, 36, 42, 48, 54, 60; 4, 5 and 6 go into 60.

—————— **EXERCISE 8** ——————

Find the LCM of the following numbers

1) 2 and 3
2) 3 and 5
3) 4 and 8
4) 2 and 6
5) 5 and 10
6) 4 and 5
7) 6 and 9
8) 2 and 5
9) 3 and 4
10) 7 and 14
11) 2, 3 and 5
12) 3, 5 and 10
13) 4, 8 and 16
14) 5, 10 and 15
15) 8 and 9
16) 6 and 7
17) 5, 10 and 20
18) 9, 18 and 2
19) 6 and 12
20) 6 and 8
21) 5 and 6
22) 2, 4 and 8
23) 2, 3, 4 and 6
24) 9 and 10

When putting fractions in order of size, make them of the same type first.

EXAMPLE

Put the following fractions in order of size with the smallest first.

(a) $\dfrac{7}{8}$ (b) $\dfrac{3}{4}$ (c) $\dfrac{1}{4}$ (d) $\dfrac{1}{2}$ (e) $\dfrac{3}{8}$

Look at the numbers underneath the fractions. The LCM of 8, 4 and 2 is 8. Change the fractions into eighths:

$$\dfrac{7}{8} \rightleftarrows \dfrac{7}{8} \qquad \dfrac{3}{4} \rightleftarrows \dfrac{6}{8} \qquad \dfrac{1}{4} \rightleftarrows \dfrac{2}{8}$$

$$\dfrac{1}{2} \rightleftarrows \dfrac{4}{8} \qquad \dfrac{3}{8} \rightleftarrows \dfrac{3}{8}$$

Now putting them in order of size we have:

$$\dfrac{2}{8}, \dfrac{3}{8}, \dfrac{4}{8}, \dfrac{6}{8}, \dfrac{7}{8}$$

and putting the fractions back into their original form:

$$\dfrac{1}{4}, \dfrac{3}{8}, \dfrac{1}{2}, \dfrac{3}{4}, \dfrac{7}{8}$$

--- **EXERCISE 9** ---

Put the following fractions in order of size, smallest first

1) (a) $\dfrac{1}{3}$ (b) $\dfrac{1}{2}$ (c) $\dfrac{5}{6}$ (d) $\dfrac{3}{4}$

2) (a) $\dfrac{3}{5}$ (b) $\dfrac{1}{2}$ (c) $\dfrac{7}{10}$ (d) $\dfrac{4}{5}$

3) (a) $\dfrac{5}{12}$ (b) $\dfrac{3}{4}$ (c) $\dfrac{1}{6}$ (d) $\dfrac{1}{4}$

4) (a) $\dfrac{1}{2}$ (b) $\dfrac{1}{8}$ (c) $\dfrac{1}{4}$ (d) $\dfrac{3}{4}$

   (e) $\dfrac{7}{8}$

— ADDITION OF FRACTIONS —

Sometimes we need to add fractions together.

EXAMPLE

I have $\frac{2}{3}$ bag of ready-mix cement left and I am given $\frac{1}{2}$ bag by my neighbour. How much do I have altogether?

We must add $\frac{2}{3} + \frac{1}{2}$.

We must change them into fractions of the *same type* before they can be added.

We do this by looking at the denominators (the numbers underneath the fractions) and finding their LCM (the smallest number that they will both go into without leaving a remainder).

The LCM of 3 and 2 is 6 so change the fractions into sixths:

$$\dfrac{2}{3} \rightleftarrows \dfrac{4}{6} \quad \text{and} \quad \dfrac{1}{2} \rightleftarrows \dfrac{3}{6}$$

So

$$\dfrac{2}{3} + \dfrac{1}{2} = \dfrac{4}{6} + \dfrac{3}{6} \quad \left(\text{sometimes written with a}\right.$$

$$= \dfrac{7}{6} \qquad \text{long line like this: } \left.\dfrac{4+3}{6}\right)$$

$$= 1\dfrac{1}{6}$$

In picture form:

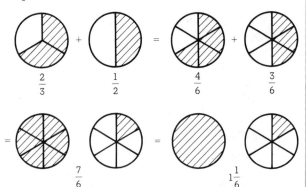

Notice that $\frac{7}{6}$ was top-heavy so it was changed to a whole number and a fraction.

42

ADDITION OF FRACTIONS WITH ─── WHOLE NUMBERS ───

If whole numbers are present we add them first.

EXAMPLE

$$5\frac{3}{4} + 6\frac{1}{2} = 11\frac{3}{4} + \frac{1}{2}$$

The smallest number that 4 and 2 go into is 4, so change the fractions to quarters.

$$= 11\frac{3}{4} + \frac{2}{4}$$

$$= 11\frac{5}{4}$$

But $\frac{5}{4}$ is top-heavy, so we turn it into a whole number and a fraction.

$$= 11 + 1\frac{1}{4}$$

$$= 12\frac{1}{4}$$

EXAMPLE

$$7\frac{1}{3} + 1\frac{1}{6} = 8\frac{1}{3} + \frac{1}{6}$$

The smallest number that 3 and 6 both go into is 6, so change the fractions to sixths.

$$= 8\frac{2}{6} + \frac{1}{6}$$

$$= 8\frac{3}{6}$$

But $\frac{3}{6}$ is not in its lowest terms.

$$= 8\frac{1}{2}$$

─── **EXERCISE 10** ───

1) $\dfrac{3}{4} + \dfrac{1}{2}$

2) $\dfrac{1}{3} + \dfrac{1}{2}$

3) $\dfrac{3}{4} + \dfrac{1}{8}$

4) $\dfrac{2}{5} + \dfrac{3}{10}$

5) $\dfrac{3}{4} + \dfrac{1}{3}$

6) $\dfrac{5}{6} + \dfrac{2}{3}$

7) $\dfrac{5}{8} + \dfrac{1}{4}$

8) $\dfrac{4}{5} + \dfrac{2}{3}$

9) $\dfrac{1}{7} + \dfrac{2}{3}$

10) $\dfrac{1}{3} + \dfrac{1}{4}$

11) $\dfrac{5}{6} + \dfrac{1}{12}$

12) $\dfrac{2}{3} + \dfrac{5}{12}$

13) $\dfrac{3}{4} + \dfrac{5}{8}$

14) $\dfrac{7}{8} + \dfrac{3}{16}$

15) $\dfrac{3}{5} + \dfrac{2}{15}$

16) $\dfrac{3}{8} + \dfrac{3}{4}$

17) $\dfrac{1}{3} + \dfrac{1}{5}$

18) $\dfrac{3}{7} + \dfrac{1}{3}$

19) $\dfrac{1}{2} + \dfrac{2}{3}$

20) $\dfrac{2}{5} + \dfrac{3}{4}$

21) $\dfrac{1}{8} + \dfrac{1}{2}$

22) $\dfrac{3}{4} + \dfrac{4}{5}$

23) $\dfrac{5}{6} + \dfrac{1}{5}$

24) $\dfrac{7}{9} + \dfrac{1}{3}$

25) $5\dfrac{1}{2} + 2\dfrac{3}{4}$

26) $3\dfrac{3}{4} + 4\dfrac{1}{8}$

27) $1\dfrac{1}{5} + 2\dfrac{1}{4}$

28) $5\dfrac{3}{8} + 4\dfrac{1}{4}$

29) $6\dfrac{3}{5} + 2\dfrac{1}{10}$

30) $4\dfrac{1}{3} + 2\dfrac{3}{5}$

31) $3\dfrac{1}{4} + 2\dfrac{3}{8}$

32) $4\dfrac{3}{5} + 1\dfrac{1}{4}$

33) $5\dfrac{3}{7} + 1\dfrac{1}{14}$

34) $3\dfrac{1}{2} + 1\dfrac{1}{2}$

35) $4\dfrac{1}{3} + \dfrac{2}{3}$

36) $12\dfrac{1}{4} + 1\dfrac{1}{2}$

37) $4\dfrac{1}{3} + 1\dfrac{5}{6}$

38) $7\dfrac{1}{3} + \dfrac{1}{12}$

39) $2\dfrac{2}{3} + 3\dfrac{4}{9}$

40) $\dfrac{1}{7} + \dfrac{1}{8}$

41) $\dfrac{5}{9} + \dfrac{4}{7}$

42) $2\dfrac{3}{7} + 3\dfrac{1}{14}$

43) $3\frac{1}{7} + 2\frac{5}{7}$ 46) $3\frac{4}{5} + \frac{2}{3}$

44) $5\frac{3}{4} + 2\frac{1}{8}$ 47) $\frac{4}{11} + \frac{3}{22}$

45) $3\frac{1}{3} + 1\frac{1}{6}$ 48) $3\frac{3}{10} + \frac{4}{5}$

– SUBTRACTION OF FRACTIONS –

If a pane of glass has to be replaced in a window then some space allowance has to be made. If the pane of glass was bought to fit the window exactly it would be under too much strain and might break. About 6 mm ($\frac{1}{8}$ inch) is subtracted from the measurements of the window.

To be able to take fractions away from each other they have to be made into fractions of the same type. (In the same way as for adding fractions.)

EXAMPLE

$\frac{2}{3} - \frac{1}{2}$ We cannot say $\frac{2-1}{3-2}$.

This is quite *wrong!*

The smallest number that 3 and 2 goes into without leaving a remainder is 6 so we turn the fractions into sixths.

$$\frac{2}{3} \;\rightleftharpoons\; \frac{4}{6}$$

and

$$\frac{1}{2} \;\rightleftharpoons\; \frac{3}{6}$$

so that

$$\frac{2}{3} - \frac{1}{2} = \frac{4}{6} - \frac{3}{6} = \frac{1}{6}$$

1) $\frac{1}{5} - \frac{1}{10}$ 6) $\frac{3}{11} - \frac{1}{22}$

2) $\frac{2}{7} - \frac{1}{14}$ 7) $\frac{5}{6} - \frac{1}{3}$

3) $\frac{1}{2} - \frac{3}{8}$ 8) $\frac{7}{12} - \frac{1}{2}$

4) $\frac{4}{5} - \frac{3}{10}$ 9) $\frac{2}{5} - \frac{1}{10}$

5) $\frac{3}{8} - \frac{1}{4}$ 10) $\frac{3}{4} - \frac{2}{3}$

SUBTRACTION OF FRACTIONS
— WITH WHOLE NUMBERS —

When there are whole numbers, we take them away first and then subtract the fractions (sometimes we have to "borrow" from the whole number).

EXAMPLE

Subtract $11\frac{5}{7}$ from $13\frac{3}{7}$.

$$13\frac{3}{7} - 11\frac{5}{7} = 2 + \frac{3}{7} - \frac{5}{7}$$

We cannot take $\frac{5}{7}$ from $\frac{3}{7}$ so "borrow" from the 2 and say

$$2 = 1 + \frac{7}{7}$$

$$= 1 + \frac{7}{7} + \frac{3}{7} - \frac{5}{7}$$

$$= 1\frac{5}{7}$$

For small numbers "borrowing" can be avoided by making the fractions top-heavy first:

EXAMPLE

$$1\frac{1}{2} - \frac{7}{8} = \frac{3}{2} - \frac{7}{8} \quad \text{but} \quad \frac{3}{2} \rightleftarrows \frac{12}{8}$$

$$= \frac{12}{8} - \frac{7}{8}$$

$$= \frac{5}{8}$$

——————— **EXERCISE 12** ———————

1) $1\frac{3}{4} - \frac{1}{4}$ 7) $2\frac{5}{12} - 1\frac{7}{12}$

2) $2\frac{5}{6} - \frac{1}{3}$ 8) $1\frac{1}{3} - \frac{3}{4}$

3) $1\frac{1}{2} - \frac{2}{3}$ 9) $1\frac{1}{4} - \frac{7}{8}$

4) $3\frac{1}{3} - \frac{1}{12}$ 10) $2\frac{1}{3} - 1\frac{11}{12}$

5) $2\frac{1}{12} - \frac{1}{6}$ 11) $2\frac{1}{5} - 1\frac{7}{10}$

6) $3\frac{3}{4} - 2\frac{7}{8}$ 12) $3\frac{1}{3} - \frac{5}{6}$

COMBINED ADDITION AND –SUBTRACTION OF FRACTIONS–

EXAMPLE

$$1\frac{1}{2} - 1\frac{1}{8} + 2\frac{1}{4} = \frac{3}{2} - \frac{9}{8} + \frac{9}{4}$$

$$= \frac{12 - 9 + 18}{8}$$

$$= \frac{21}{8}$$

$$= 2\frac{5}{8}$$

——————— **EXERCISE 13** ———————

1) $2\frac{1}{4} - 1\frac{1}{8} + 1\frac{1}{2}$

2) $3\frac{1}{12} - 2\frac{11}{12} + \frac{5}{6}$

3) $\frac{1}{5} - \frac{1}{10} + \frac{1}{20}$

4) $1\frac{1}{3} - \frac{5}{6} + \frac{5}{12}$

5) $1\frac{1}{4} + 1\frac{1}{8} - 1\frac{1}{2}$

6) $2\frac{1}{3} - 1\frac{5}{6} + \frac{7}{12}$

7) $\frac{5}{12} - \frac{1}{6} + \frac{1}{3}$

8) $1\frac{1}{3} + 1\frac{1}{6} - 2\frac{1}{12}$

9) $1\frac{3}{4} + \frac{1}{2} - \frac{5}{8}$

MULTIPLICATION OF FRACTIONS

If there are whole numbers make them top-heavy first. Then multiply the numbers on top together and multiply the numbers on the bottom together.

EXAMPLE

$$1\frac{2}{3} \times 2\frac{1}{2} = \frac{5}{3} \times \frac{5}{2} = \frac{5 \times 5}{3 \times 2} = \frac{25}{6} = 4\frac{1}{6}$$

Sometimes you can make the working easier by *cancelling*. You may cancel any number on top with any number underneath.

EXAMPLE

Multiply $\frac{2}{3}$ by $\frac{3}{5}$.

$$\frac{2}{3} \times \frac{3}{5} = \frac{2 \times \cancel{3}^{1}}{\cancel{3}_{1} \times 5} = \frac{2 \times 1}{1 \times 5} = \frac{2}{5}$$

If a number on top is the same as any number underneath, they can be *cancelled*.

Cross out 3 at the top and put 1.
Cross out 3 at the bottom and put 1.

Cancelling can only be done after the numbers have been made top-heavy.

EXAMPLE

Multiply $1\frac{3}{11}$ by $\frac{3}{7}$.

The 3s cannot be cancelled.

$$1\frac{3}{11} \times \frac{3}{7} = \frac{\cancel{14}^{2}}{11} \times \frac{3}{\cancel{7}_{1}} = \frac{2 \times 3}{11 \times 1} = \frac{6}{11}$$

If a number on top has a *common factor* with a number underneath, then divide both numbers by it.

7 is a common factor because it goes into both 14 and 7. $14 \div 7 = 2$ and $7 \div 7 = 1$

Cross out 14 and put 2. Cross out 7 and put 1.

EXAMPLE

$$1\frac{1}{2} \times 3\frac{1}{3} = \frac{\cancel{3}^{1}}{\cancel{2}_{1}} \times \frac{\cancel{10}^{5}}{\cancel{3}_{1}} = \frac{1 \times 5}{1 \times 1} = 5$$

The 3s have been *cancelled*. Also 2 and 10 are divided by 2 leaving 1 and 5. If there are whole numbers, do not cancel until the fractions have been made top-heavy.

EXERCISE 14

1) $\frac{2}{3} \times \frac{3}{4}$

2) $\frac{3}{5} \times \frac{4}{7}$

3) $\frac{3}{4} \times \frac{1}{5}$

4) $\frac{3}{5} \times \frac{5}{9}$

5) $2\frac{1}{2} \times \frac{4}{5}$

6) $\frac{3}{5} \times 1\frac{1}{3}$

7) $3\frac{1}{2} \times 4\frac{1}{7}$

8) $1\frac{1}{2} \times \frac{2}{3}$

9) $4\frac{1}{2} \times \frac{1}{9}$

10) $5\frac{2}{5} \times \frac{5}{9}$

11) $\frac{7}{9} \times \frac{3}{14}$

12) $\frac{4}{15} \times \frac{5}{8}$

13) $1\frac{1}{5} \times \frac{2}{3}$

14) $1\frac{1}{4} \times \frac{1}{5}$

15) $3\frac{1}{2} \times \frac{4}{7}$

16) $\frac{4}{5} \times \frac{25}{32}$

17) $\frac{1}{3} \times \frac{2}{5}$

18) $1\frac{1}{4} \times \frac{3}{5}$

19) $1\frac{1}{2} \times \frac{1}{2}$

20) $\frac{1}{4} \times \frac{1}{8}$

21) $3\frac{1}{2} \times \frac{8}{21}$

22) $\frac{7}{8} \times \frac{32}{63}$

23) $\frac{6}{7} \times \frac{7}{12}$

24) $\frac{5}{11} \times \frac{33}{50}$

25) $2\frac{1}{2} \times \frac{2}{5}$

26) $7\frac{1}{7} \times \frac{21}{25}$

27) $1\frac{1}{2} \times 1\frac{1}{3}$

28) $\frac{5}{7} \times \frac{14}{15}$

— DIVISION OF FRACTIONS —

$\frac{2}{3} \div \frac{1}{2}$ is a division sum which asks the question: "how many halves in two-thirds?"

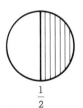  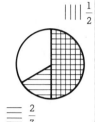

$||||\, \frac{1}{2}$

$\frac{1}{2}$ $\frac{2}{3}$ $\equiv \frac{2}{3}$

If we draw a picture you can see that there is more than *one* half in two-thirds but less than *two* halves. In fact there are $1\frac{1}{3}$ halves in two-thirds.

The rule for division of fractions is:

If there are mixed fractions make them top-heavy first. Then turn the second fraction upside down and multiply.

EXAMPLES

$$\frac{2}{3} \div \frac{1}{2} = \frac{2}{3} \times \frac{2}{1} = \frac{2 \times 2}{3 \times 1} = \frac{4}{3} = 1\frac{1}{3}$$

$$1\frac{3}{4} \div 1\frac{2}{3} = \frac{7}{4} \div \frac{5}{3} = \frac{7}{4} \times \frac{3}{5} = \frac{7 \times 3}{4 \times 5} = \frac{21}{20}$$

$$= 1\frac{1}{20}$$

$$1\frac{1}{2} \div 2\frac{1}{2} = \frac{3}{2} \div \frac{5}{2} = \frac{3}{2} \times \frac{2}{5}$$

it is not a good idea to try to squash these two steps into one

$$= \frac{3 \times \overset{1}{\cancel{2}}}{\underset{1}{\cancel{2}} \times 5} = \frac{3 \times 1}{1 \times 5} = \frac{3}{5}$$

You can cancel any number on the top with any number on the bottom if they have a common factor at the multiplication stage. *Do not cancel across a division sign.*

1) $\frac{3}{4} \div \frac{1}{2}$

2) $\frac{5}{6} \div \frac{2}{3}$

3) $\frac{4}{5} \div \frac{2}{7}$

4) $\frac{2}{5} \div \frac{3}{5}$

5) $2\frac{1}{2} \div 1\frac{1}{4}$

6) $3\frac{3}{4} \div \frac{5}{16}$

7) $5\frac{1}{3} \div \frac{4}{9}$

8) $3\frac{1}{3} \div 1\frac{2}{3}$

9) $\frac{5}{9} \div \frac{2}{3}$

10) $\frac{6}{11} \div \frac{3}{22}$

11) $\frac{5}{12} \div 2\frac{1}{2}$

12) $\frac{7}{8} \div 1\frac{5}{16}$

13) $6\frac{1}{2} \div 4\frac{1}{3}$

14) $3\frac{1}{4} \div 4\frac{1}{3}$

15) $2\frac{1}{3} \div \frac{7}{8}$

16) $1\frac{1}{2} \div \frac{3}{7}$

17) $2\frac{1}{2} \div \frac{1}{2}$

18) $1\frac{3}{4} \div \frac{1}{8}$

19) $\frac{2}{3} \div 1\frac{1}{3}$

20) $\frac{3}{5} \div \frac{9}{20}$

21) $1\frac{1}{4} \div 1\frac{1}{2}$

22) $1\frac{1}{6} \div 4\frac{2}{3}$

23) $3\frac{1}{3} \div \frac{5}{9}$

24) $1\frac{7}{8} \div 1\frac{1}{4}$

To make a whole number top-heavy divide by 1.

For example:

$$6 = \frac{6}{1} \qquad 9 = \frac{9}{1} \qquad 20 = \frac{20}{1} \quad \text{etc.}$$

EXAMPLE

Find $5 \div 1\frac{1}{3}$.

$$5 \div 1\frac{1}{3} = \frac{5}{1} \div \frac{4}{3}$$

$$= \frac{5}{1} \times \frac{3}{4}$$

$$= \frac{5 \times 3}{1 \times 4}$$

$$= \frac{15}{4}$$

$$= 3\frac{3}{4}$$

1) $4 \div \frac{3}{4}$ 5) $2\frac{1}{2} \div 10$

2) $3 \div 1\frac{1}{2}$ 6) $3\frac{3}{4} \div 5$

3) $7 \div \frac{7}{9}$ 7) $3\frac{1}{3} \div 15$

4) $11 \div 2\frac{3}{4}$ 8) $3\frac{1}{5} \div 8$

CALCULATING FRACTIONS OF A
─── WHOLE NUMBER ───

EXAMPLE

Find $\frac{3}{4}$ of 360.

Method 1

$$\frac{3}{4} \text{ of } 360 = \frac{3}{4} \times 360$$

$$= \frac{3}{\cancel{4}_1} \times \frac{\cancel{360}^{90}}{1}$$

$$= \frac{3 \times 90}{1 \times 1}$$

$$= 270$$

Method 2

To find $\frac{3}{4}$ of 360 *divide by 4 and multiply by 3.*

$$4\overline{)360} \quad 90$$

$$\begin{array}{r} 90 \\ \underline{3\times} \\ 270 \end{array} \quad \text{so that } \frac{3}{4} \text{ of } 360 = 270$$

1) $\frac{3}{4}$ of 16   5) $\frac{5}{7}$ of 84

2) $\frac{4}{5}$ of 15   6) $\frac{3}{4}$ of 240

3) $\frac{5}{8}$ of 24   7) $\frac{1}{7}$ of 630

4) $\frac{3}{8}$ of 56   8) $\frac{3}{5}$ of 200

1) Here are some drawings of fractions of a disc. What fractions have been shaded?

(a)   (b)

(c)   (d)

2) Draw three sketches of a circle and shade in

(a) $\frac{1}{4}$ (b) $\frac{7}{12}$ (c) $\frac{3}{8}$

3) Draw a sketch to show that
$$\frac{5}{6} = \frac{10}{12}$$

4) What fraction of this circle is shaded?

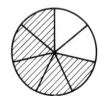

5) Fill in the missing numbers

(a) $\dfrac{3}{4} = \dfrac{}{8}$ (b) $\dfrac{1}{2} = \dfrac{}{20}$

(c) $\dfrac{1}{6} = \dfrac{}{12}$ (d) $\dfrac{3}{5} = \dfrac{}{15}$

6) Reduce these fractions to their lowest terms

(a) $\dfrac{5}{10}$ (b) $\dfrac{3}{15}$ (c) $\dfrac{9}{12}$ (d) $\dfrac{7}{56}$

7) Turn these top-heavy fractions into mixed fractions

(a) $\dfrac{27}{5}$ (b) $\dfrac{29}{8}$ (c) $\dfrac{31}{6}$ (d) $\dfrac{43}{6}$

8) Make these mixed fractions top-heavy

(a) $1\dfrac{2}{3}$ (b) $6\dfrac{2}{3}$ (c) $7\dfrac{3}{8}$ (d) $11\dfrac{1}{2}$

9) Find the LCM of 2, 4, 5 and 8.

10) Put these fractions in order of size with the smallest first

$\dfrac{3}{4},\ \dfrac{5}{8},\ \dfrac{1}{2},\ \dfrac{3}{5}.$

11) (a) $3\dfrac{1}{2} + 2\dfrac{1}{4}$ (b) $\dfrac{4}{5} + \dfrac{3}{10}$

(c) $3\dfrac{1}{3} + 1\dfrac{1}{6} + 2\dfrac{5}{12}$

12) (a) $2\dfrac{1}{5} - 1\dfrac{3}{10}$ (b) $3\dfrac{1}{12} - 1\dfrac{1}{6}$

(c) $\dfrac{5}{8} - \dfrac{3}{16}$ (d) $\dfrac{7}{20} - \dfrac{1}{5}$

13) $1\dfrac{3}{4} + 2\dfrac{1}{2} - 3\dfrac{1}{8}$

14) $3\dfrac{1}{2} - 1\dfrac{1}{4} + 1\dfrac{1}{8} - \dfrac{5}{8}$

15) (a) $\dfrac{3}{5} \times \dfrac{5}{6}$ (b) $\dfrac{4}{7} \times \dfrac{7}{8}$

(c) $1\dfrac{1}{2} \times 1\dfrac{1}{3}$ (d) $1\dfrac{1}{4} \times 1\dfrac{1}{5}$

16) (a) $\dfrac{4}{5} \div \dfrac{5}{6}$ (b) $1\dfrac{1}{3} \div 2\dfrac{1}{2}$

17) Work out the following, remembering that brackets must be worked out first and that multiplication and division must be done before addition and subtraction.

(a) $\left(1\dfrac{1}{4} + \dfrac{5}{8}\right) \div 2\dfrac{1}{2}$ (b) $1\dfrac{1}{7} \times \left(2\dfrac{1}{3} - 1\dfrac{1}{6}\right)$

(c) $\dfrac{5}{8} + \dfrac{3}{4} \times \dfrac{2}{3}$ (d) $\dfrac{5}{6} - \dfrac{1}{3} \div \dfrac{1}{2}$

18) Find (a) $\dfrac{3}{4}$ of 560 (b) $\dfrac{4}{5}$ of 350.

USING A CALCULATOR
——— FOR FRACTIONS ———

Some calculators have a fraction key which is usually marked $a\dfrac{b}{c}$.

EXAMPLE

Enter the fraction $\dfrac{3}{4}$.

Input	Display
3	3.
$a\dfrac{b}{c}$	3⌐.
4	3⌐4.

EXAMPLE

Enter the mixed number $3\dfrac{2}{5}$.

Input	Display
3	3.
$a\dfrac{b}{c}$	3⌐.
2	3⌐2.
$a\dfrac{b}{c}$	3⌐2⌐.
5	3⌐2⌐5.

EXAMPLE

Add $\dfrac{3}{4}$ and $3\dfrac{1}{2}$

Input	Display
3	3.
$a\dfrac{b}{c}$	3⌐.
4	3⌐4.
+	3⌐4.
3	3.
$a\dfrac{b}{c}$	3⌐.
1	3⌐1.
$a\dfrac{b}{c}$	3⌐1⌐.
2	3⌐1⌐2.
=	4⌐1⌐4.

which stands for $4\dfrac{1}{4}$.

EXAMPLE

Find $\dfrac{7}{8}$ of £216

Input	Display
7	7.
$a\dfrac{b}{c}$	7⌐.
8	7⌐8.
×	7⌐8.
216	216.
=	189.

So $\dfrac{7}{8}$ of £216 is £189.

EXAMPLE

Write $\dfrac{35}{56}$ in its lowest terms

The calculator will automatically reduce the fraction to its lowest terms.

Input	Display
35	35.
$a\dfrac{b}{c}$	35⌐.
56	35⌐56.
=	5⌐8.

So $\dfrac{35}{56}$ in its lowest terms is $\dfrac{5}{8}$.

EXAMPLE

Change $\dfrac{37}{5}$ into a mixed number

Input	Display
37	37.
$a\dfrac{b}{c}$	37⌐.
5	37⌐5.
=	7⌐2⌐5.

So $\dfrac{37}{5}$ is equivalent to $7\dfrac{2}{5}$.

Try Exercise 18, questions 11 to 18, using a calculator.

──────────── **EXERCISE 19** ────────────

1) Which is bigger, $\frac{3}{4}$ of £20 or $\frac{1}{2}$ of £32?

2) A nurse is told to give a patient $\frac{3}{5}$ of a bottle of medicine which contains 25 ml. How much should she give the patient?

3) A man's salary of £16 000 is reduced by $\frac{1}{4}$. How much does he get now?

4) It takes $1\frac{1}{2}$ minutes to make a machine part. How many parts can be made in 15 minutes?

5) $\frac{3}{4}$ of the biscuits in a tin are chocolate. If there are 24 biscuits altogether, how many chocolate ones are there?

6) A cake is divided into 8 equal slices. Sita eats 2 of them. What fraction of the cake has been eaten?

7) When the profits of a business are shared between two people, one receives $\frac{5}{8}$. What fraction does the other receive?

8) Ten people share £240 between them equally. What fraction does each receive and how much is this?

EXERCISE 20

The questions in this exercise are all of a practical nature. They all depend on fractions for their solution.

1) A girl spends $\frac{3}{4}$ of her pocket money and has 45 pence left. How much did she have to start with?

2) What is $\frac{7}{8}$ of £1600?

3) A reel of ribbon contains $31\frac{1}{2}$ yards. Lengths of $3\frac{1}{4}$, $5\frac{3}{8}$ and $8\frac{7}{8}$ yards are cut from it. How much ribbon remains on the reel?

4) A brand of tea is sold in packs containing $\frac{1}{8}$ kilogram. How many packs can be obtained from a box containing $51\frac{1}{2}$ kilograms?

5) A cask of lemonade contains 32 litres. It is poured into glasses each containing $\frac{1}{5}$ of a litre. How many glasses can be obtained?

6) A plank of wood is $31\frac{1}{2}$ metres long. It is cut into 6 equal lengths. What is the length of each piece?

7) A snail crawls 1 metre in $1\frac{1}{2}$ minutes. How long will it take to crawl $29\frac{1}{4}$ metres?

8) A man can make $2\frac{1}{2}$ articles per hour. How many articles can he make in 38 hours?

9) A watering can holds $11\frac{1}{2}$ litres. It is filled 9 times from a tank containing 250 litres. How much is left in the tank?

10) A school has 800 pupils, $\frac{1}{5}$ are in the upper school, $\frac{3}{8}$ are in the middle school and the remainder are in the lower school. How many pupils are there in the lower school?

11) An alloy consists of $\frac{7}{10}$ of copper, $\frac{9}{50}$ of tin and $\frac{3}{25}$ of zinc. How much of each metal is there in 900 grams of the alloy?

12) Calculate the total weight of coal carried by a train of 38 trucks if 20 carry $19\frac{1}{2}$ tonnes and the remainder carry $18\frac{3}{4}$ tonnes.

13) A motorcyclist covers $484\frac{1}{2}$ metres in $12\frac{3}{4}$ seconds. How far does he travel in 1 second?

14) A factory produces 660 engines per week. It increases its output by $\frac{5}{12}$. How many engines per week does it then produce each week?

15) The profits of a business are £14 500. It is shared between two partners A and B. If A receives $\frac{2}{5}$, how much does B receive?

4

The Four Rules for Decimals

We are familiar with "addition" (adding on) and "subtraction" (taking away) of whole numbers.

We need to be able to do *decimal* addition and subtraction in order to be able to add up money or to work out change.

All money can be written in decimal form, for example

£20 can be written as £20.00

with a *decimal point* separating the pounds and the pence.

The decimal point always goes *after* the whole number.

Figures *before* the decimal point are the *whole number*	Figures *after* the decimal point are a *fraction* (part) of a *whole number*
56 .	24

56.24 is a number between 56 and 57.

£56.24 is an amount of money between £56 and £57. The 24 p is *part* of a whole pound.

£6.01 is six pounds and one penny. Since there are one hundred pence in a pound then one penny is *one hundredth* of a pound.

Pounds	Pence
£6 .	01

Units		Hundredths
6 .	0	1

The second column after the point in decimals means hundredths and the number in the column tells you how many hundredths to take.

£6.10 is six pounds ten pence. Since there are ten 10 ps in one pound then 10 p is *one tenth* of a pound.

Pounds	Pence
£6 .	10

Units	Tenths	Hundredths
6 .	1	0

The first column after the decimal point means tenths and the number in the column tells you how many tenths to take.

The *first* place after the decimal point represents *tenths*

The *second* place after the decimal point represents *hundredths*

This can be extended:

The *third* place after the decimal point represents *thousandths*

The *fourth* place after the decimal point represents *ten thousandths*

The *fifth* place after the decimal point represents *hundred thousandths*

The *sixth* place after the decimal point represents *millionths*

etc.

Each column represents a number ten times smaller than the one before it.

Units		Tenths	Hundredths	Thousandths	Ten thousandths	Hundred thousandths
4	.	2	5	4	1	6

↑   ↑   ↑

These figures represent *very small* quantities compared to the others.

6.2 means 6 units 2 tenths

0.04 means 4 hundredths

13.01 means 1 ten 3 units 0 tenths
     1 hundredth

3.12 means 3 units 1 tenth 2 hundredths

0.728 means 7 tenths 2 hundredths
     8 thousandths

──────────── **EXERCISE 1** ────────────

What do these decimal numbers mean?

1) 0.2
2) 3.4
3) 10.04
4) 16.02
5) 8.24
6) 15.16
7) 3.817
8) 0.082
9) 50.207
10) 0.304

11) What does the 6 in 10.63 mean?
12) What does the 2 in 1.02 mean?
13) What does the 7 in 71.3 mean?
14) What does the 5 in 12.135 mean?

──────────────────────

MARKING DECIMALS ON A SCALE

──────────────────────

──────────── **EXERCISE 2** ────────────

Make a copy of the scale below.

Draw arrows and label these points. The first one has been done for you.

1) 6.3
2) 5.8
3) 2.5
4) 4.4
5) 0.7
6) 5.3
7) 7.0
8) 6.8
9) 2.1
10) 1.4
11) 0.2
12) 5.2

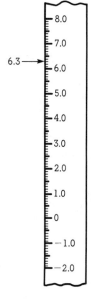

Say which is larger

13) 0.2 or 0.7
14) 5.3 or 5.8
15) 6.8 or 7.0
16) 0.7 or 7.0
17) 2.5 or 5.2
18) 2.5 or 2.1
19) 6.3 or 6.8
20) 4.4 or 1.4
21) 0.2 or 2.1
22) 5.3 or 5.2

THE DECIMAL POINT AND FOLLOWING ZEROS

Notice that:

7 means 7 units

7.0 means 7 units 0 tenths

7.00 means 7 units 0 tenths 0 hundredths

7.000 means 7 units 0 tenths 0 hundredths 0 thousandths

so that

7 = 7.0 = 7.00 = 7.000, etc.

To make a whole number into a decimal, put a decimal point after the whole number and add as many zeros as required.

EXERCISE 3

1) Write (a) 5 (b) 8 (c) 9 as a decimal with (i) one zero (ii) two zeros (iii) three zeros after the decimal point.

2) Does 4 equal (a) 4.00 (b) 4.000 (c) 4.0 (d) 40?

3) Does 15 equal (a) 15.000 (b) 15.0 (c) 15.00 (d) 150?

4) Write the following numbers as whole numbers without zeros and without a decimal point
(a) 2.000   (b) 56.0   (c) 342.000
(d) 15.00

UNNECESSARY ZEROS

One source of confusion is which noughts are needed and which are not.

1) You do not need noughts in front of whole numbers (though they may be put in when dividing to make sure the numbers are lined up correctly), for example:

$$\begin{array}{r} 05.2 \\ 3\overline{)15.6} \end{array} \quad \text{and} \quad 05.2 = 5.2$$

2) Noughts are not needed if they come after the decimal point at the end of the figures (but they are sometimes put in to show how accurate a measurement is) e.g. 0.700 m would mean that the measurement has been taken correct to one millimetre.

6.300 = 6.3

4.70 = 4.7

3.00 = 3

3) The noughts *are* needed in these numbers:

70 .203 500 .3005

Notice that in 70 the nought is put in to keep the place for the missing tens, in .203 the noughts keeps the place for the missing hundredths, etc.

.203 is *not* equal to .23

4) One zero is generally left *before* the point for tidiness but it is not essential.

.827 is generally written 0.827

.5 is generally written 0.5 etc.

00.5 would be the same as 0.5 but it is unusual to write more than one nought before the decimal point.

EXAMPLES

0067.50 is the same as 67.5

3.000   is the same as 3

004.0   is the same as 4

0.700   is the same as 0.7 or .7

00.023   is the same as 0.023 or .023
         (but *not* .23)

60.0   is the same as 60 (but *not* 6)

──────── **EXERCISE 4** ────────

Copy these numbers and then write them without the unnecessary zeros. Leave one zero before the decimal point for tidiness.

1) 0048.80
2) 4.0000
3) 006.00
4) 08
5) 0.8000
6) .70
7) 00.045
8) 1.1030
9) 5070.00
10) 300.000
11) 0000.2720
12) 11.000 000 0
13) 002.0200
14) 00.055
15) 6.0
16) Does 6.05 equal 6.50?
17) Does 6.5 equal 6.50?
18) Does £6.50 equal £6.05?
19) On a calculator would £6.50 appear as 6.05 or 6.5?

EXAMPLE

Write the following fractions as decimals

(a) $\dfrac{4}{10}$ (b) $\dfrac{5}{100}$ (c) $\dfrac{12}{100}$ (d) $\dfrac{2}{1000}$

(e) $\dfrac{145}{1000}$ (f) $\dfrac{12}{1000}$.

	Units	Tenths	Hundredths	Thousands
		$\dfrac{1}{10}$	$\dfrac{1}{100}$	$\dfrac{1}{1000}$
(a) $\dfrac{4}{10} =$	0 .	4		
(b) $\dfrac{5}{100} =$	0 .	0	5	
(c) $\dfrac{12}{100} =$	0 .	1	2	
(d) $\dfrac{2}{1000} =$	0 .	0	0	2
(e) $\dfrac{145}{1000} =$	0 .	1	4	5
(f) $\dfrac{12}{1000} =$	0 .	0	1	2

Look closely at example (c).

The 2 goes in the hundredths column and the 1 spills over into the tenths column.

For example (f) the 2 goes into the thousandths column and the 1 spills over into the hundredths column.

──────── **EXERCISE 5** ────────

Write the following as decimals

1) $\dfrac{6}{10}$ 7) $\dfrac{8}{100}$

2) $\dfrac{3}{100}$ 8) $\dfrac{9}{1000}$

3) $\dfrac{4}{1000}$ 9) $\dfrac{14}{100}$

4) $\dfrac{25}{100}$ 10) $\dfrac{628}{1000}$

5) $\dfrac{127}{1000}$ 11) $\dfrac{5}{10}$

6) $\dfrac{3}{10}$ 12) $\dfrac{7}{100}$

13) $\dfrac{6}{1000}$ 19) $\dfrac{15}{100}$

14) $\dfrac{75}{100}$ 20) $\dfrac{205}{1000}$

15) $\dfrac{954}{1000}$ 21) $\dfrac{11}{1000}$

16) $\dfrac{2}{10}$ 22) $\dfrac{15}{1000}$

17) $\dfrac{6}{100}$ 23) $\dfrac{62}{1000}$

18) $\dfrac{3}{1000}$ 24) $\dfrac{58}{1000}$.

─── DECIMAL ADDITION ───

The important thing to remember when doing decimal addition is to *line up the decimal points*. If whole numbers are present then put a decimal point at the *end* of the whole number and add zeros as required.

EXAMPLE

Add $62.1 + 0.53 + 215$.

215 is the same as 215.00.

```
  62.1
   0.53
 215.00 +
 ──────
 277.63
```

The decimal point in the answer goes underneath the other decimal points.

─────── EXERCISE 6 ───────

1) $62.34 + 41.11$
2) $1.3 + 4.2 + 3.1$
3) $12.11 + 8.71 + 29.11$
4) $89.71 + 49.1 + 98$

5) $761.3 + 291.1 + 497.4$
6) $12.3 + 0.56 + 1.12$
7) $0.3 + 7 + 6.25$
8) $17 + 11 + 29.4$
9) $18 + 16.23 + 0.25$
10) $141 + 20 + 17.1$
11) $9.125 + 0.726$
12) $17.63 + 0.129 + 11$
13) $18.251 + 11.257$
14) $6.28 + 12 + 18.19$
15) $17.1 + 3.05 + 25.1$
16) $7.12 + 0.67 + 0.2$
17) $11.6 + 12.8 + 1.9$
18) $14.52 + 0.05 + 6.1$
19) $17.56 + 2.14 + 0.006$
20) $32.4 + 14 + 16.5$
21) $67.104 + 18.029 + 7.315$
22) $17.51 + 1.105 + 0.069 + 0.37$

23)
```
191.103
206.040
 15.900 +
───────
```

24)
```
21.09
 2.56
 0.18
25.07 +
─────
```

25)
```
671.32
 19.41
151.5  +
──────
```

26)
```
726.13
 13.5
  0.462 +
──────
```

27)
```
175.68
 13.5
  0.462 +
──────
```

28)
```
39.76
21.95
 0.275 +
─────
```

29) 165.4
     1.98
    17.2 +
———

30) 318.1
    0.12
   67.9 +
———

31) Add together 53.124, 0.761 and 24.128.
32) Add together 72.1, 11.81 and 14.29.
33) Add together 26.03, 0.15 and 2.91.
34) Add together 7.21, 3.48 and 11.57.

DECIMAL SUBTRACTION

The decimal points must be lined up underneath each other.

The answer can be *checked* by adding the bottom two lines which should give the top line. The actual subtraction is carried out in the same manner as for whole number subtraction.

EXAMPLE

$7.2 - 3.4$

The first number goes on top. Start at the right hand end.

  7.2
  3.4 −
  ———
  3.8

Check by adding 3.4 and 3.8 to get 7.2.

EXAMPLE

$500 - 0.14$

Remember that 500 is the same as 500.00.

  500.00
    0.14 −
  ———
  499.86

1) $15.38 - 11.06$
2) $2.75 - 1.36$
3) $14.2 - 9.1$
4) $70.08 - 13.59$
5) $315.75 - 0.93$
6) $17.2 - 15.8$
7) $1.42 - 1.38$
8) $11.5 - 7.7$
9) $142.8 - 29.7$
10) $12.5 - 7$
11) $13 - 0.12$
12) $800 - 0.16$
13) $49.5 - 13.8$
14) $7000 - 52.5$
15) $6003.5 - 729.9$
16) $218.6 - 204.1$
17) $198.3 - 4.27$
18) $502 - 34.508$
19) $62.46 - 50$
20) $157.2 - 3.898$

21) 17.58
   11.24 −
   ———

22) 77.51
   36.90 −
   ———

23) 159.87
    27.16 −
   ———

24) 103.12
    27.06 −
   ———

25) 61.19
   51.07 −
   ———

26) 300.0
    1.5 −
   ———

27) 418.26
    29.38 −
    ‾‾‾‾‾

28) 51.123
    38.762 −
    ‾‾‾‾‾

29) 0.0152
    0.0098 −
    ‾‾‾‾‾

30) 0.607
    0.509 −
    ‾‾‾‾‾

31) Subtract 17.38 from 29.72.
32) Subtract 15.1 from 207.5.
33) Subtract 0.82 from 0.97.
34) Subtract 7.285 from 15.34.
35) Subtract 65.2 from 158.3.
36) Subtract 218.6 from 527.5.

USING A CALCULATOR FOR ── DECIMAL NUMBERS ──

With a calculator the addition shown in the previous example is quite straightforward. As shown below we simply enter the numbers and signs in the order given.

Input	Display
62.5	62.5
+	62.5
33.14	33.14
−	95.64
0.67	0.67
+	94.97
11.5	11.5
−	106.47
19.81	19.81
−	86.66
8.21	8.21
=	78.45

COMBINED ADDITION AND ── SUBTRACTION OF DECIMALS ──

When working out the answer to a question such as:

$62.5 + 33.14 - 0.67 + 11.5 - 19.81 - 8.21$

first add the numbers with a plus sign (or no sign, such as the first one), then add the numbers with a minus sign and take away the second sum from the first:

```
  62.5          0.67
 33.14         19.81
 11.5  +        8.21 +
‾‾‾‾‾‾         ‾‾‾‾‾
107.14         28.69
```

```
107.14
 28.69 −
‾‾‾‾‾‾
 78.45
```

── EXERCISE 8 ──

1) $14.3 + 25.8 - 0.51 - 2.42$
2) $6.54 + 0.29 + 3.12 - 0.67$
3) $14.21 - 0.28 + 5.32 - 7.11$
4) $152.6 + 104.12 - 0.253 - 0.45$
5) $53.125 + 120.61 - 72.188 - 27.1 + 3.1$
6) $57.6 + 106.1 - 15 - 11.3$
7) $23.4 - 34.5 + 67.91 - 3.1 - 6.2$
8) $0.92 - 0.64 + 2.1 - 4.1 + 11.4$
9) $23.45 + 34.89 - 14.5 - 13.8$
10) $13.5 + 98.1 - 12.1 - 0.3 - 5.09$

MULTIPLICATION OF DECIMALS ── BY POWERS OF 10 ──

Decimals may easily be multiplied by 10, 100, 1000, etc.

58

EXAMPLE

Find (a) 6.2×10 (b) 6.2×100 (c) 6.2×1000.

(a) $6.2 \times 10 = 62$ since

$$6.2 = 6\frac{2}{10} \text{ and } 6\frac{2}{10} \times 10 = \frac{62}{\cancel{10}} \times \frac{\cancel{10}}{1} = 62$$

(b) $6.2 \times 100 = 620$ since

$$6.2 = 6\frac{2}{10} \text{ and } 6\frac{2}{10} \times 100 = \frac{62}{\cancel{10}} \times \frac{\overset{10}{\cancel{100}}}{1}$$
$$= 620$$

(c) $6.2 \times 1000 = 6200$ since

$$6.2 = 6\frac{2}{10} \text{ and } 6\frac{2}{10} \times 1000 = \frac{62}{\cancel{10}} \times \frac{\overset{100}{\cancel{1000}}}{1}$$
$$= 6200$$

Notice *the effect of multiplying by 10 is to move the decimal point one place to the right.*

The effect of multiplying by 100 is to move the decimal point two places to the right.

The effect of multiplying by 1000 is to move the decimal point three places to the right.

EXAMPLE

Multiply these numbers by (i) 10 (ii) 100 (iii) 1000:

(a) 0.53   (b) 62.73   (c) 0.6.

(a) (i) 5.3 (ii) 53 (iii) 530
(b) (i) 627.3 (ii) 6273 (iii) 62 730
(c) (i) 6 (ii) 60 (iii) 600

In some of these cases zero has to be put in to make up the missing places.

――――――― **EXERCISE 9** ―――――――

Multiply these numbers by (i) 10 (ii) 100 (iii) 1000.

1) 5.7	6) 0.002 854
2) 6.21	7) 0.5823
3) 0.023	8) 0.007
4) 0.46	9) 160.07
5) 0.2178	10) 0.147 29

MULTIPLICATION OF DECIMALS

First multiply the numbers as if there were no decimal point. Then add the number of decimal places in the first number to the number of decimal places in the second number, and count back this number of places from the *right* to find where to put the decimal point.

EXAMPLE

6.2	*one* decimal place
3 ×	*no* decimal place
18.6	*one* decimal place in the answer

4.41	*two* decimal places
3 ×	*no* decimal place
13.23	*two* decimal places in the answer

0.087	*three* decimal places
1.1 ×	*one* decimal place
870	*four* decimal places in the answer
87	
0.0957	

When multiplying with decimal numbers a rough estimate using rounded numbers should be made before attempting to multiply.

EXAMPLE

Find the product of 5.63 and 7.21

Rough estimate $= 6 \times 7 = 42$

Input	Display
5.63	5.63
×	5.63
7.21	7.21
=	40.5923

So the product of 5.63 and 7.21 is 40.5923.

1) 6.7×4
2) 3.5×3
3) 4.8×7
4) 12.2×6
5) 17.8×7
6) 35.4×9
7) 1.25×6
8) 3.41×7
9) 5.83×8
10) 30.5×15
11) 25.6×24
12) 3.05×17
13) 0.052×1.1
14) 0.02×0.4
15) 0.357×0.9
16) 10.07×0.2
17) 43.5×0.4
18) 59.6×0.7
19) 0.72×0.8
20) 6.781×0.12
21) 9.32×2.5
22) 10.6×3.8
23) 3.21×5.1
24) 4.03×2.7
25) 83.6×4.8
26) 7.25×2.1
27) 0.06×1.2
28) 35.8×2.3
29) 4.25×2.7
30) 2.06×1.9
31) Multiply 2.251 by 9.
32) Multiply 3.02 by 0.08.
33) Multiply 0.013 by 1.8.
34) Multiply 34.2 by 7.
35) Multiply 8.7 by 0.003.
36) Multiply 6.1 by 0.12.
37) Multiply 7.2 by 1.2.
38) Multiply 8.4 by 9.5.
39) Multiply 7.3 by 0.87.
40) Multiply 0.56 by 0.27.

DIVISION OF DECIMALS BY POWERS OF 10

EXAMPLE

Divide 2.5 by (a) 10 (b) 100 (c) 1000.

(a) $2.5 \div 10 = 0.25$

since

$$2.5 = 2\frac{5}{10} \text{ and } 2\frac{5}{10} \div 10 = \frac{25}{10} \div \frac{10}{1}$$

$$= \frac{25}{10} \times \frac{1}{10}$$

$$= \frac{25}{100}$$

$$= 0.25$$

Similarly

(b) $2.5 \div 100 = 0.025$

and

(c) $2.5 \div 1000 = 0.0025$

The effect of dividing by 10 is to move the decimal point one place to the left.

The effect of dividing by 100 is to move the decimal point two places to the left.

The effect of dividing by 1000 is to move the decimal point three places to the left.

EXAMPLE

Divide these numbers by (i) 10 (ii) 100 (iii) 1000:

(a) 0.6 (b) 5.78 (c) 5671.2

(a) (i) 0.06 (ii) 0.006 (iii) 0.0006
(b) (i) 0.578 (ii) 0.0578 (iii) 0.005 78
(c) (i) 567.12 (ii) 56.712 (iii) 5.6712

Divide these numbers by (i) 10 (ii) 100 (iii) 1000

1) 6.1
2) 72.5
3) 0.05
4) 260.05
5) 0.469
6) 1.28
7) 0.0026
8) 325.5
9) 4628.9
10) 713.598

DIVISION OF DECIMALS BY WHOLE NUMBERS

EXAMPLE

$72.12 \div 6$

$$\begin{array}{r} 12.02 \\ 6\overline{)72.12} \end{array}$$

6s into 7 go 1 remainder 1. Put 1 above the 7 and the remainder 1 by the 2 to make 12.

6s into 12 go 2. Put 2 above the 2.

Put a decimal point above the other decimal point. 6s into 1 won't go so *put a nought* above the 1. Carry the 1 by the 2 to make 12. 6s into 12 go 2 so put 2 above the line above the other 2.

Make sure you line up your figures carefully and also line up the decimal points. This is very important.

1) $52.5 \div 3$
2) $48.5 \div 5$
3) $8.12 \div 4$
4) $0.552 \div 3$

5) $2.55 \div 5$
6) $45.32 \div 4$
7) $16.32 \div 2$
8) $1.561 \div 7$
9) $2.16 \div 6$
10) $8.127 \div 9$
11) $216.3 \div 3$
12) $804.4 \div 4$
13) $715.6 \div 4$
14) $3.824 \div 8$
15) $0.819 \div 9$
16) $2.2 \div 11$
17) $4.632 \div 3$
18) $8.664 \div 4$
19) $9.333 \div 9$
20) $4.284 \div 12$
21) $8.565 \div 5$
22) $0.352 \div 2$
23) $1.89 \div 9$
24) $32.45 \div 5$
25) $62.10 \div 6$
26) $5.532 \div 3$
27) $72.36 \div 9$
28) $15.4 \div 7$
29) $5.232 \div 3$
30) $4.824 \div 4$

A calculator can be used to find the answer to a division as a whole number and remainder.

EXAMPLE

Using a calculator, work out $27 \div 4$ giving your answer as a whole number and a remainder.

Input	Display	
27.	27.	
÷	27.	
4	4.	
=	6.75	
−	6.75	
6	6.	This is the whole number
=	0.75	
×	0.75	
4	4.	
=	3.	This is the remainder

The whole number before the decimal point is 6. There are six fours in 27.

We now have to find the remainder as a whole number. We can do this by multiplying the decimal part by 4. We see that the remainder is 3.

Hence

$$27 \div 4 = 6 \text{ remainder } 3$$

EXAMPLE

Find the remainder when 92 579 is divided by 24.

Input	Display
92579	92579.
÷	92579.
24	24.
=	3857.4583
−	3857.4583
3857	3857.
=	0.4583
×	0.4583
24	24.
=	10.9992

The calculator gives the remainder as 10.9992.

However, the remainder must be a whole number. Because the calculator is limited to 8 digits calculations of this type are seldom exactly right and we expect small errors to occur. 10.9992 is so near to 11 that this must be the remainder. Hence.

$$92579 \div 24 = 3857 \text{ remainder } 11$$

Note that:

$$3857 \times 24 + 11 = 92\,579$$

which can be checked using the calculator.

──────── EXERCISE 13 ────────

Try Exercises 16 and 17 in Chapter 2 using a calculator, giving the answers as a whole number and a remainder.

DIVISION OF A DECIMAL BY A
──── DECIMAL ────

EXAMPLE

$0.255 \div 0.5$

Move the decimal point the *same number of places to the right* in both numbers until the second number is a whole number. Then divide as before:

$$0.255 \div 0.5 = 2.55 \div 5$$

$$\begin{array}{r} 0.51 \\ 5\overline{)2.55} \end{array} \quad \text{so } 0.255 \div 0.5 = 0.51$$

EXAMPLE

Divide 87.64 by 0.07.

$$87.64 \div 0.07 = 8764 \div 7$$

$$\begin{array}{r} 1252 \\ 7\overline{)8764} \end{array} \quad \text{so } 87.64 \div 0.07 = 1252$$

EXAMPLE

Divide 3 by 0.06.

$$3 \div 0.06 = 300 \div 6 \quad \textit{two zeros must be}$$
$$\textit{written after the 3}$$

$$\begin{array}{r} 50 \\ 6\overline{)300} \end{array} \quad \text{so } 3 \div 0.06 = 50$$

EXAMPLE

Divide 336.6 by 16.5

$$336.6 \div 16.5 = 3366 \div 165$$

$$\begin{array}{r} 20.4 \\ 165\overline{)3366.0} \\ 330 \\ \hline 66\,0 \\ 66\,0 \end{array}$$

Use long division. We can put in a decimal point and zero since 3366 = 3366.0

EXAMPLE

Divide 931 by 17.85, using a calculator, stating the answer correct to three decimal places.

Rough estimate = $900 \div 15 = 60$

Input	Display
931	931.
÷	931.
17.85	17.85
=	52.15685677

So $931 \div 17.85 = 52.157$ correct to 3 decimal places (see below for exercises on decimal places).

EXERCISE 14

1) $4.24 \div 0.4$
2) $5.25 \div 0.5$
3) $62.1 \div 0.9$
4) $34.5 \div 0.005$
5) $5.6 \div 0.07$
6) $64.8 \div 0.008$
7) $0.231 \div 1.1$
8) $0.42 \div 0.7$
9) $0.65 \div 0.05$
10) $205.5 \div 0.003$
11) $4.9 \div 0.07$
12) $0.032 \div 0.08$
13) $4.25 \div 0.5$
14) $7.21 \div 0.7$
15) $62.82 \div 0.0002$
16) $51.63 \div 0.3$
17) $78.8 \div 0.04$
18) $9.216 \div 0.9$
19) $8.8 \div 0.11$
20) $45.09 \div 0.09$
21) $515.5 \div 0.5$
22) $0.3372 \div 0.003$
23) $0.1561 \div 0.7$
24) $14.4 \div 0.2$
25) $9.612 \div 0.09$
26) $12.12 \div 0.6$
27) $36.54 \div 0.9$
28) $0.72 \div 0.8$
29) $0.2414 \div 0.02$
30) $7.35 \div 0.5$

31) Divide 1.428 by 0.35.
32) Divide 0.007 263 by 0.807.
33) Divide 0.0196 by 0.14.
34) Divide 0.002 55 by 0.015.
35) Divide 10.8 by 1.25.

DECIMAL PLACES

Sometimes the numbers do not divide exactly. In this case the division is usually only carried out to two or three decimal places.

Decimal places (often shortened to "d.p.") means the *number of figures after the decimal point*:

6.253   has three decimal places

0.25   has two decimal places

15.6142 has four decimal places

EXERCISE 15

How many decimal places do these numbers have?

1) 6.23
2) 7.1298
3) 6.414
4) 4.155
5) 0.000 562 03
6) 0.5
7) 3.128 62
8) 25.554
9) 109.1
10) 19 087.1651

The number 6.22 is marked on the ruler below. It is nearer to 6.2 than 6.3:

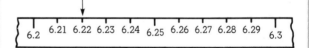

We would say that 6.22 = 6.2 correct to one decimal place.

The number 6.28 is marked on the ruler below. It is nearer to 6.3 than 6.2:

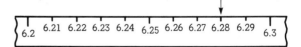

We would say that 6.28 = 6.3 correct to one decimal place.

4.82 = 4.8 correct to one decimal place

4.89 = 4.9 correct to one decimal place

3.71 = 3.7 correct to one decimal place

3.76 = 3.8 correct to one decimal place

By convention, if the last figure is a 5 (or more) we go up to the next figure so that 6.85 = 6.9 correct to one decimal place.

EXAMPLE

Put (a) 5.291 (b) 5.297 correct to two decimal places.

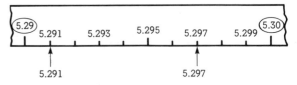

From the diagram 5.291 is nearer to 5.29 and 5.297 is nearer to 5.30:

5.291 = 5.29 correct to two decimal places

5.297 = 5.30 correct to two decimal places

EXERCISE 16

Put these numbers correct to one decimal place

1)	2.21	6)	2.28
2)	5.79	7)	5.71
3)	9.35	8)	9.32
4)	8.12	9)	8.19
5)	1.23	10)	1.26

EXERCISE 17

Put these numbers correct to two decimal places

1)	0.251	7)	0.799
2)	0.357	8)	0.791
3)	0.683	9)	5.231
4)	0.786	10)	5.237
5)	0.112	11)	7.216
6)	0.119	12)	7.218

We said that 6.22 = 6.2 correct to one decimal place and 6.28 = 6.3 correct to one decimal place (because 6.28 is nearer to 6.3 than 6.2).

We could carry this accuracy further by giving a number correct to *two*, *three* or *four* decimal places. (Remember decimal places are the number of figures after the decimal point.)

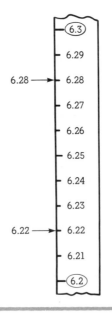

Here are some more difficult examples:

0.955 = 0.96   correct to 2 d.p.

0.683 71 = 0.68   correct to 2 d.p.

2.995 = 3.00   correct to 2 d.p.

5.179 82 = 5.18   correct to 2 d.p.

107.198 = 107.20 correct to 2 d.p.

0.683 71
=====

we ignore figures in the fourth or later decimal places

0.955 is halfway between 0.95 and 0.96 but for a 5 or over in the third decimal place we increase the previous figure by one

Put these numbers correct to two decimal places

1)	0.855	9)	3.695
2)	0.755	10)	4.595
3)	0.555	11)	8.295
4)	5.655	12)	11.195
5)	0.2487	13)	102.183
6)	0.1925	14)	84.1932
7)	0.3856	15)	15.7811
8)	0.4889	16)	14.285 67

For *three decimal places* we count three figures after the decimal point and look at the figure in the fourth decimal place. If it is a 5 or more we increase the previous figure by 1.

EXAMPLE

Put 0.7786 correct to three decimal places.

There is a 6 in the fourth place so we increase 8 by 1

0.7786 = 0.779 correct to 3 d.p.

EXAMPLE

Put 2.4195 correct to three decimal places.

$$
\begin{array}{r}
2.419 \\
1 + \\
\hline
2.420
\end{array}
$$

There is a 5 in the fourth place, increasing 9 by 1 causes a difficulty so we increase 19 by 1 to make 20

2.4195 = 2.420 correct to 3 d.p.

EXAMPLE

Put 3.9997 correct to three decimal places.

This is a difficult one because the numbers change so completely. However since 7 is more than 5 we have to add 1

$$
\begin{array}{r}
3.999 \\
1 + \\
\hline
4.000
\end{array}
$$

3.9997 = 4.000 correct to 3 d.p.

EXAMPLE

Put 0.8812 correct to three decimal places.

This is much more straight forward. We do not need to increase the previous figure because 2 is less than 5

0.8812 = 0.881 correct to 3 d.p.

Put these numbers correct to three decimal places

1)	0.6528	11)	0.7799
2)	0.6522	12)	0.7791
3)	0.6529	13)	0.7795
4)	0.6525	14)	2.9995
5)	0.6521	15)	2.9999
6)	5.5123	16)	2.9991
7)	5.5129	17)	2.9993
8)	5.5120	18)	2.9998
9)	5.5127	19)	2.9997
10)	5.5125	20)	2.9992

EXAMPLE

Work out $2.51 \div 0.3$ correct to three decimal places.

$2.51 \div 0.3 = 25.1 \div 3$ (make the second number a *whole* number by moving the decimal point in *both* numbers)

$$\begin{array}{r} 8.3666 \\ 3\overline{)25.1000} \end{array}$$

work to four decimal places and correct to three:

$$2.51 \div 0.3 = 8.367$$

EXERCISE 20

Carry out the following correct to three decimal places. (Hint: remember to make the second number a *whole* number before dividing by moving the point the same number of places to the right in both numbers.)

1) $5.2 \div 3$
2) $6.2 \div 0.09$
3) $27.8 \div 7$
4) $7.1 \div 0.9$
5) $0.51 \div 0.7$
6) $352 \div 3$
7) $8.2 \div 0.6$
8) $0.28 \div 3$
9) $42.1 \div 0.6$
10) $5.4 \div 0.07$
11) $4.51 \div 0.06$
12) $5.15 \div 0.9$

CHANGING FRACTIONS TO DECIMALS

Suppose we wanted to share £5 between eight people because an outing cost £5 and eight people each wanted to pay their share.

We should have to divide £5 by 8.

We can write this as $£5 \div 8$ or $£\frac{5}{8}$ but what is $£\frac{5}{8}$?

(It wouldn't be much good to cut a £1 note into eight equal parts and take five!) So we need to convert $£\frac{5}{8}$ into decimal form. The method is to write 5 as a decimal and to divide by 8.

5 is the same as 5.000

$$\begin{array}{r} 0.625 \\ 8\overline{)5.000} \end{array}$$

so that

$$\frac{5}{8} = 0.625$$

and

$$£\frac{5}{8} = £0.625$$

which gives an answer exactly half way between 62p and 63p.

To turn a fraction into a decimal divide the number on top by the number underneath.

EXAMPLE

Turn $\frac{3}{4}$ into a decimal.

3 is the same as 3.00

$$\begin{array}{r} 0.75 \\ 4\overline{)3.00} \end{array}$$

so that

$$\frac{3}{4} = 0.75$$

EXAMPLE

Use a calculator to turn $\frac{5}{8}$ into a decimal number.

Input	Display
5	5.
÷	5.
8	8.
=	0.625

So $\frac{5}{8} = 0.625$.

66

Turn these fractions into decimal form

1) $\dfrac{1}{4}$ 6) $\dfrac{2}{5}$

2) $\dfrac{1}{2}$ 7) $\dfrac{3}{5}$

3) $\dfrac{1}{8}$ 8) $\dfrac{1}{20}$

4) $\dfrac{7}{8}$ 9) $\dfrac{1}{25}$

5) $\dfrac{1}{5}$ 10) $\dfrac{3}{20}$

3) $1\dfrac{1}{2}$ 12) $3\dfrac{4}{5}$

4) $1\dfrac{1}{4}$ 13) $2\dfrac{2}{5}$

5) $2\dfrac{3}{4}$ 14) $2\dfrac{1}{8}$

6) $4\dfrac{3}{4}$ 15) $5\dfrac{3}{5}$

7) $5\dfrac{1}{2}$ 16) $2\dfrac{1}{5}$

8) $1\dfrac{3}{8}$ 17) $6\dfrac{1}{4}$

9) $2\dfrac{5}{8}$ 18) $10\dfrac{1}{2}$

10) $3\dfrac{1}{2}$ 19) $11\dfrac{3}{4}$

11) $15\dfrac{1}{2}$ 20) $4\dfrac{7}{8}$

If a whole number is included then *the whole number goes before the decimal point.*

EXAMPLE

Turn $1\frac{3}{4}$ into a decimal.

First we deal with the fraction:

$3 = 3.00$

$$\begin{array}{r} 0.75 \\ 4\overline{)3.00} \end{array}$$

so that

$$\dfrac{3}{4} = 0.75$$

Then we put in the whole number so that

$$1\dfrac{3}{4} = 1.75$$

RECIPROCALS

The reciprocal of a number is 1 divided by that number.

Thus the reciprocal of 5 is $\frac{1}{5}$ as a fraction or $1 \div 5 = 0.2$ as a decimal.

EXAMPLE

Find the reciprocal of 93 correct to 4 decimal places.

Input	Display
1	1.
÷	1.
93	93.
=	0.010752688

So the reciprocal of 93 is 0.0108 correct to 4 decimal places.

Write these mixed fractions as decimal numbers

1) $3\dfrac{3}{4}$ 2) $2\dfrac{1}{2}$

EXAMPLE

Find the reciprocal of 0.039 correct to 2 decimal places.

Input	Display
1	1.
÷	1.
0.039	0.039
=	25.64102564

So the reciprocal of 0.039 is 25.64 correct to 2 decimal places.

To find the reciprocal of a fraction turn the fraction upside down.

EXAMPLE

Find the reciprocal of $1\frac{1}{3}$

$$1\frac{1}{3} = \frac{4}{3}$$

The reciprocal is $3 \div 4 = 0.75$ as a decimal.

--- EXERCISE 23 ---

Find the reciprocals of these whole numbers

(a) as a fraction   (b) as a decimal.

1) 2   2) 4   3) 10
4) 50   5) 100   6) 200
7) 1000   8) 1 000 000

Find the reciprocals of these decimal numbers correct to the number of decimal places stated.

9) 1.35 correct to 4 decimal places
10) 23.6 correct to 3 decimal places
11) 498 correct to 5 decimal places
12) 0.07 correct to 3 decimal places
13) 0.009 correct to 2 decimal places
14) 0.0768 correct to 3 decimal places
15) 0.316 correct to 4 decimal places.

Find the reciprocals of these numbers giving your answer as a whole number or a decimal.

16) 1.25   17) 2.5   18) 0.25
19) 0.5   20) 0.1   21) 0.01
22) 0.001   23) 0.0001

Find the reciprocals of these fractions as a whole number.

24) $\frac{1}{2}$ 25) $\frac{1}{5}$ 26) $\frac{1}{8}$

27) $\frac{1}{10}$ 28) $\frac{1}{40}$ 29) $\frac{1}{75}$

Find the reciprocals of these fractions giving your answer (a) as a fraction (b) as a decimal.

30) $\frac{5}{2}$ 31) $\frac{2}{3}$ 32) $\frac{10}{7}$

33) $\frac{100}{11}$ 34) $\frac{8}{7}$ 35) $\frac{8}{3}$

CHANGING DECIMALS TO FRACTIONS

To change a decimal to a fraction it is necessary to remember the table headings:

Units		Tenths	Hundredths	Thousandths	Ten thousandths	etc.
0	.	5				

So

$$0.5 = \frac{5}{10} \text{ (five tenths)}$$

68

Here are some more examples:

$$0.03 = \frac{3}{100} \text{ (three hundredths)}$$

$$0.42 = \frac{4}{10} + \frac{2}{100} = \frac{40}{100} + \frac{2}{100} = \frac{42}{100}$$

$$0.625 = \frac{6}{10} + \frac{2}{100} + \frac{5}{1000} = \frac{625}{1000}$$

Sometimes it is convenient to reduce these fractions to their lowest terms (by dividing top and bottom by the same number):

$$0.5 = \frac{5}{10} = \frac{1}{2}$$

$$0.03 = \frac{3}{100}$$

which is in its lowest terms already.

$$0.42 = \frac{42}{100} = \frac{21}{50}$$

$$0.625 = \frac{625}{1000} = \frac{125}{200} = \frac{25}{40} = \frac{5}{8}$$

(dividing each time by 5).

──────────── EXERCISE 24 ────────────

Change these decimals to fractions in their lowest terms

1) 0.6	9) 0.62
2) 0.7	10) 0.55
3) 0.09	11) 0.15
4) 0.02	12) 0.23
5) 0.04	13) 0.725
6) 0.05	14) 0.123
7) 0.08	15) 0.513
8) 0.17	

When there are whole numbers they go in front of the fraction.

EXAMPLE

Change 4.5 to a mixed fraction.

$$0.5 = \frac{5}{10} = \frac{1}{2}$$

$$4.5 = 4\frac{1}{2}$$

──────────── EXERCISE 25 ────────────

Change these numbers to mixed fractions

1) 2.5	6) 27.5
2) 3.75	7) 3.3
3) 5.25	8) 5.25
4) 6.2	9) 14.5
5) 18.1	10) 22.2

The following should be learnt by heart:

$$0.1 = \frac{1}{10} \qquad 0.5 = \frac{1}{2}$$

$$0.25 = \frac{1}{4} \qquad 0.75 = \frac{3}{4}$$

and it is also helpful to remember that:

$$0.2 = \frac{1}{5}$$

── RECURRING DECIMALS ──

When changing $\frac{1}{3}$ to a decimal it is found that the figures repeat themselves:

1 is the same as 1.000 000

$$\begin{array}{r} 0.333\,333 \\ 3)\overline{1.000\,000} \end{array}$$

so that

$$\frac{1}{3} = 0.333\,333\,3\ldots$$

This is an example of a *recurring decimal.* 0.333 333 33 . . . may be written as $0.\dot{3}$. Recurring decimals are not often of practical use. It is generally more useful to work to so many decimal places.

0.666 66 . . . is written as $0.\dot{6}$

and is equal to $\frac{2}{3}$. ($\frac{2}{3} = 0.67$ correct to 2 d.p.)

0.121 212 12 . . . is written as $0.\dot{1}\dot{2}$

0.124 581 245 812 458 . . . is written as $0.\dot{1}24 5\dot{8}$

14.504 154 154 154 15 . . . is written as $14.50\dot{4}1\dot{5}$

This is called "dot notation".

——————— EXERCISE 26 ———————

Write these recurring decimals using dot notation (i.e. dots over the figures as in the examples just given).

1) 0.444 444 . . .
2) 0.343 434 34 . . .
3) 0.467 534 675 346 753 . . .
4) 7.888 888 . . .
5) 30.787 878 . . .
6) 12.503 093 093 093 09 . . .
7) 0.222 222 2
8) 0.191 919 . . .
9) 2.314 531 453 145 3145 . . .

— SEQUENCE OF OPERATIONS —

Brackets must be worked first then multiplication and division before subtraction and addition.

EXAMPLE

$8.6 \times 2 + 0.5 \times 0.3$

We do the multiplication first:

8.6	0.5	17.2
2 ×	0.3 ×	0.15 +
17.2	0.15	17.35

$= 17.2 + 0.15$

$= 17.35$

EXAMPLE

$(1.5 + 2.5) \times (3.2 + 0.8)$
This time the brackets must be done first:

1.5	3.2	4.0 = 4	4 × 4 = 16
2.5 +	0.8 +		
4.0	4.0		

$= 4.0 \times 4.0$

$= 16$

——————— EXERCISE 27 ———————

1) $4.3 \times 5 + 5.1 \times 2$
2) $0.4 \times 0.5 + 0.2 \times 0.3$
3) $(1.5 + 1.5) \times (3.5 + 3.5)$
4) $0.5 \times (5.4 + 6.6)$
5) $5.1 \times 0.5 + 2.7 \times 0.9$
6) $(0.3 + 0.4 + 0.5) \times 0.05$
7) $5.5 \div 0.5 + 4.5 \div 0.5$
8) $(52.5 - 49.3) \div 4$
9) $(15.6 - 4.9) - (0.3 \times 3)$
10) $(1.335 + 23.45 + 13.1) - (15.7 + 0.3)$
11) $0.2 \times 0.3 - 0.15 \times 0.2$
12) $0.3 + 0.4 \times 5$

1) What does the number 4 mean in the following numbers?
   (a) 42   (b) 1.4   (c) 1.04   (d) 34.131

2) Write the number 17 as a decimal with (a) one zero (b) three zeros after the decimal point.

3) Does (a) 7.06 equal 7.60 (b) 7.6 equal 7.60?

4) Write these numbers leaving out unnecessary zeros
   (a) 000 72.5   (b) 63.100
   (c) 402.0500.

5) Write these fractions as decimals
   (a) $\dfrac{3}{10}$ (b) $\dfrac{6}{100}$ (c) $\dfrac{7}{1000}$ (d) $\dfrac{35}{100}$.

6) (a) Add 2.1, 3.4, 2.5 and 1.3.
   (b) Take 7.04 from 8.16.

7) (a)   2.56   (b) 100.00
        13.21   1.96 −
        0.568 +   ‾‾‾‾‾
        ‾‾‾‾‾

8) $62.51 + 12.72 - 4.38$

9) Multiply 0.72 by (a) 10 (b) 100.

10) Multiply (a) 0.65×4 (b) 1.3×8.

11) (a) 16.5   (b) 6.03   (c) 10.32
        0.9 ×   0.04 ×   0.15 ×
        ‾‾‾‾   ‾‾‾‾   ‾‾‾‾

12) Divide 17.56 by (a) 100 (b) 1000.

13) Divide 217.8 by 3.

14) (a) $5\overline{)25.60}$ (b) $8\overline{)0.976}$

15) $35.84 \div 0.7$

16) Divide 415.2 by 0.4.

17) $0.005\overline{)2.500}$

18) Write (a) 4.28 (b) 4.21 correct to one decimal place.

19) Write (a) 14.6295 (b) 14.6291 correct to three decimal places.

20) Write 59.341 56 correct to (a) the nearest whole number (b) one decimal place (c) two decimal places (d) three decimal places (e) four decimal places.

21) Divide 5 by 3 and give your answer correct to three decimal places.

22) Write (a) $\dfrac{4}{5}$ (b) $\dfrac{7}{10}$ (c) $\dfrac{1}{3}$ as decimals.

23) Write the following numbers as decimals
   (a) $2\dfrac{1}{4}$ (b) $5\dfrac{1}{10}$ (c) $4\dfrac{1}{5}$.

24) Write these decimals as fractions
   (a) 0.2   (b) 0.08   (c) 0.45.

25) Change these numbers to mixed fractions in their lowest terms
   (a) 6.5   (b) 4.25   (c) 9.1   (d) 1.75.

26) Write these recurring numbers using dot notation
   (a) 0.111 11 . . .   (b) 0.292 929 29 . . .
   (c) 0.501 250 125 012 . . .

27) (a) $0.1 \times 21.2 + 0.3 \times 0.5$
   (b) $(6.2 + 3.1) \times (0.1 + 0.2)$
   (c) $(5.60 - 2.79) - (3.41 - 2.38)$

This exercise is for further practice in the correct use of signs +, −, × and ÷. Show all working and show clearly which signs you have used.

1) 0.3 kilograms is removed from 1.1 kilograms of cheese. How much does the cheese weigh now?

2) A piece of wood 1.54 metres long is divided into 7 equal lengths. How long is each piece?

3) 3.5 metres of cloth cost £7. How much would 1 metre cost?

4) In a sale a woman buys 4 items costing £2.50, £0.37, £6.50 and £4.99. How much change would she get from a £20 note?

5) A litre of petrol costs £0.52. If the bill was £5.20, how many litres were bought?

6) Wages are £4.50 an hour. How much is paid for $2\frac{1}{2}$ hours work?

7) £1 buys 1.6 dollars. How many dollars can be bought for £15?

8) Water flows from a tap at a rate of 2.5 litres a minute. How many litres would flow in 6 minutes?

9) A lorry weighs 2.5 tonnes when loaded with 12 boxes, each weighing 0.025 tonnes. What is the weight of the lorry when unloaded?

10) A metal bar that is 3.5 metres long is divided into 0.5 metre lengths. How many pieces are there?

11) A box of cornflakes weighs 0.44 kilograms. How heavy would a box with only $\frac{3}{4}$ of this weight be?

12) In a test, a child was asked to work out $\frac{3}{8}$ of 5.6. What should his answer have been?

13) A $2\frac{1}{2}$ litre bottle of Cola costs £2.30. How much is this per litre?

14) All of Sanjay's year group are going on a school holiday. The holiday costs £54.50 for each pupil. There are 80 pupils in the year group. Work out the total amount to be collected.

15) If a large pot of speciality jam costs £1.20 and $\frac{1}{3}$ is to be taken off this price in a sale, how much will the jam now cost?

16) A group of business people decide to spend £6000 on a racehorse. They each contribute an equal amount. There are 18 people in the group. How much will each have to contribute?

17) A café owner is furnishing her café with 4 large tables which seat 6 people and 9 smaller tables which seat 4 people. How many chairs will she need? The larger tables cost £59.50, the smaller tables £44.50 and the chairs £28.50. How much does she need to spend on the tables and chairs?

18) (a) A 500 millilitre bottle of shampoo costs 92 p. How much is this per millilitre?
    (b) The 250 millilitre bottle costs 65 p. How much is this per millilitre?

19) 5 miles are about 8 kilometres. How much is 1 mile in kilometres?

20) There are 2.54 centimetres in 1 inch. How long is 30 centimetres in inches, correct to two decimal places?

——————— EXERCISE 30 ———————

The questions in this exercise are all of a practical nature. They all depend on decimals for their solution.

1) A lorry weighs 3.5 tonnes when empty. It is loaded with 48 metal ingots each weighing 0.075 tonnes. What is the total weight of the lorry when loaded?

2) What is the total weight carried by a train consisting of 40 trucks if 20 of them carry 19.75 tonnes per truck and the remainder carry 18.5 tonnes per truck?

3) A herd of cows yields 19.8 litres of milk per cow. If the herd consists of 28 cows, what is the total amount of milk obtained?

4) A farmer feeds his cattle at the rate of 1.6 kilograms of concentrate per 5 litres of milk produced per day. If he has a herd of 24 cows, each producing 19.25 litres per day, calculate the amount of concentrate required per day.

5) The diagram shows a shaft for a diesel engine. Calculate its overall length.

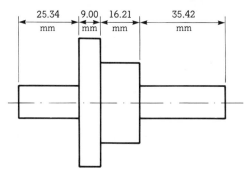

6) Rivets are placed 35.2 millimetres apart. If 28 rivets are placed in an assembly, calculate the distance between the first and last rivets.

7) An operator takes 0.74 minutes to make a soldered joint. Work out the time required to make 379 such joints.

8) The resistance of a copper wire 27.6 metres long is 1515.24 ohms. What is the resistance per metre?

9) An oil tank contains 1750 litres. How many full tins each containing 0.6 litres can be filled from this tank?

10) It takes 29 strips of wallpaper to paper a room, each piece being 2.35 metres long. What length of wallpaper is needed?

11) The cost of telephone calls is 4.2 pence per call. Work out the cost of making 204 calls.

12) 125 sheets of metal, all of the same thickness, are in a stack which is 37.500 millimetres high. What is the thickness of each sheet of metal?

13) Which is the greater, $\frac{7}{8}$ or 0.882, and by how much?

14) A consignment of screws has a total weight of 1757 grams. If each screw weighs 2.51 grams, calculate the number of screws in the consignment.

5

— Measurement —

We weigh light objects using milligrams or grams. Heavier objects are weighed in kilograms or metric tonnes.

There are:

    1000 milligrams in 1 gram

    1000 grams in 1 kilogram

    1000 kilograms in 1 metric tonne

Milligrams are shortened to mg

Grams are shortened to g

Kilograms are shortened to kg

It is easy to change from one unit to another in the metric system. We move the decimal point to the left when changing to larger units and to the right when changing to smaller units.

EXAMPLE

Change 4000 g to kg.

$$1000\,g = 1\,kg \quad so$$
$$4000\,g = 4\,kg$$

To change grams to kilograms divide by 1000 (move the decimal point three places to the left).

EXAMPLE

Change 5285 g to kg.

$$1000\,g = 1\,kg$$
$$5285\,g = 5285 \div 1000\,kg$$
$$= 5.285\,kg$$

Notice that it is more usual to write 5.285 kg rather than 5 kg 285 g.

EXAMPLE

Change 700 g to kg.

$$1000\,g = 1\,kg$$
$$700\,g = 700 \div 1000\,kg$$
$$= 0.7\,kg$$

EXAMPLE

Change 15 000 g to kg.

$$1000\,g = 1\,kg$$
$$15\,000\,g = 15\,000 \div 1000\,kg$$
$$= 15\,kg$$

— EXERCISE 1 —

Change to kilograms

1)	3000 g	7)	3850 g
2)	5000 g	8)	5670 g
3)	9000 g	9)	4500 g
4)	1500 g	10)	1654 g
5)	6500 g	11)	3045 g
6)	7500 g	12)	6723 g

13) 423 g	16) 25 000 g
14) 560 g	17) 40 000 g
15) 305 g	18) 50 000 g

To change kilograms to grams multiply by 1000 (move the decimal point three places to the right).

EXAMPLE

Change (a) 3 kg (b) 3.84 kg (c) 0.2 kg to grams.

$$1 \text{ kg} = 1000 \text{ g}$$
$$3 \text{ kg} = 3 \times 1000 \text{ g} = 3000 \text{ g}$$
$$3.84 \text{ kg} = 3.84 \times 1000 \text{ g} = 3840 \text{ g}$$
$$0.2 \text{ kg} = 0.2 \times 1000 \text{ g} = 200 \text{ g}$$

--- **EXERCISE 2** ---

Change to grams

1) 5 kg	10) 5.1 kg
2) 7 kg	11) 4.15 kg
3) 9 kg	12) 3.2 kg
4) 2.34 kg	13) 25 kg
5) 9.16 kg	14) 50 kg
6) 2.08 kg	15) 60 kg
7) 0.5 kg	16) 0.15 kg
8) 0.9 kg	17) 0.35 kg
9) 0.4 kg	18) 0.95 kg

Very small objects are weighed in milligrams (mg). There are 1000 milligrams in 1 gram.

Very heavy objects are weighed in metric tonnes. There are 1000 kg in 1 metric tonne.

--- **EXERCISE 3** ---

How many mg are there in

1) 2 g	4) 0.5 g
2) 4 g	5) 3.5 g?
3) 10 g	

How many kg are there in

6) 3 metric tonnes
7) 5 metric tonnes
8) 0.5 metric tonnes
9) 8.5 metric tonnes
10) 0.25 metric tonnes?

Imperial units of measurement are ounces, pounds, stones, hundredweights and tons.

There are:

16 ounces in 1 pound
14 pounds in 1 stone
112 pounds in 1 hundredweight
20 hundredweights in 1 ton

Light objects are weighed in pounds and ounces and heavier objects in tons and hundredweights.

An ounce is shortened to 1 oz
A pound is shortened to 1 lb
A stone is shortened to 1 st
A hundredweight is shortened to 1 cwt

Approximate imperial/metric conversions:

2 pounds is slightly less than 1 kilogram
4 ounces is slightly more than 100 grams

Medicines might be weighed in mg.

Sweets are weighed in oz (or grams).

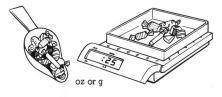

oz or g

Vegetables are weighed in pounds (or kilograms).

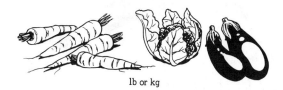

lb or kg

A man can carry 1 cwt of coal.

cwt

Lorries are weighed in tons (or metric tonnes).

Tons or tonnes

EXERCISE 4

1) How many ounces are there in
   (a) $\frac{1}{4}$ lb   (b) $2\frac{1}{2}$ lb   (c) 5 lb?

2) How many cwt are there in
   (a) $\frac{1}{2}$ ton   (b) 3 tons   (c) 5.5 tons?

3) Is 1 oz nearly equal to
   (a) 3 g   (b) 30 g   (c) 300 g   (d) 3 kg?

4) Is 1 kg more or less than 2 lb?

5) Is 1 cwt approximately equal to
   (a) 500 kg   (b) 50 kg   (c) 5 kg
   (d) 500 g?

LENGTH

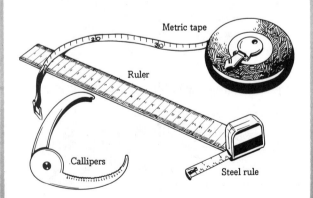

Metric tape

Ruler

Callipers

Steel rule

The "length" of a line gives us a measure of how far a line stretches out from end to end. Some instruments for measuring length are pictured above.

The length of this line is 8 cm.

|←————————— 8 —————————→|

The units of length are:

kilometres, metres, centimetres or millimetres, in metric measurement

or

miles, yards, feet or inches in imperial measurement

We measure short distances in millimetres, centimetres or inches and longer distances in metres, kilometres, yards or miles.

Wood might be measured in millimetres, centimetres or inches.

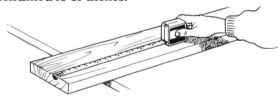

Curtain and dress materials are sold by the metre or the yard.

A metre is about 3 inches longer than 1 yard.

A metre is about the depth of the shallow end of a swimming pool.

You could walk 1 kilometre in about 10 minutes.

8 kilometres = 5 miles (approx.)

Kilometres are shortened to km
Metres are shortened to m
Centimetres are shortened to cm
Millimetres are shortened to mm

There are:

　10 mm in 1 cm
　100 cm in 1 m
　1000 m in 1 km

EXAMPLE

How many mm are there in

(a) 5 cm　　　(b) 2.5 cm　　(c) 125 cm?

　1 cm = 10 mm

so *to change cm to mm we multiply by 10:*

　5 cm = 5 × 10 mm = 50 mm
　2.5 cm = 2.5 × 10 mm = 25 mm
　125 cm = 125 × 10 mm = 1250 mm

―――――――― EXERCISE 5 ――――――――

How many mm are there in

1) 2 cm	7) 5.5 cm
2) 4 cm	8) 9.5 cm
3) 6 cm	9) 240 cm
4) 9 cm	10) 350 cm
5) 1.5 cm	11) 500 cm?
6) 0.5 cm	

EXAMPLE

How many cm are there in　(a) 8 m　(b) 0.5 m
(c) 25 m?

　1 m = 100 cm

so *to change m to cm we multiply by 100:*

　8 m = 8 × 100 cm = 800 cm
　0.5 m = 0.5 × 100 cm = 50 cm
　25 m = 25 × 100 cm = 2500 cm

―――――――― EXERCISE 6 ――――――――

How many cm are there in

1) 2 m	5) 0.4 m	9) 40 m
2) 4 m	6) 0.6 m	10) 50 m
3) 7 m	7) 0.8 m	11) 85 m
4) 9 m	8) 0.9 m	12) 95 m?

We write 1 cm 7 mm as 1.7 cm.
This line is 1.7 cm long measured to the nearest mm.

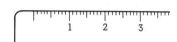

―――――――― EXERCISE 7 ――――――――

Measure the lines below and give their length in cm to the nearest mm.

1) _____
2) _____
3) _____
4) _____
5) _____
6) _____
7) _____
8) ____
9) _____
10) _____

EXAMPLE

Change to km:
(a) 5000 m　　(b) 1500 m　　(c) 400 m.

　1000 m = 1 km

so to change to km we divide by 1000

　5000 m = 5000 ÷ 1000 km = 5 km
　1500 m = 1500 ÷ 1000 km = 1.5 km
　　400 m = 400 ÷ 1000 km = 0.4 km

(Reminder: *To divide by 1000 move the decimal point three places to the left.*)

Change to km

1) 2000 m
2) 4000 m
3) 7000 m
4) 9000 m
5) 2500 m
6) 3550 m
7) 4500 m
8) 7500 m
9) 500 m
10) 650 m
11) 700 m
12) 900 m

There are:

12 inches in 1 foot

3 feet in 1 yard

1760 yards in 1 mile

Inches are shortened to in or "

Feet are shortened to ft or '

Yards are shortened to yd

Approximate imperial/metric conversions:

1 inch is slightly more than $2\frac{1}{2}$ cm

1 foot is approximately equal to 30 cm

1 yard is slightly less than 1 metre

$\frac{5}{8}$ mile is approximately equal to 1 km
(or 5 miles = 8 km approx.)

---

1) How many feet are there in 3 yards?

2) How many yards are there in $\frac{1}{2}$ mile?

3) How many inches are there in 3′6″?

4) Change 4 yd 2 ft into feet.

5) Is $\frac{1}{4}$ mile more or less than 800 yd?

6) How many $\frac{1}{16}″$ are there in 1″?

7) Is 1 metre longer or shorter than 1 yard?

8) How many inches are approximately equal to (a) 10 cm (b) 5 cm (c) 20 cm?

9) How many cm are approximately equal to (a) 3 ft (b) 1′6″ (c) 6 in?

10) How many km are approximately equal to (a) 10 miles (b) 15 miles?

11) Is 30 miles approximately equal to (a) 20 km (b) 48 km or (c) 60 km?

12) Are 20 miles more or less than 50 km?

CAPACITY

Capacity is a measure of the amount of space that a liquid takes up.

The metric units in common use are the millilitre and the litre which are shortened to ml and ℓ.

There are:

1000 mℓ in 1 litre

Sometimes the centilitre is used. This is shortened to cℓ.

There are:

100 cℓ in 1 litre

The imperial units of capacity are the fluid ounce (fl oz), the pint (pt), the quart (qt) and the gallon (gall). There are others but they are not used so often.

5 mℓ plastic medicine spoon

1 pt milk bottle

Petrol is sold in gallons or litres

Cream is sold in fluid ounces

78

There are:

  20 fl oz in 1 pt
  2 pts in 1 qt
  8 pts in 1 gall

Approximate imperial/metric conversions:

  1 litre is approximately equal to $1\frac{3}{4}$ pints
  30 mℓ is approximately equal to 1 fl oz
  1 gallon is approximately equal to $4\frac{1}{2}$ litres

──────── **EXERCISE 10** ────────

1) How many cℓ are there in $\frac{1}{2}$ litre?

2) How many 5 mℓ spoonfuls would make up 1 cℓ?

3) Are 2 pints more or less than 1 litre?

4) Are 120 mℓ approximately equal to (a) 1 (b) 2 (c) 3 or (d) 4 fluid ounces?

5) Are 3 gallons of petrol more or less than 15 litres?

6) How many pints are there in (a) $2\frac{1}{2}$ gallons (b) 3 quarts?

7) Change 0.75 litres to cℓ.

8) Is there more cream in a 5 fl oz pot than a 135 mℓ pot?

9) Change 2500 mℓ to litres.

10) I want to buy about 5 gallons of petrol. How many litres should I ask for to make sure of getting at least 5 gallons?

11) How many pints are there in 5 gallons?

12) How many fluid ounces are there in $\frac{1}{4}$ pint?

13) How many mℓ are there in 0.5 litres?

14) Change 2.5 litres to mℓ.

15) Are 2 gallons approximately equal to (a) 3 (b) 9 (c) 12 or (d) 15 litres?

In some countries milk is sold in litre packs or 50 cℓ packs

─────────── TIME ───────────

In the winter, time in the United Kingdom is measured using Greenwich Mean Time (GMT). In summer the clocks are put forward by 1 hour and this is called British Summer Time.

  60 seconds (s) = 1 minute (min)
  60 minutes = 1 hour (h)
  24 hours = 1 day
  7 days = 1 week (wk)
  31/30/28 days = 1 month
  365 days = 1 year
  366 days = 1 leap year
  13 weeks = 1 quarter
  52 weeks = 1 year
  12 months = 1 year
  p.a. = per annum = per year
  a.m. = ante meridiem = morning
  p.m. = post meridiem = afternoon

The time of day can be expressed *either* using the 12 hour clock *or* the 24 hour clock. Some clocks and watches use the 12 hour clock and we tell morning from afternoon by the use of a.m. or p.m.

Timetables are usually written using the 24 hour clock which runs from 0000 hours (midnight) through to 2400 hours (midnight again). 8 a.m. would be 0800 hours ("eight hundred hours"), noon would be 1200 hours ("twelve hundred hours") and 5.30 p.m. would be 1730 hours ("seventeen thirty hours").

Here is a simplified portion of a railway timetable using the 24 hour clock:

Padlow	Tyford	Maxby	Bixley
0830	0835	0855	0910
0930	⟶	⟶	1000
1030	1035	1055	1110
1130	1137	1159	1215
1200	⟶	⟶	1230
1330	1335	1355	1410
1530	1545	1610	1620
1700	1705	1725	1740

EXERCISE 11

Using the timetable above, answer these questions

1) Which train is the slowest?
2) I wish to catch the next train from Padlow after 4.30 p.m. Which train must I catch?
3) How long must I wait for the first train from Maxby after 9 a.m.?
4) How long does it take the fastest trains to get from Padlow to Bixley?
5) Does the last train run after 6 p.m.?

EXERCISE 12

1) How many minutes are in $2\frac{1}{4}$ hours?
2) How many seconds are in $3\frac{1}{2}$ minutes?
3) Change 150 min to hours and minutes.
4) A 4 lb chicken is to be cooked for 5×20 min. How many hours and minutes is this?
5) How many days are there in
   (a) January   (b) November
   (c) February in a leap year?
6) How many weeks are there in 2 quarters?
7) If a man earns £6000 per annum how much is this per month?

MONEY

There are 100 pence in one pound:

100 p = £1   £1

£5.50 is read as "five pounds fifty" and £6.02 as "six pounds two pence". The zero is important as it keeps the place for the missing tens.

EXAMPLE

Add £6.25 and £14.03.

```
    6.25
   14.03 +
   ─────
   20.28
```

so that £6.25 + £14.03 = £20.28

EXAMPLE

Subtract £4.07 from £5.

```
   5.00
   4.07 −
   ────
   0.93
```

so that £5 − £4.07 = 93p.

EXERCISE 13

Add

1) £3.71 + £2.03
2) £2.90 + £1.07
3) £54.10 + £3.30
4) £0.65 + £4.18
5) £0.75 + £1.30
6) £0.05 + £1.07
7) £10.95 + £19.13
8) £35.47 + £14
9) £23.45 + £7.08
10) £107 + £235

EXERCISE 14

Subtract

1) £97.53 − £25.22
2) £357.65 − £23.47
3) £405.50 − £203.60
4) £75 − £15.50
5) £34.91 − £14.80
6) £567.10 − £25.45
7) £45.75 − £15.80
8) £30.50 − £29.75
9) £55.73 − £35.25
10) £40 − £6.50

EXAMPLE

Multiply £2.51 by 3.

```
  2.51
   3 ×
 7.53
```

so that £2.51 × 3 = £7.53.

EXAMPLE

Divide £6.75 by 9.

```
    0.75
9)6.75
```

or change to p first

£6.75 = 675 p

```
     75
9)675  = 75 p
```

so that £6.75 ÷ 9 = 75 p.

EXERCISE 15

Multiply

1) £4.50 × 2
2) £3.30 × 3
3) £5.72 × 4
4) £1.05 × 5
5) £16.60 × 5
6) £34.10 × 2
7) £10.05 × 3
8) £15 × 7
9) £135.50 × 8
10) £0.45 × 9
11) £40 × 7
12) £56.35 × 3
13) £25 × 5
14) £35.50 × 7
15) £0.50 × 8
16) £1.50 × 9

EXERCISE 16

Divide

1) £3.36 ÷ 3
2) £4.80 ÷ 4
3) £14.21 ÷ 7
4) £15.60 ÷ 5
5) £7.20 ÷ 6
6) £10.05 ÷ 3
7) £7.10 ÷ 2
8) £1.35 ÷ 5
9) £8.32 ÷ 8
10) £10.50 ÷ 7
11) £2.20 ÷ 4
12) £10.20 ÷ 5
13) £5.01 ÷ 3
14) £9.04 ÷ 8
15) £14 ÷ 4
16) £13.50 ÷ 2
17) £31 ÷ 2
18) £35.70 ÷ 7
19) £25.55 ÷ 5
20) £2.14 ÷ 2

ADDITION AND SUBTRACTION OF METRIC QUANTITIES

EXAMPLE

Add 250 g, 700 g and 325 g giving your answer in kg.

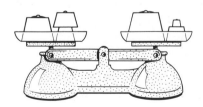

```
 250
 700
 325 +
1275
```
1275 g = 1.275 kg

EXAMPLE

Subtract 2.8 m from 3.0 m giving your answer in cm.

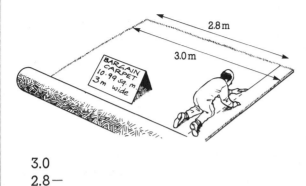

$$
\begin{array}{r}
3.0 \\
2.8 - \\
\hline
0.2
\end{array}
\qquad 0.2\,\text{m} = 20\,\text{cm}
$$

EXERCISE 17

1) Add 500 g, 950 g and 750 g giving your answer in kg.
2) Add 200 m, 500 m and 350 m giving your answer in km.
3) Add 450 mℓ, 500 mℓ and 750 mℓ giving your answer in litres.
4) Add 3.2 kg, 4.5 kg and 5 kg giving your answer in kg.
5) Add 50 cm, 20 cm and 140 cm giving your answer in metres.
6) Subtract 500 g from 1750 g giving your answer in kg.
7) Subtract 7.5 km from 8 km giving your answer in metres.
8) Subtract 750 cℓ from 1500 cℓ giving your answer in litres.
9) Subtract 8.25 kg from 9.50 kg giving your answer in grams.
10) Subtract 25 mm from 75 mm giving your answer in cm.

In the next exercise make the units the *same* before adding or subtracting.

EXAMPLE

Add 2.5 kg, 500 g and 0.75 kg.

$$2.5\,\text{kg} = 2500\,\text{g} \quad (\text{change to g})$$

$$0.75\,\text{kg} = 750\,\text{g} \quad (\text{change to g})$$

$$
\begin{array}{r}
2500 \\
500 \\
750 + \\
\hline
3750
\end{array}
\qquad 3750\,\text{g}
$$

EXERCISE 18

1) 1.5 kg + 500 g + 0.25 kg (answer in grams).
2) 20 cm + 1 m + 5.5 m (answer in cm).
3) 50 g + 2 kg + 1.5 kg (answer in grams).
4) 15 mm + 2 cm + 0.3 cm (answer in mm).
5) Add 5 km, 200 m and 1500 m giving your answer in km.
6) Subtract 400 m from 1 km giving your answer in metres.
7) Subtract 550 g from 4 kg giving your answer in grams.
8) 7 m − 600 cm (answer in cm).
9) 1.5 kg − 750 g (answer in grams).
10) 5.5 cm − 25 mm (answer in mm).

MULTIPLICATION AND DIVISION OF METRIC QUANTITIES

EXAMPLE

Multiply 250 g by 8 giving your answer in kg.

$$
\begin{array}{r}
250 \\
8 \times \\
\hline
2000
\end{array}
\qquad 2000\,\text{g} = 2\,\text{kg}
$$

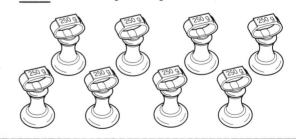

EXAMPLE

Multiply 2.5 km by 4 giving your answer in m.

$$\begin{array}{r} 2.5 \\ 4\times \\ \hline 10.0 \end{array}$$ 10.0 km = 10 000 m

or change to metres first

2.5 km = 2500 m

2500 × 4 = 10 000 m

EXAMPLE

Divide 750 km by 3.

750 ÷ 3 = 250   250 km

EXAMPLE

Divide 5 kg by 4 giving your answer in g.

$$\begin{array}{r} 1.25 \\ 4\overline{)5.00} \end{array}$$ (take care to line up the figures and the decimal point)

1.25 kg = 1250 g

or change to grams first

5 kg = 5000 g

5000 ÷ 4 = 1250 g

—————— **EXERCISE 19** ——————

1) Multiply 200 g by 4 giving your answer in grams.
2) Multiply 500 m by 2 giving your answer in km.
3) Multiply 25 mm by 6 leaving your answer in mm.
4) Multiply 3.5 kg by 5 leaving your answer in kg.
5) Multiply 50 mℓ by 10 giving your answer in litres.
6) Divide 450 km by 3 leaving your answer in km.
7) Divide 6 kg by 4 giving your answer in g.
8) Divide 750 mm by 5 giving your answer in cm.
9) Divide 28 kg by 7 leaving your answer in kg.
10) Divide 3.5 m by 7 giving your answer in cm.
11) 250 g × 3 (answer in g).
12) 450 g ÷ 9 (answer in g).
13) 700 mℓ × 5 (answer in litres).
14) 5 km ÷ 10 (answer in m).
15) 1500 m × 4 (answer in km).
16) 56 mm ÷ 8 (answer in mm).
17) 3.5 kg × 3 (answer in kg).
18) 7.2 m ÷ 9 (answer in cm).
19) 350 cm × 8 (answer in m).
20) 1.8 litres ÷ 6 (answer in cℓ).

—————————————————————

PRACTICAL PROBLEMS USING METRIC AND IMPERIAL —————— MEASUREMENT ——————

EXAMPLE

I bought 2 loaves at 42 p each, $\frac{1}{2}$ kg butter at 65 p per 250 g and $\frac{1}{4}$ lb cheese at £1.20 per pound. How much did I pay?

2 loaves
@ 42 p each = 2 × 42 p =   84 p

500 g butter
@ 65 p per 250 g = 2 × 65 p =   130 p

$\frac{1}{4}$ lb cheese

    @ £1.20 a lb = $\frac{1}{4}$ × £1.20 =   30 p

              I paid   £2.44

EXAMPLE

A man receives "time and a half" for overtime on Sundays and "time and a third" for overtime during the week. If his basic wage is £6 per hour how much per hour does he get for overtime (a) on Sunday (b) on Friday?

For Sunday he gets

$£6 + \frac{1}{2} × £6 = £6 + £3 = £9$ per hour

For Friday he gets

$£6 + \frac{1}{3} × £6 = £6 + £2 = £8$ per hour

EXAMPLE

An electricity bill was £65 for 1 quarter (which is 13 weeks). How much was it per week?

    £65 ÷ 13 = £5

It was £5 per week.

EXAMPLE

Time in New York is 5 hours behind Greenwich Mean Time (GMT). If it is 1300 hours in New York what time is it according to Greenwich Mean Time?

    1300 hours + 0500 hours = 1800 hours

It is 1800 hours or 6 p.m. (GMT)

             EXERCISE 20 ————

1) Find the cost of 8 tubes of sweets at 7p each.
2) How many days in (a) 2 weeks (b) 5 weeks (c) 7 weeks (d) 9 weeks?
3) Three people share the rent for a flat. It is £120 per week. If they all pay an equal share how much does each person pay?
4) I buy 7 yards of material at £2.50 per yard. How much does this cost?

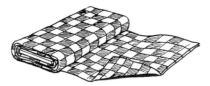

5) Fudge costs 39 p for 2 oz. How much do 10 oz cost?
6) What is the cost of 3 loaves of bread at 42 p per loaf?

7) 1 dozen = 12. What is the cost of 2 dozen eggs at 62 p for half a dozen?
8) Taking 4 weeks in 1 month how many weeks are there in (a) 2 months (b) 5 months (c) 7 months?
9) A man receives "time and a quarter" for overtime on weekdays. If his basic wage is £6.40 per hour how much does he get for overtime on Wednesdays?

84

10) An electricity bill was £52 in 1 quarter (13 weeks). How much was it per week?

11) Three people share the rent for a flat. One pays $\frac{1}{2}$ and a second pays $\frac{1}{3}$ of the total weekly rent of £60. The third pays the rest. How much do each of the three people pay?

12) Find the cost of 3 lb of carrots at 20 p per lb, 4 lb eating apples at 40 p per lb, 1 lb cooking apples at 52 p per lb and 2 lb potatoes at 35 p for a 2 lb bag.

13) Rolls cost 16 p each. Find the cost of 1 dozen.

14) Some dress material costs £5.50 for a metre length. Find the cost of 3 metres of the material.

15) How much is $3\frac{1}{2}$ yards of material at £3.50 per yard?

16) Find the cost of 1.6 kg cheese at £2.40 per kg.

17) Time in New York is 5 hours behind Greenwich Mean Time. Find the time (GMT) when New York time is 1700 hours.

18) When milk cost 20 p per pint I had 2 pints a day. How much did I pay per week?

19) Find the cost of $2\frac{1}{2}$ lb of beans at 84 p per lb, 10 lb potatoes at 18 p per lb, $3\frac{1}{4}$ lb cabbage at 40 p per lb and 7 oranges at 22 p each.

20) 100 bottles each holding $\frac{2}{5}$ litre are filled from a container holding 150 litres. How much is left in the container?

21) A radio programme of songs, poetry and stories is 1 h 30 mins long. $\frac{3}{5}$ of the time is devoted to songs, $\frac{1}{4}$ to poetry and the rest to stories. How much time is spent on stories?

6

Percentages

% = per cent

per cent = out of 100

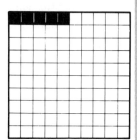

5 out of 100 squares have been shaded.

5% of the squares have been shaded.

A percentage is another way of writing a fraction (or a decimal):

$$10\% = \frac{1}{10} = 0.1$$

$$25\% = \frac{1}{4} = 0.25$$

$$33\frac{1}{3}\% = \frac{1}{3} = 0.3\dot{3}$$

$$50\% = \frac{1}{2} = 0.5$$

$$66\frac{2}{3}\% = \frac{2}{3} = 0.6\dot{6}$$

$$75\% = \frac{3}{4} = 0.75$$

These should be learnt by heart.

An increase on wages of 10% means an increase of $\frac{1}{10}$ (for example, £20 wage rises to £22).

A 25% surcharge means paying $\frac{1}{4}$ extra (for example, a £4 bill becomes £5).

A 50% discount means the price is reduced by $\frac{1}{2}$ (for example, a £15 article is sold for £7.50).

To find 10% just divide by 10.

EXAMPLE

Find 10% of £75.

   £75 ÷ 10 = £7.50

so that 10% of £75 = £7.50.

Most calculators possess a percentage key which is used in the way shown in the following example.

EXAMPLE

Find 12% of £80

Input	Display
80	80.
×	80.
12	12.
%	9.6

So 12% of £80 is £9.60.

Find 10% of
1) £30
2) 40 cm
3) 20 p
4) 60 min
5) £5.50
6) 75 km
7) 90 m
8) £3.20

CHANGING A PERCENTAGE TO A FRACTION

$$5\% = \frac{5}{100} = \frac{1}{20}$$

(dividing top and bottom by 5).

To change a percentage to a fraction divide by 100 and leave out the % sign.

Harder examples

$$12\frac{1}{2}\% = \frac{12\frac{1}{2}}{100} = \frac{25}{200} \quad \text{(doubling top and bottom)}$$
$$= \frac{1}{8}$$

$$33\frac{1}{3}\% = \frac{33\frac{1}{3}}{100} = \frac{100}{300} \quad \text{(multiplying top and bottom by 3)}$$
$$= \frac{1}{3}$$

$$66\frac{2}{3}\% = \frac{66\frac{2}{3}}{100} = \frac{200}{300} \quad \text{(multiplying top and bottom by 3)}$$
$$= \frac{2}{3}$$

Change these percentages to fractions (cancel where possible)

1) 20%		14) 4%	
2) 30%		15) 2%	
3) 40%		16) 1%	
4) 50%		17) $2\frac{1}{2}\%$	
5) 60%			
6) 70%		18) $\frac{1}{2}\%$	
7) 80%			
8) 90%		19) $37\frac{1}{2}\%$	
9) 55%			
10) 35%		20) $62\frac{1}{2}\%$	
11) 15%			
12) 75%		21) 150%	
13) 85%			

CALCULATING PERCENTAGES OF A QUANTITY

EXAMPLE

A coat has been reduced by 15% in a sale. The old price was £40. How much reduction is there?

We have to find 15% of £40:

Method 1

10% of £40 is £4

5% of £40 is £2

15% of £40 is £6 so there is a £6 reduction.

87

Method 2

$$15\% = \frac{15}{100}$$

so 15% of £40

$$= \frac{15}{100} \times £40 = £\frac{\overset{3}{15} \times 4\cancel{0}}{\underset{2}{\cancel{100}}} = £\frac{\overset{12}{6\cancel{0}}}{\underset{2}{\cancel{10}}} = £6$$

Method 1 is suitable when working in your head but Method 2 can be used for more complicated numbers.

To calculate a percentage of a quantity, multiply by the number and divide by 100.

EXERCISE 3

1) 15% of £60
2) 25% of 36 cm
3) 50% of 4 km
4) 5% of 20 p
5) 20% of 30 mm
6) 60% of 25 m
7) 40% of £50
8) $33\frac{1}{3}$% of £6
9) $12\frac{1}{2}$% of £72
10) 75% of 12 kg
11) 50% of £36.56
12) 20% of £300
13) 150% of £40
14) 38% of £1
15) $\frac{1}{2}$% of £200

SIMPLE INTEREST

If you save money with a Building Society or the Post Office, or a Bank then *interest* is paid to you for your lending them the money.

EXAMPLE

I save £300 at 10% interest p.a. with a Building Society. How much interest is paid at the end of the year?

10% of £300 is £30

£30 interest will be paid at the end of the year.

If you *borrow* money then you have to *pay* interest for the loan.

EXAMPLE

I borrow £500 at 13% interest p.a. How much do I have to pay back at the end of the year?

$$13\% \text{ of } £500 = \frac{13}{100} \times £500 = £\frac{13 \times 500}{100}$$

$$= £65$$

I have to pay back £500 + £65 = £565 at the end of the year.

EXAMPLE

Find the simple interest on £300 for 4 years at 17% per annum.

Insert	Display
300	300.
×	300.
17	17.
%	51.
×	51.
4	4.
=	204.

So the amount of simple interest is £204.

EXERCISE 4

Find the interest for 1 year on

1) £600 at 3% p.a.
2) £400 at 5% p.a.
3) £500 at 8% p.a.

4) £1000 at 11% p.a.
5) £500 at 7% p.a.
6) £650 at 9% p.a.
7) £20 at 8% p.a.
8) £75 at 10% p.a.
9) £350 at 8% p.a.
10) £45.70 at 10% p.a.
11) £525 at 10% p.a.
12) £3000 at 15% p.a.

For *simple interest* the interest is the same every year. For 2 years the interest will be doubled. For 3 years it will be multiplied by 3. For 6 months it would be multiplied by $\frac{1}{2}$, for 3 months by $\frac{1}{4}$ and so on. This situation occurs when the interest earned is taken out of the Building Society each year.

Find the simple interest on

13) £100 for 2 years at 15% p.a.
14) £50 for 3 years at 8% p.a.
15) £600 for 6 months at 10% p.a.
16) £200 for 3 months at 10% p.a.

COMPOUND INTEREST

For compound interest the savings and interest are added together at the end of the year. The following year there will be more interest (since there is more capital). This situation occurs when the interest earned is left in the Building Society.

EXAMPLE

A man saves £200 at compound interest of 10% for 3 years. How much will he have at the end of the third year?

1st year
10% interest on £200 = £20
New capital (savings + interest) = £220

2nd year
10% interest on £220 = £22
New capital (savings + interest) = £242

3rd year
10% interest on £242 = £24.20
New capital (savings + interest) = £266.20

He will have £266.20 at the end of the third year.

Find the compound interest on

1) £300 at 10% for 2 years
2) £400 at 10% for 3 years
3) £500 at 10% for 3 years
4) £600 at 10% for 2 years
5) £600 at 5% for 2 years
6) £800 at 15% for 2 years

Tables of compound interest are available for more difficult numbers (but this is outside the scope of this book).

CHANGING A FRACTION TO A PERCENTAGE

$$\frac{3}{5} = \frac{3}{5} \times 100\% \quad \left(\text{since } 100\% = \frac{100}{100} = 1\right)$$

$$= \frac{3 \times \overset{20}{\cancel{100}}}{\underset{1}{\cancel{5}}}\%$$

$$= 60\%$$

$\frac{3}{5}$ of £100 = £60

60% of £100 = £60.

To change a fraction to a percentage multiply by 100 and put in a % sign.

A calculator will convert a fraction into a percentage automatically.

EXAMPLE

Convert $\frac{7}{8}$ into a percentage.

Input	Display
7	7.
÷	7.
8	8.
%	87.5

So $\frac{7}{8}$ is equivalent to 87.5%.

--------- **EXERCISE 6** ---------

Change these fractions to percentages

1) $\frac{1}{5}$

2) $\frac{4}{5}$

3) $\frac{1}{4}$

4) $\frac{1}{2}$

5) $\frac{3}{4}$

6) $\frac{1}{3}$

7) $\frac{2}{3}$

8) $\frac{1}{8}$

9) $\frac{3}{8}$

10) $\frac{5}{8}$

11) $\frac{2}{5}$

12) $\frac{7}{8}$

13) $\frac{1}{20}$

14) $\frac{1}{40}$

15) $\frac{7}{10}$

16) $\frac{9}{10}$

17) $\frac{1}{10}$

18) $\frac{1}{100}$

19) $\frac{3}{100}$

20) $\frac{1}{50}$

Reminder: To change a fraction to a decimal divide by the number underneath:

$$\frac{3}{4} = 4\overline{)3.00}^{0.75}$$

90

To change a decimal to a fraction remember that:

the first place is $\dfrac{1}{10}$ ths

the second place is $\dfrac{1}{100}$ ths

the third place is $\dfrac{1}{1000}$ ths etc.

$$0.125 = \dfrac{1}{10} + \dfrac{2}{100} + \dfrac{5}{1000}$$

$$= \dfrac{125}{1000} \text{ and } cancel$$

$$= \dfrac{1}{8}$$

--- **EXERCISE 7** ---

Copy and complete this table

Percentage	Fraction	Decimal
		0.5
	$\frac{3}{4}$	
$33\frac{1}{3}\%$		
	$\frac{2}{3}$	
		0.125
25%		
		0.1
20%		
		0.3
$62\frac{1}{2}\%$		

EXPRESSING ONE NUMBER AS A PERCENTAGE OF ANOTHER NUMBER

To express 36 as a percentage of 50 we say that

$$36 \text{ out of } 50 = \dfrac{36}{50}$$

and then change this fraction to a percentage:

$$\dfrac{36}{50} = \dfrac{36}{\cancel{50}} \times \overset{2}{\cancel{100}}\%$$

$$= 36 \times 2\%$$

$$= 72\%$$

EXAMPLE

A girl gains 14 marks out of 20 in a test. What percentage is this?

$$14 \text{ out of } 20 = \dfrac{14}{20} = \dfrac{14}{\cancel{20}} \times \overset{5}{\cancel{100}}\%$$

$$= 14 \times 5\% = 70\%$$

--- **EXERCISE 8** ---

Express the first number as a percentage of the second

1) 10, 50
2) 15, 25
3) 15, 150
4) 20, 125
5) Express 60 as a percentage of 200.
6) Express 75 as a percentage of 150.
7) A boy gains 8 marks out of 10. What percentage is this?
8) 5 out of 30 bolts are defective. What percentage are defective?
9) 15 minutes out of an hour and a quarter TV programme is spent on commercials. What percentage is spent on commercials?
10) There are 12 women in an evening class of 30 adults. What percentage are women? What percentage are men?

11) 3 out of 20 apples are bad. What percentage of apples are bad?

12) There are 50 red tablets, 25 green tablets, 20 yellow tablets and 105 white tablets of soap in a cardboard carton. What percentage are green. What percentage are white? What percentage are coloured?

13) Two girls gain 25 and 30 marks out of 40 in a test, respectively. What percentage does the girl with the higher mark get? What percentage does the other girl obtain?

14) In a survey it is found that out of 400 swans, 350 are Mute swans, 34 are Bewick swans and the rest are Whooper swans. What percentage are Whooper swans?

15) Express $32\frac{1}{2}$ marks out of 50 as a percentage.

COMPARING PERCENTAGES, — FRACTIONS AND DECIMALS —

Since percentages and fractions are different ways of writing decimal numbers, it is possible to see which are smaller or larger by first making them all into decimals.

EXAMPLE

Write in order of size 0.375, 30%, $\frac{1}{3}$.

Changing the percentage first to a fraction and then a decimal we have:

$$30\% = \frac{30}{100} = 0.3$$

Changing the fraction to a decimal (by dividing the bottom number into the top number) we have:

$$\frac{1}{3} = 0.333 \ldots \text{ or } 0.3\dot{3}$$

Now, putting them in order of size, smallest first, we have:

$$0.3, \quad 0.3\dot{3}, \quad 0.375$$
$$\text{or} \quad 30\%, \quad \tfrac{1}{3}, \quad 0.375$$

Remember to change the percentage and fraction back to their original forms when you write the answer.

Write in order of size, smallest first

1) $\frac{1}{2}$, 60%, 0.4

2) 25%, $\frac{1}{5}$, 0.22

3) $\frac{4}{5}$, 45%, 0.54

4) 0.67, 66%, $\frac{2}{3}$

5) 0.75, $\frac{3}{4}$, 75%

6) $\frac{3}{20}$, $12\frac{1}{2}$%, 0.2

7) $\frac{1}{9}$, 9%, 0.9

8) 35%, 0.33, $\frac{3}{8}$

9) 85%, 0.875, $\frac{5}{6}$

10) 150%, 1.75, $1\frac{1}{4}$

PERCENTAGES IN EVERYDAY —————— SITUATIONS ——————

VAT

The tax collected on most goods and services is known as Value Added Tax or VAT for short. Some goods, such as food and books, are exempt (no tax is paid). VAT is added on to the end of a bill. In 1995 the VAT on fuel was 8% and on most other goods was $17\frac{1}{2}$%.

Income Tax

Income tax is collected to pay for large building projects, roads, education, health services, etc. In 1996/1997 the rates were 20% (lower) 24% (basic) and 40% (higher) rate.

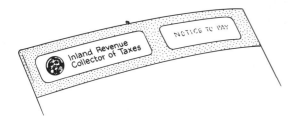

Earnings

Wage increases are often made in terms of percentages. An employee might get an 8%, 10% or 12% increase on his or her earnings. In harder times only a 2% or 3% increase might be paid.

Discounts

If a shop wishes to promote a sale then it will offer a discount of, say, 12%. 12% will be taken off the price. Often regular customers will be allowed a special discount.

Compound Interest

Banks and Building Societies offer compound interest on savings if the interest is left in the account together with the savings at the end of the year. The interest rate is in terms of a percentage.

Rate of Inflation

If goods cost £100 one year and £108 the next year we say the rate of inflation is 8%. If goods then go up by another 10% the following year we say the rate of inflation has increased, but if they only go up by another 6% we say the rate of inflation has decreased (even though the actual price has increased!).

Index Numbers

Index numbers are a way of showing how the *cost of living* rises or falls. If an index number goes up from 50 to 52 we would say it has increased by 4% (2 on 50 being equivalent to 4 on 100).

Population Statistics

These are kept to show by how much the number of people in the country (or a particular town or district) increases or decreases each year and they are usually expressed as a percentage.

Marks in Examinations

These are often given in percentages. If a student has 35 marks out of 50 this is equivalent to 70 out of a 100, or 70%.

To work out the percentage first work out the fraction:

$$\frac{35}{50}$$

and then multiply by 100%:

$$\frac{35}{50} \times 100\%^{2} = 35 \times 2\%$$
$$= 70\%$$

Tolerances and Efficiency

These are engineering terms. If a bolt must be made to a tolerance of 2% then all measurements must be within 2% of the "standard" bolts. This is often written as ±2%. (Usually the measurement must be much more accurate than this.)

The efficiency of a machine is:

$$\frac{\text{Output}}{\text{Input}} \times 100\%$$

Profit and Loss

If goods are bought for £5 and sold for £7 we say the *profit* is £2. £5 is called the *cost price* and £7 the *selling price*.

The percentage profit is:

$$\frac{\text{Profit}}{\text{Cost price}} \times 100\%$$

In this example the percentage profit is $\frac{2}{5} \times 100\% = 40\%$.

If the selling price is less than the cost price then a loss has been made and the percentage loss is given by:

$$\frac{\text{Loss}}{\text{Cost price}} \times 100\%$$

────── **EXERCISE 10** ──────

Simple problems involving percentage.

1) 8% VAT is charged on a £300 gas bill. Find the cost of the bill after VAT has been added.

2) A woman working part-time earns £6600 p.a. She is allowed £4000 tax free and pays 20% income tax on the rest. How much income tax does she pay?

3) Wages of £4.50 per hour are increased by 8%. Find the new hourly wage.

4) A shop offers 15% discount on a £120 piece of furniture. How much discount is allowed? What is the furniture sold for?

5) A credit card company loans £500 at 2% interest per month. Give the amount of interest payable at the end of the first month.

6) A $12\frac{1}{2}$% reduction is made on a £24 coat.

   How much does the coat cost?

7) Goods costing £100 one year cost £113 the following year. What is the rate of inflation?

8) A carrycot is bought for £15 and sold for £20. Give the percentage profit.

9) A car is bought for £4000 and sold for £3500. Give the percentage loss.

10) An index number is raised by 10%. The index number is 480. What will the new index number be?

11) The population of Tadington is 10 000. Find the population the following year if it is increased by 5%.

12) A boy earns 23 marks out of 50 in a test. What percentage is this?

13) Bolts must be made to a tolerance of ± 2%. If the standard measurement is 10 mm what are the least and most allowable measurements?

14) What is the efficiency of a machine with a 5 kW input and a 3 kW output?

More difficult problems involving percentage will be found in later chapters.

—————— **EXERCISE 11** ——————

1) Find 10% of (a) £40 (b) £200 (c) £36.50 (d) £18.20 (e) 400 g (f) 2.5 kg (g) 340 m (h) 20 s.

2) Write $12\frac{1}{2}$% (a) as a fraction (b) as a decimal.

3) Find $12\frac{1}{2}$% of (a) £16 (b) £32.

4) Write $33\frac{1}{3}$% (a) as a fraction (b) as a decimal.

5) Find $33\frac{1}{3}$% of (a) £6 (b) £15.

6) Change 95% to (a) a fraction (b) a decimal.

7) Find the simple interest for 1 year on (a) £200 at 3% p.a. (b) £950 at 7% p.a.

8) A man saves £150 at compound interest of 10% for 2 years. Find how much he has at the end of the second year.

9) For savings of £250 at compound interest of 10% for 3 years find (a) the *total* interest earned and (b) the amount of savings (including interest) at the end of the third year.

10) Change (a) $\frac{17}{100}$ (b) $\frac{19}{20}$ to percentages.

11) Write $\frac{3}{10}$ (a) as a decimal (b) as a percentage.

12) $17\frac{1}{2}\%$ VAT is charged on a £45 table. Find the total cost of the table including the VAT.

13) A woman earns £7000 p.a. and is allowed £2000 tax free. How much income tax does she pay if the rate of tax is 30%?

14) Find the new wage when wages of £75 per week are increased by 12%.

15) A price reduction of 17% is made on a three piece suite costing £625. How much does the three piece suite cost?

16) A credit card company makes a loan of £480 at 5% interest for 3 months. Find the amount of interest to be paid at the end of 3 months.

17) An index number is 25. What will the new index number be if it is raised by 20%?

18) The population of a town is 50 000. It is increased by 8% the following year. The year after that the population increases by 10%. What is the new population after 2 years?

19) 3 pupils earn 20, 22 and 24 marks out of 25 in a test. What percentage marks are these?

20) Screws of length 2 cm have to be made to a tolerance of ± 2%. Give the maximum and minimum permissible measurements.

21) Find the efficiency of a heater with a 10 kW input and an 8 kW output.

22) Goods are bought for £50 and sold for £55. Give the percentage profit.

23) (a) Write down the calculator buttons you would press to work out $17\frac{1}{2}\%$ of £350.
    (b) Give the answer to this calculation correct to the nearest penny.
    (c) Work out $17\frac{1}{2}\%$ of 80 p.

24) $17\frac{1}{2}\%$ VAT is charged on a £500 bill.
    (a) Find the amount of VAT charged.
    (b) What is the total cost of the bill?

25) How much VAT at 17.5% would be charged on a toaster costing £40?

26) Work out $17\frac{1}{2}\%$ of (a) £100 (b) £200 (c) £50 (d) £400

27) Complete this bill.

LABOUR	£300
PARTS	£150
VAT @ $17\frac{1}{2}\%$	_____
TOTAL COST	_____

7

Squares and Square Roots

SQUARES OF WHOLE NUMBERS

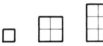 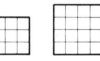

The square of 4 is $4 \times 4 = 16$. (This can be written as 4^2.)

The square of a number is the number multiplied by itself.

EXERCISE 1

Work out

1) 2^2 4) 7^2
2) 5^2 5) 8^2
3) 6^2 6) 9^2

Long multiplication is needed for the squares of larger numbers.

EXAMPLE

Work out the square of 75.

$75^2 = 75 \times 75$

```
    75
    75 ×
   ───
   375
  5250
  ────
  5625
```

Multiply by the *units* figure. Put a 0 and multiply by the *tens* figure.

EXERCISE 2

Work out

1) 13^2 7) 72^2
2) 26^2 8) 76^2
3) 35^2 9) 77^2
4) 48^2 10) 81^2
5) 57^2 11) 88^2
6) 69^2 12) 95^2

SQUARES OF FRACTIONS

EXAMPLE

Find the square of $\frac{2}{3}$.

$$\frac{2}{3} \times \frac{2}{3} = \frac{2 \times 2}{3 \times 3} = \frac{4}{9}$$

EXAMPLE

Find the square of $1\frac{4}{5}$.

$$1\frac{4}{5} \times 1\frac{4}{5} = \frac{9}{5} \times \frac{9}{5} = \frac{81}{25} = 3\frac{6}{25}$$

Mixed fractions must be made *top-heavy* before multiplying.

---------- EXERCISE 3 ----------

Find the squares of these fractions

1) $\dfrac{1}{4}$ 9) $1\dfrac{2}{3}$

2) $\dfrac{5}{7}$ 10) $1\dfrac{1}{3}$

3) $\dfrac{4}{9}$ 11) $2\dfrac{1}{2}$

4) $\dfrac{2}{5}$ 12) $2\dfrac{1}{4}$

5) $\dfrac{5}{8}$ 13) $4\dfrac{1}{2}$

6) $\dfrac{3}{7}$ 14) $5\dfrac{1}{2}$

7) $\dfrac{4}{11}$ 15) $3\dfrac{1}{3}$

8) $\dfrac{7}{12}$

—— SQUARES OF DECIMALS ——

EXAMPLE

Find the square of 7.5.

First multiply as if there were no decimal point:

$$\begin{array}{r} 75 \\ 75\times \\ \hline 375 \\ 5250 \\ \hline 5625 \end{array}$$ then

To find where to put the decimal point in the answer:

Count the number of figures after the point in the first number. Add this to the number of figures after the point in the second number. Count back this number of places from the *right* and put in the point.

7.5 one figure after the point

7.5 one figure after the point

56.25 count back two figures and put in the point

---------- EXERCISE 4 ----------

Work out

1) $(1.5)^2$ 6) $(5.5)^2$
2) $(3.5)^2$ 7) $(5.8)^2$
3) $(4.2)^2$ 8) $(6.3)^2$
4) $(4.9)^2$ 9) $(0.5)^2$
5) $(5.1)^2$ 10) $(0.7)^2$

The squares of numbers from 1 to 12 should be learnt by heart:

1 4 9 16 25 36 49 64 81 100 121 144

SQUARE ROOTS OF WHOLE —— NUMBERS ——

The *square* of 6 is $6 \times 6 = 36$.

The *square root* of 36 is 6

$$\sqrt{36} = 6$$

$\sqrt{\ }$ is the square root sign and it means "the square root of".

We read $\sqrt{25}$ as "the square root of 25" and it has a value of 5.

---------- EXERCISE 5 ----------

Write down the square roots of these numbers

1) 64   5) 16
2) 49   6) 25
3) 81   7) 100
4) 9   8) 36

SQUARE ROOTS OF FRACTIONS

To find the square root of a fraction we find the square roots of the two numbers separately:

EXAMPLE

Find $\sqrt{\dfrac{9}{16}}$.

$$\sqrt{\frac{9}{16}} = \frac{\sqrt{9}}{\sqrt{16}} = \frac{3}{4}$$

--- **EXERCISE 6** ---

Find

1) $\sqrt{\dfrac{9}{25}}$

2) $\sqrt{\dfrac{16}{49}}$

3) $\sqrt{\dfrac{49}{64}}$

4) $\sqrt{\dfrac{36}{121}}$

5) $\sqrt{\dfrac{49}{100}}$

6) $\sqrt{\dfrac{25}{36}}$

7) $\sqrt{\dfrac{25}{81}}$

8) $\sqrt{\dfrac{9}{100}}$

9) $\sqrt{\dfrac{16}{25}}$

To find the square root of a decimal, say 50.4, we either make an estimate or use a calculator.

SQUARE ROOTS OF DECIMALS
─── BY ESTIMATION ───

EXAMPLE

Estimate $\sqrt{50.4}$ correct to 1 decimal place.

50.4 lies between the whole number squares 49 and 64. 49 is the *square* of 7 and 64 is the *square* of 8, so the $\sqrt{50.4}$ will be a decimal number which lies between 7 and 8.

Try 7.1:

7.1	*one* decimal place
7.1×	*one* decimal place
7 1	*two* decimal places in the answer
49 7 0	
50.4 1	

Estimate means "make a good guess".

So 7.1 is slightly too big.
However $7.0 \times 7.0 = 49.0$, so 7.1 is a better estimate than 7.0.

$\sqrt{50.4} = 7.1$ correct to 1 decimal place.

--- **EXERCISE 7** ---

Estimate (correct to 1 decimal place) the square roots of

1)	10.5	7)	83.8
2)	30.2	8)	86
3)	45.1	9)	90.3
4)	50.9	10)	35
5)	62.0	11)	29
6)	75	12)	60

Since this process is rather laborious a calculator will usually prove more convenient. It is useful to estimate first so you know your calculator answer is right.

USING A CALCULATOR FOR
- SQUARES AND SQUARE ROOTS -

EXAMPLE

Use a calculator to find the square of 18.7.

Input	Display
18.7	18.7
×	18.7
=	349.69

So $(18.7)^2 = 349.69$.

EXAMPLE

Use a calculator to find the square root of 93.21 correct to 3 decimal places.

Input	Display
93.21	93.21
$\sqrt{}$	9.6545326

So

So $\sqrt{93.21} = 9.655$ correct to 3 decimal places.

EXERCISE 8

Use a calculator for this exercise.

Find the square of

1) 17   2) 255   3) 650
4) 0.25   5) 2.56   6) 21.8.

Find the square root (correct to 2 decimal places) of

7) 1.4   8) 14.4   9) 14
10) 895   11) 89.5   12) 0.895.

Work out:

13) $(1.2)^2$ 14) $(24)^2$ 15) $(42)^2$
16) $\sqrt{0.64}$ 17) $\sqrt{0.01}$ 18) $\sqrt{0.16}$

EXERCISE 9
---

1) Work out (a) 3^2 (b) 10^2.
2) Write down the squares of whole numbers from 7 to 12.
3) Work out (a) 14^2 (b) 20^2 (c) 150^2.
4) Find (a) $\left(\frac{1}{2}\right)^2$ (b) $\left(\frac{3}{5}\right)^2$ (c) $\left(1\frac{1}{2}\right)^2$.
5) Find (a) $(0.2)^2$ (b) $(1.1)^2$ (c) $(0.04)^2$.
6) In these examples you must square before adding or subtracting. Find (a) $3^2 + 4^2$ (b) $11^2 - 9^2$.
7) Write down the square roots of (a) 121 (b) 144.
8) Estimate, correct to one decimal place the square root of (a) 12 (b) 28.
9) Find (a) $\sqrt{\frac{4}{9}}$ (b) $\sqrt{\frac{25}{64}}$.
10) In these examples you must add or subtract before finding the square root. Find (a) $\sqrt{9+16}$ (b) $\sqrt{25-9}$.
11) Use a calculator to work out (a) 54^2 (b) $\sqrt{0.54}$ correct to 2 d.p.
12) Using a calculator find correct to three decimal places

    (a) 0.568^2 (e) $\sqrt{12}$
    (b) 1.7^2 (f) $\sqrt{24}$
    (c) 3.56^2 (g) $\sqrt{5.08}$
    (d) 0.78^2 (h) $\frac{1}{3}\sqrt{81}$.

13) Write down what appears in the display of a scientific calculator as the answer to
    (a) $(200\,000)^2$ (b) $(0.000\,02)^2$.

Note: *When using a calculator to find the square of a very large number or the square of a very small number, the result may appear in the display in standard form (see pp. 103–6).*

8

— Approximation and Estimation —

If you ask a builder to put an extension on your house then you would expect him to give you an estimate of the final cost. (If he tells you exactly how much the work will cost this is called a "quotation".) An estimate may be a little more or a little less than the actual cost.

To make his estimate the builder would have to work out roughly what he would charge for building materials, electrical work, plumbing, labour, etc.

```
Building
  materials  . £2000
Plumbing
  work  . . . .  £400
Electrical
  work  . . . .  £200
Labour . . . . £2400
Final          _____
  estimate . . .£5000
```

His costs would only be approximate (i.e. not exactly correct but nearly correct). The costs might be rounded to the nearest £100, £50 or the nearest £10.

ROUNDING

£298 rounded to the nearest £100 would be £300

£445 rounded to the nearest £50 would be £450

£179 rounded to the nearest £10 would be £180

Round these numbers to the nearest £100
1) £598
2) £602
3) £730
4) £890
5) £1520
6) £23 890

Round these numbers to the nearest £50
7) £45
8) £135
9) £190
10) £240
11) £780
12) £15 595

Round these numbers to the nearest £10
13) £82
14) £176
15) £181
16) £292.50
17) £475.85

Round these numbers to the nearest £1
18) £45.69
19) £72.85
20) £31.10
21) £173.55
22) £273.99

ROUGH CHECKS FOR CALCULATIONS

Always make sure your answers are sensible!

This can be achieved by rounding the numbers and doing a rough check.

EXAMPLE

How much money would I need to be able to buy 19 small presents each costing 98 p?

Rough estimate:

$20 \times 100\,p$

$= 2000\,p$

$= £20$

Accurate calculation:

```
    19
    98 ×
   ‾‾‾‾
   152
  1710
  ‾‾‾‾
  1862
```

$1862\,p = £18.42$

I would need £18.62.

EXAMPLE

What is the weight of 28 nails at 1.9 g each?

Rough estimate:

$30 \times 2\,g$

$= 60\,g$

Accurate calculation:

```
    28
   1.9 ×
   ‾‾‾‾
   252
   280
   ‾‾‾‾
  53.2
```

The weight of the 28 nails is 53.2 g.

—————————— **EXERCISE 2** ——————————

First make a rough estimate and then perform these calculations accurately.

1) Find the total length of 9 planks of wood each of length 1.9 metres.

2) Calculate the cost of 21 notebooks at 9 p each.

3) What is the weight of 39 books each weighing 102 g?

4) A length of rope 801 cm long is cut into 9 equal pieces. Find the length of each piece.

5) I pay for 11 items each costing 9 p with a £10 note. Find the change I should receive.

6) I buy goods at £11.99, £4.02, £6.03 and £2.98. What is the total bill?

7) 399 sweets are divided equally between nine children. How many sweets does each child get and how many are left over?

8) How heavy are 401 books each weighing 2.1 kg?

9) If the diameter of a coin is 2.9 cm, how far would 99 coins stretch if laid end to end?

10) The weight of 11 boxes is 319 g. How much does each box weigh?

————————————————————————————————

EXAMPLE

$$\frac{20.1 \times 1.9}{0.9}$$

Rough estimate:

$$\frac{20 \times 2}{1}$$

$= 40$

Accurate calculation:

```
   20.1
    1.9 ×
   ‾‾‾‾‾
   1809
   2010
   ‾‾‾‾‾
  38.19
```

$38.19 \div 0.9 = 381.9 \div 9$

```
     42.433 ...
   9)381.900
```

Accurate answer 42.4$\dot{3}$ or 42.4 to one decimal place.

102

First perform a rough estimate and then an accurate calculation

1) $\dfrac{11 \times 9}{3}$

2) 149×9

3) $99 \div 1.1$

4) $64.8 \div 9$

5) 10.5×0.9

6) $\dfrac{30.1 \times 1.9}{0.7}$

7) $\dfrac{11 \times 19}{9}$

8) $99 \div 11 + 11 \times 9$

9) 2.1×1.1

10) 7.01×3.02

11) $11 \times 9\dfrac{1}{2}$

12) 4.05×1.1

INDICES

$2 \times 2 \times 2 \times 2$ may be written as 2^4.

This is known as "index form". The small raised number, 4, is called the *index*. Another word that may be used instead of index is *power*.

We read 2^4 as "2 to the power 4" or "2 to the fourth".

Brackets may be used for clarity so 2^4 and $(2)^4$ are the same.

Numbers to the power 2 are known as *square* numbers so 8^2 is called 8 *squared*. Numbers to the power 3 are *cubic* numbers so 7^3 is called 7 *cubed* or the cube of 7.

EXAMPLE

(a) Write $9 \times 9 \times 9$ in index form.
(b) Write out $(8)^4$ as a product.
(c) Evaluate $(7)^3$.

(a) $9 \times 9 \times 9 = 9^3$
(b) $(8)^4 = 8 \times 8 \times 8 \times 8$
(c) $(7)^3 = 7 \times 7 \times 7 = 49 \times 7 = 343$

Notice that "evaluate" means "work out to a single number".

Similarly $3^5 = 3 \times 3 \times 3 \times 3 \times 3$, with 3 being multiplied by itself 5 times and $2^7 = 2 \times 2 \times 2 \times 2 \times 2 \times 2 \times 2$, and so on.

EXAMPLE

Using a calculator find the value of 5^4.

Input	Display
5	5.
× ×	5.
=	25.
=	125.
=	625.

So $5^4 = 625$. Note that the number of equals signs is one less than the power to which the number is raised.

Write in index form

1) 3×3
2) $4 \times 4 \times 4$
3) $6 \times 6 \times 6 \times 6$
4) $5 \times 5 \times 5 \times 5 \times 5 \times 5$
5) 10×10
6) $9 \times 9 \times 9 \times 9 \times 9 \times 9 \times 9 \times 9 \times 9 \times 9$
7) $11 \times 11 \times 11$

Write out as a product

8) $(3)^4$
9) 7^2
10) 8^3
11) 10^6
12) 12^3

Evaluate

13) 2^3
14) 5^4
15) 6^3
16) 3^4
17) 8^3

18) Add together 4 squared and 2 cubed.
19) What is twice the cube of 3?
20) Subtract the square of 4 from the cube of 3.

WAYS OF WRITING VERY LARGE AND VERY SMALL NUMBERS

In atomic physics measurements may be very small whereas in astronomy measurements may be very large. It is inconvenient to write very large or very small numbers out in full.

The short way of writing these numbers is known as *standard index form* (or *standard form*). Scientific calculators are able to store standard form numbers and they do this by storing two numbers: one that gives the figures and the other that shows where the decimal point should go.

To understand standard form you need to know the meaning of 10^0, 10^1, 10^2, 10^3, etc., and also 10^{-1}, 10^{-2}, etc.

$$10^3 = 10 \times 10 \times 10 = 1000$$

$$10^2 = 10 \times 10 = 100$$

$$10^1 = 10$$

$$10^0 = 1$$

$$10^{-1} = 0.1$$

$$10^{-2} = 0.01$$

etc.

so that $10^5 = 10 \times 10 \times 10 \times 10 \times 10 = 100\,000$

and $10^{-5} = 0.000\,01$

EXERCISE 5

Convert from index form to single numbers

1) 10^2	6) 10^1
2) 10^0	7) 10^{-2}
3) 10^{-1}	8) 10^{-3}
4) 10^3	9) 10^{-4}
5) 10^6	10) 10^4

Standard Form Numbers

The first number has a value between 1 and 10. (It is either a single figure or a decimal point follows the first figure.)

It is followed by "$\times 10^3$" or "$\times 10^{-2}$", etc., to show the *size* of the number. The small raised number is called *the index*.

Here are some examples of standard form numbers:

	Calculator display
2.56×10^2	2.56 02
3.7×10^3	3.7 03
5×10^{-1}	5. -01
2.7×10^{-2}	2.7 -02
3.56×10^0	3.56 00
5.3×10^1	5.3 01

The index is stored in the calculator by means of a special key, usually labelled EXP or EE

EXAMPLE

Write 256 in standard form.

$$256 = 2.56 \times 100 = 2.56 \times 10^2$$

EXAMPLE

Write 3700 in standard form.

$$3700 = 3.7 \times 1000 = 3.7 \times 10^3$$

EXERCISE 6

Write these numbers in standard form

1) 248	6) 4100
2) 6800	7) 5200
3) 781	8) 343
4) 563	9) 230
5) 7800	10) 4560

EXAMPLE

Write 0.5 in standard form.

$0.5 = 5 \times 0.1 = 5 \times 10^{-1}$

EXAMPLE

Write 0.34 in standard form.

$0.34 = 3.4 \times 0.1 = 3.4 \times 10^{-1}$

EXAMPLE

Write 0.06 in standard form.

$0.06 = 6 \times 0.01 = 6 \times 10^{-2}$

EXAMPLE

Write 0.027 in standard form.

$0.027 = 2.7 \times 0.01 = 2.7 \times 10^{-2}$

—————— **EXERCISE 7** ——————

Write these numbers in standard form

1) 0.6	7) 0.054
2) 0.2	8) 0.061
3) 0.09	9) 0.1
4) 0.05	10) 0.087
5) 0.45	11) 0.08
6) 0.56	12) 0.3

The two numbers which are stored in a calculator are, of course, the first number (between 1 and 10) and the index. If the index is large then the number is large and if the index is low (negative) then the number is small:

5.6×10^9 is a large number

3.28×10^{-8} is a small number

Quick Rule for Finding the Index

Count the number of places you need to move the decimal point until it is just after the first figure. (If you move to the *right* the index will be negative.)

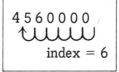

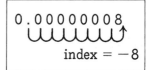

So from the diagrams above:

$3638 = 3.638 \times 10^3$

$0.0075 = 7.5 \times 10^{-3}$

$853.2 = 8.532 \times 10^2$

$4\,560\,000 = 4.56 \times 10^6$

$0.000\,000\,08 = 8 \times 10^{-8}$

—————— **EXERCISE 8** ——————

Write these numbers in standard form

1) 58 000	8) 0.000 452
2) 3400	9) 567 000 000
3) 156 000	10) 0.000 000 59
4) 63	11) 497 000
5) 597	12) 0.000 35
6) 0.0078	13) 0.000 026
7) 0.0543	14) 635 124 000

It is important to be able to work out *the value* of standard form numbers.

EXAMPLES

Evaluate (a) 5.2×10^4 (b) 6.11×10^{-3}.

$$5.2 \times 10^4 = 5.2 \times 10\,000$$
$$= 52\,000$$
$$6.11 \times 10^{-3} = 6.11 \times 0.001$$
$$= 0.006\,11$$

--- **EXERCISE 9** ---

Evaluate these numbers

1) 5.3×10^3
2) 3.4×10^2
3) 6.9×10^5
4) 9.2×10^{-1}
5) 9.43×10^{-1}
6) 3.11×10^{-2}

7) 4.7×10^0
8) 5.66×10^1
9) 4.505×10^{-1}
10) 3.2×10^6
11) 1.4×10^{-2}
12) 8×10^{-3}

--- ## SIGNIFICANT FIGURES ---

Sometimes a number has far too many figures for practical use. This is overcome by putting the number "correct to 3 significant figures" (or "correct to 2 significant figures" or "correct to 4 significant figures" or however many figures are required).

For example

$70.196\,754\,02$

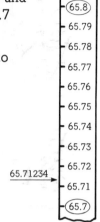

is roughly equal to 70.1 but if you look at the scale it is closer to 70.2.

We say it is equal to 70.2 correct to 3 significant figures.

"Significant figures" is generally shortened to "sig.fig." or to "S.F.".

EXAMPLE

Put 65.712 34 correct to 3 S.F.

The first three figures are 6, 5 and 7. The number is closer to 65.7 than 65.8, so:

$65.712\,34 = 65.7$ correct to 3 S.F.

(Since all figures after the fourth are ignored, the following examples only have 4 figures to start with.)

EXAMPLE

Put (a) 3.268 (b) 3.265 (c) 3.264 correct to 3 S.F.

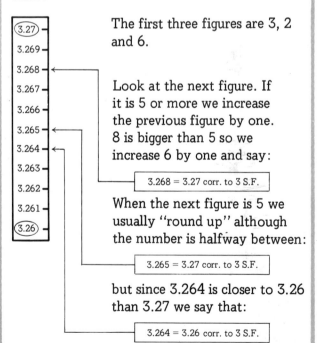

The first three figures are 3, 2 and 6.

Look at the next figure. If it is 5 or more we increase the previous figure by one. 8 is bigger than 5 so we increase 6 by one and say:

$3.268 = 3.27$ corr. to 3 S.F.

When the next figure is 5 we usually "round up" although the number is halfway between:

$3.265 = 3.27$ corr. to 3 S.F.

but since 3.264 is closer to 3.26 than 3.27 we say that:

$3.264 = 3.26$ corr. to 3 S.F.

Put these numbers correct to 3 S.F.

1) 3.261
2) 3.269
3) 3.263
4) 3.262
5) 3.266
6) 3.260
7) 3.267
8) 62.89
9) 62.83
10) 62.87
11) 62.86
12) 62.85
13) 62.81
14) 62.84
15) 62.88
16) 62.82

Scale:
62.9
62.89
62.88
62.87
62.86
62.85
62.84
62.83
62.82
62.81
62.8

EXAMPLE

Put 4571 correct to 3 S.F. Count three figures and then look at the figure in the next place:

4   5   7   1
① ② ③   ↑

It is a 1 so we leave the figure before alone.

It may be tempting to say 4571 = 457 correct to 3 S.F. but this is quite wrong as you can see if you look at the scale. It is obvious then that 4571 is nowhere near 457!

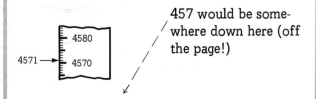

457 would be somewhere down here (off the page!)

The correct answer is 4571 = 4570 correct to 3 S.F. We have to put in a nought. It is called a "trailing zero".

Put these numbers correct to 3 S.F. (Don't forget the trailing zeros – sometimes there will be more than one.)

1) 1351
2) 6732
3) 5674
4) 6789
5) 7128
6) 24 561
7) 145 881
8) 3009
9) 4028
10) 5421
11) 5563
12) 4325
13) 6072
14) 4516
15) 3399
16) 1199
17) 4599
18) 23 410
19) 567 889
20) 4561

Further examples:

12 151 = 12 200 correct to 3 S.F.

5009 = 5010 correct to 3 S.F.

6699 = 6700 correct to 3 S.F.

Reminder: *Putting a number correct to 3 significant figures is simply a way of getting a rough value of the number which is good enough for most purposes.*

When the number has a value less than 1 we start counting at the first non-zero figure.

EXAMPLE

Put (a) 0.5121 (b) 0.5128 (c) 0.5125 correct to 3 S.F.

(a) 0 . 5 1 2 1 = 0.512 correct to 3 S.F.
    ①②③ ↑

(b) 0 . 5 1 2 8 = 0.513 correct to 3 S.F.
    ①②③ ↑

(c) 0 . 5 1 2 5 = 0.513 correct to 3 S.F.
    ①②③ ↑

EXAMPLE

Put (a) 0.008 127 (b) 0.008 025 (c) 0.008 199 correct to 3 S.F.

(a) 0.0 0 8 1 2 7 = 0.008 13 correct to 3 S.F.
       ①②③↑

(b) 0.0 0 8 0 2 5 = 0.008 03 correct to 3 S.F.
       ①②③↑

We do count the zero "inside" the number.

(c) 0.0 0 8 1 9 9 = 0.008 20 correct to 3 S.F.
       ①②③↑

When 1 is added to 819 it becomes 820.

EXERCISE 12

Put these numbers correct to 3 S.F.

1) 0.4121	9) 0.089 13
2) 0.4129	10) 0.089 19
3) 0.4125	11) 0.080 91
4) 0.4123	12) 0.080 99
5) 0.006 112	13) 0.005 199
6) 0.006 118	14) 0.005 191
7) 0.006 115	15) 0.006 099
8) 0.006 111	16) 0.006 092

EXAMPLES

3.61 = 3.6 correct to 2 S.F.

4.92 = 4.9 correct to 2 S.F.

389 = 390 correct to 2 S.F.

EXERCISE 13

Put these numbers correct to 2 S.F.

1) 5.61	7) 279
2) 3.92	8) 0.369
3) 6.15	9) 1499
4) 7.29	10) 0.512
5) 2.31	11) 63.9
6) 5.38	12) 5.81

EXERCISE 14

Round

1) £43.98 to the nearest £

2) £780 to the nearest £100

3) £6.72 to the nearest 10p

4) 1.95 m to the nearest metre

5) 485 g to the nearest 100 g

6) 3 h 55 min to the nearest hour

7) 25.41 cm to the nearest cm

8) 7950 ml to the nearest litre (1000 mℓ = 1 litre)

9) 5.7 kg to the nearest kg

10) 3959 m to the nearest km (1000 m = 1km).

First make a rough estimate and then work out accurately

11) the weight of 9 boxes each weighing 190 g

12) the total length of 2 ropes each 1.95 m long

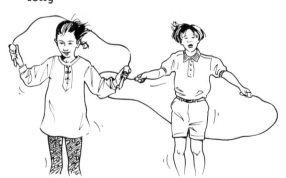

13) a shopping bill of 5 items at £3.96, £4.02, £2.10, £1.98 and £0.96.

108

Make a rough estimate and then perform these calculations accurately.

14) A length of wood is sawn into 7 equal pieces. The wood is 490 cm long. How long is each piece?

15) $\dfrac{21 \times 9}{3}$

16) 119×9

17) $810 \div 8$

18) Write out (a) 10^2 (b) 10^{-1} (c) 10^0 (d) 10^1.

19) Write these numbers in standard form
(a) 372 (b) 45 (c) 0.65
(d) 0.09.

20) Write (a) 57 000 (b) 0.000 05 in standard form.

21) Write these numbers correct to 3 significant figures
(a) 2.561 (b) 0.6592 (c) 8.079
(d) 5723.

22) Write these numbers correct to 2 significant figures
(a) 0.125 (b) 7280 (c) 0.0366
(d) 4.5152.

ROUNDING THE NUMBERS
── ON A CALCULATOR ──

Using a calculator takes all the hard work out of calculations but new problems arise. Errors can be made keying in, so *check that the answer is sensible:* make a rough estimate; key in twice.

If there are too many digits in the answer then it should be rounded. It could be rounded to the nearest whole number or to two decimal places or to three significant figures, etc. Which you choose will depend on the problem you are solving. If the numbers in a problem are all given to three significant figures then it would be sensible to round the answer to three significant figures.

Sometimes rounding errors occur. For example on some calculators you might input

$$\boxed{7} \; \boxed{\div} \; \boxed{3} \; \boxed{\times} \; \boxed{3} \; \boxed{=} \text{ and get}$$

6.9999999 in the display. A rounding error will have occurred and 6.9999999 should be taken to mean 7.

If you are working out a problem with a calculator don't round the numbers until the end. If you round too soon then all the errors reinforce each other and the answer can be wrong by quite a significant amount.

────────── **EXERCISE 15** ──────────

Use a calculator to solve these problems. Show all your working.

1) A bill of £15.17 is to be split between 7 people. Discuss what happens when 15.17 is divided by 7 using a calculator. What does the answer mean? How much should each person pay? How could it be made fair?

2) Say why $\dfrac{45}{3 \times 5}$ cannot be work out by

pressing $\boxed{4} \; \boxed{5} \; \boxed{\div} \; \boxed{3} \; \boxed{\times} \; \boxed{5}$

$\boxed{=}$. What keys would you press to work this out correctly?

3) The area of a triangle is found by working out a half the length of the base times the height. If the length of the base is 12.7 cm and the height is 8.5 cm use a calculator to work out the area of the triangle correct to three significant figures.

4) (a) Add £4.50, 65 p and £4.25, giving your answer correct to the nearest pound.

(b) Add £4.50, 65 p, £4.25 and 40 p giving your answer correct to the nearest pound. Why is your answer wrong if you just add 40 p to the answer of part (a)?

5) The number in a calculator display is 3.141592654. Write this
(a) to the nearest whole number
(b) to two decimal places
(c) to four significant figures.

6) Find correct to the nearest 1 p
(a) 8% of £55   (b) 6% of £42.15
(c) 9% of 90 p.

7) Evaluate $\sqrt{(43.7)^2 + (59.1)^2}$ correct to three significant figures.

8) $1\frac{1}{2}$ metres are cut from a length of cloth that is 4.3 metres long.

Write the length that remains as a decimal. How many whole metres are left? Why would it not make sense to round the number of metres left to the nearest whole metre?

9) Find (a) $(0.000\,04)^2$ (b) $(400\,000)^2$ using a calculator, giving your answers in standard form.

(c) Draw diagrams to show the answers in the display of your calculator.

(d) How are the answers to (a) and (b) different from those of (c)?

10) Find $\dfrac{2.46}{1.08 \times 1.15}$ correct to 3 S.F.

11) Experiment to find the largest and smallest numbers your calculator will accept.

Typing $\boxed{1}$ $\boxed{\div}$ $\boxed{0}$ $\boxed{=}$ gives an error.

Find three more instructions that give an error and try to explain why.

12) Write $19.999\,999\,99$ correct to the nearest whole number.

9

Ratio and Proportion

Concrete is made by mixing sand and cement in the ratio 3:1 (three to one) and then mixing this with water. You could put three *bags* of sand to one *bag* of cement:

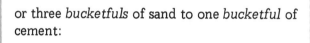

or three *bucketfuls* of sand to one *bucketful* of cement:

or three *cups* of sand to one *cup* of cement:

These ratios are all 3:1.

As long as sand and cement are kept "in the same proportion", when mixed with water they will make concrete. The actual amounts of sand and cement only affect the total *amount* of concrete made.

Jam is made by mixing fruit to sugar in a given ratio according to a recipe:

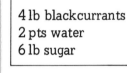

Recipe
4 lb blackcurrants
2 pts water
6 lb sugar

To make *twice* as much jam you would *double* the amounts given in the recipe. To make *half* the amount of jam you would *halve* the amounts in the recipe. The recipe gives the proportion of fruit to water to sugar. Here three quantities are involved, fruit, water and sugar.

Model trains are usually made to a "scale of 1:72". Every measurement on the model is 1/72th of the real measurement. They are in a ratio of 1:72.

The ratio shows how many times bigger one quantity is than another. The real measurements are 72 times bigger.

Sometimes problems are made easier by first putting ratios into "simpler terms". These ratios are already in their *simplest terms*:

3:1   5:7   10:9   1:100   500:3

The numbers in the ratios are all *whole numbers*. *No* number divides exactly into both sides of the ratio.

These ratios are *not* in their simplest terms:

6:4   15:3   100:50   40 cm:50 cm

1 m:1 cm $1\frac{1}{4}:\frac{1}{3}$

2 will divide into both 6 and 4. We can simplify the ratio 6:4 by dividing both sides by 2:

6:4 is the same ratio as 3:2

In the second example, 3 divides into both 15 and 3:

15:3 is the same ratio as 5:1

(If you put 15 cups to 3 cups you would get the same "mix" as if you put 5 cups to 1 cup.)

In the third example, 50 divides into both 100 and 50 (50 is a "common factor"):

100:50 is the same ratio as 2:1

To simplify a ratio you may divide both sides by the same number.

EXERCISE 1

Simplify the following ratios

1) 3:6		19) 45:81	
2) 5:20		20) 33:44	
3) 8:16		21) 48:60	
4) 4:12		22) 72:84	
5) 4:10		23) 50:60	
6) 7:21		24) 45:50	
7) 3:9		25) 100:150	
8) 3:15		26) 200:300	
9) 42:49		27) 1400:2100	
10) 25:30		28) 1500:3000	
11) 35:42		29) 175:200	
12) 16:24		30) 5000:7500	
13) 25:20		31) 900:1000	
14) 36:24		32) 350:400	
15) 40:50		33) 50:60:70	
16) 15:18		34) 25:30:35	
17) 21:24		35) 40:50:60:70	
18) 56:64			

The above exercise should bring out the similarities between ratios and fractions.

Compare 3:6 = 1:2 with $\dfrac{3}{6} = \dfrac{1}{2}$

and   5:20 = 1:4 with $\dfrac{5}{20} = \dfrac{1}{4}$

SIMPLIFYING RATIOS WITH UNITS

We can simplify the ratio 40 cm:50 cm by

(a) removing the units (because they are the same) and

(b) dividing both sides by 10

40 cm : 50 cm is the same ratio as 4 : 5

The ratio 1 m : 1 cm can be simplified by making the units the same and then removing them

1 m : 1 cm is the same ratio as 100 cm : 1 cm (because 1 m = 100 cm)

100 cm : 1 cm is the same ratio as 100 : 1

To simplify a ratio make the units the same and then remove them. Then simplify as before if possible.

EXAMPLE

Simplify (a) 5 cm : 1 mm (b) 5 km : 1 m (c) £2.50 : 50 p.

(a) 5 cm : 1 mm is the same as 50 mm : 1 mm which is the same as 50 : 1.

(b) 5 km : 1 m is the same as 5000 m : 1 m which is the same as 5000 : 1.

(c) £2.50 : 50 p is the same as 250 p : 50 p which is 250 : 50 or 5 : 1 (dividing by 50).

——————— **EXERCISE 2** ———————

Simplify the following ratios

1)   300 cm : 100 cm
2)   500 m : 1 km
3)   40 g : 1 kg
4)   £1.50 : 50 p
5)   6 kg : 60 g
6)   2 km : 500 m
7)   100 g : 0.5 kg
8)   30 p : 90 p
9)   75 p : 25 p
10) 75 p : £1.00
11) 400 m : 4 km
12) $2\frac{1}{2}$ km : 1000 m
13) 75 p : £1.50
14) 40 cm : 4 m

15) 1 cm : 5 mm
16) £2.80 : £3.50
17) 5.5 cm : 20 mm
18) 60 g : 100 g
19) £5.00 : £1.00
20) 3 km : 2.5 km
21) 2 kg : 500 g
22) 30 m : 27 m
23) 12 p : 8 p
24) 3 hours : 30 min
25) £12.50 : £2.50
26) 16 kg : 8 kg
27) 5 g : 50 g
28) 2 hours : 45 min
29) £3.00 : 30 p
30) 3 kg : 300 g

To simplify a ratio such as $1\frac{1}{4} : \frac{1}{3}$ change mixed fractions to top-heavy fractions and make them fractions of the same type.

The smallest number that 4 and 3 will go into without leaving a remainder is 12, so change the fractions to twelfths:

$$1\frac{1}{4} : \frac{1}{3} = \frac{5}{4} : \frac{1}{3} = \frac{15}{12} : \frac{4}{12} = 15 : 4$$

You may multiply both sides of a ratio by the same number.

(To change $\frac{15}{12} : \frac{4}{12}$ to 15 : 4 we have multiplied both sides by 12.) A picture may help to make this clearer so look at this diagram:

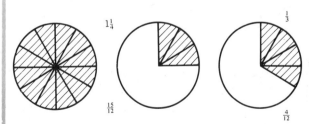

There are 15 shaded parts

There are 4 parts which are shaded

113

EXAMPLE

Simplify (a) $\dfrac{3}{5}:\dfrac{7}{10}$ (b) $1\dfrac{1}{2}:3$ (c) $1\dfrac{2}{3}:\dfrac{1}{3}$.

(a) $\dfrac{3}{5}:\dfrac{7}{10} = \dfrac{6}{10}:\dfrac{7}{10} = 6:7$

(b) $1\dfrac{1}{2}:3 = 3:6$ (doubling both sides)

$\qquad = 1:2$ (dividing both sides by 3)

(c) $1\dfrac{2}{3}:\dfrac{1}{3} = \dfrac{5}{3}:\dfrac{1}{3} = 5:1$

EXERCISE 3

Simplify these ratios

1) $\dfrac{3}{4}:\dfrac{1}{3}$

2) $\dfrac{1}{5}:\dfrac{1}{10}$

3) $\dfrac{1}{2}:\dfrac{3}{8}$

4) $\dfrac{1}{6}:\dfrac{1}{12}$

5) $2\dfrac{1}{4}:\dfrac{1}{4}$

6) $1\dfrac{4}{5}:\dfrac{3}{10}$

7) $1\dfrac{1}{2}:\dfrac{1}{3}$

8) $1\dfrac{1}{2}:\dfrac{1}{2}$

9) $3\dfrac{1}{2}:1\dfrac{1}{2}$

10) $\dfrac{5}{6}:1\dfrac{2}{3}$

11) $10:2\dfrac{1}{2}$

12) $5:2\dfrac{1}{2}$

13) $2\dfrac{3}{4}:1\dfrac{1}{2}$

14) $3\dfrac{1}{4}:2\dfrac{1}{4}$

15) $\dfrac{1}{2}:5$

16) $1\dfrac{1}{3}:1\dfrac{1}{4}$

17) Multiply the quantities in this recipe by
(a) 2 (b) 6 (c) $\frac{1}{2}$.

> *Recipe*
>
> 3 oz butter
> $2\frac{1}{2}$ oz caster sugar
> 1 egg separated
> 6 oz SR flour
> pinch salt
> $1\frac{1}{2}$ oz currants
> $\frac{1}{2}$ oz mixed peel

18) Multiply the quantities in this recipe by
(a) 4 (b) 10 (c) $\frac{1}{2}$.

> *Recipe*
>
> 1 onion
> 2 carrots
> 2 sticks celery
> 250 g cooked beef
> a 340 g can
>   oxtail soup
> salt and pepper
> 4 tsp curry
>   powder
> 50 g raisins

114

Measure the lines below.

Copy the table and put your measurements into it. Then simplify the ratio. (The first one has been done for you.)

1) (a) _____
   (b) _____
2) (a) _____
   (b) _____
3) (a) _____
   (b) _____
4) (a) _____
   (b) _____
5) (a) _____
   (b) _____
6) (a) _____
   (b) _____
7) (a) _____
   (b) _____
8) (a) _____
   (b) _____
9) (a) _____
   (b) _____
10) (a) _____
    (b) _____

	(a)	(b)	Simplified ratio of Length (a): Length (b)
1)	1.5 cm	3 cm	1 : 2
2)			
3)			
4)			
5)			
6)			
7)			
8)			
9)			
10)			

CHANGING RATIOS TO FRACTIONS AND PERCENTAGES

A ratio may be expressed as either a fraction or a percentage.

EXAMPLE

Express the ratio $4:5$ (a) as a fraction, (b) as a percentage.

(a) $4:5 = \dfrac{4}{5}$

(b) changing this to a percentage:

$$\dfrac{4}{5} = \dfrac{4}{5} \times 100\% = 80\%$$

EXAMPLE

Express the ratio $9:27$ first as a fraction in its lowest terms and then as a percentage.

$$9:27 = \dfrac{9}{27} \text{ or } \dfrac{1}{3} \text{ in its lowest terms}$$

$$\dfrac{1}{3} = \dfrac{1}{3} \times 100\% = 33\tfrac{1}{3}\%$$

Express these ratios (a) as a fraction in its lowest terms, (b) as a percentage

1) $1:5$
2) $8:16$
3) $6:9$
4) $7:10$
5) $12:15$
6) $5:8$
7) $3:20$
8) $100:300$
9) $8:64$
10) $100:150$
11) $11:20$
12) $6:120.$

13) Two steel bars have lengths 15 cm and 25 cm respectively. Give the ratio of the two lengths in its simplest form. What fraction is this? Convert this fraction to a percentage.

14) The area of a lawn is 16 m² and the area of a surrounding flowerbed is 40 m². Write down the ratio of the two areas and simplify it. Write down in its simplest form the ratio of the area of the lawn to the *total* area of the lawn and the flowerbed. What percentage is this?

15) Two girls share £3 pocket money with the younger girl getting less than the older. What percentage of the *total* pocket money does the younger girl receive if they share the pocket money in the ratio 2:3? How much does the younger girl receive? How much does the older girl get?

PROPORTIONAL PARTS

EXAMPLE

Divide £20 in the ratio 3:2.

$3 + 2 = 5$ so there are 5 parts altogether.

Divide £20 by 5 to find each part:

$20 \div 5 = 4$ so each part is £4

3 parts is $3 \times £4 = £12$

2 parts is $2 \times £4 = £8$

£20 in the ratio 3:2 is £12:£8.

EXERCISE 6

1) Divide £15 in the ratio 2:1.
2) Divide £16 in the ratio 3:1.
3) Divide £125 in the ratio 1:4.
4) Divide £20 in the ratio 3:1.
5) Divide 300 g in the ratio 3:2.
6) Divide 3000 g in the ratio 1:2.
7) Divide 400 g in the ratio 5:3.
8) Divide £35 in the ratio 4:3.
9) Divide 60 min in the ratio 3:1.
10) Divide 45 min in the ratio 4:5.
11) Divide 6 km in the ratio 5:1.
12) Divide 42 p in the ratio 2:1.
13) Divide 50 p in the ratio 1:4.
14) Divide 40 g in the ratio 5:3.
15) A man divides £500 between his two children in the ratio 2:3. How much does each child get?
16) 700 g of alloy is made up of copper, tin and nickel in the ratio 2:3:2. How many grams of each metal are there?
17) If 75 cm is divided in the ratio 4:1 how long is each part?

DIRECT PROPORTION

Two quantities are in *direct proportion* if an increase (or decrease) in one quantity is matched by an increase (or decrease) in the same ratio in the second quantity. Thus if 4 pens cost 56 p then 8 pens will cost 112 p (we have doubled the number of pens and we have also doubled the cost). Also 2 pens will cost 28 p (we have halved the number of pens and we have also halved the cost).

EXAMPLE

If 2 pencils cost 28 p how much will 5 pencils cost?

The cost is in *direct proportion* to the number of pencils. There are two methods for solving this problem:

Method 1 (the method of unity)

2 pencils cost 28 p
1 pencil costs $28 \div 2 = 14$ p
5 pencils cost $5 \times 14 = 70$ p

Method 2 (fractional method)

2 pencils cost 28 p
5 pencils cost $\dfrac{5}{2} \times 28 = 70$ p

──────── **EXERCISE 7** ────────

1) 5 toffees cost 15 p. What do 6 cost?

2) 3 oranges cost 30 p. What do 7 cost?

3) 9 paper cups cost 27 p. What do 8 cost?

4) 2 kg potatoes cost £1.04. What do 12 kg potatoes cost?

5) 3 bananas cost 24 p. What do 6 cost?

6) 5 hooks cost 25 p. Find the cost of 8 hooks.

7) 9 bolts cost 36 p. Find the cost of 27 bolts.

8) 2 televisions cost £500. What do 3 cost?

9) 4 bowls cost £3.00. What do 5 cost?

10) 3 rubbers cost 33 p. What do 9 cost?

11) 2 bottles of ink cost £1.20. What do 7 bottles of ink cost?

12) 6 sweets cost 12 p. Find the cost of 18 sweets.

13) 5 razor blades cost 35 p. Find the cost of 15 razor blades.

14) 8 plates cost £8.80. What do 3 cost?

15) A car travels 60 miles in 2 hours. How far will it travel in (a) 4 hours (b) 1 hour?

16) 3 boxes weigh 12 kg. Find the weight of 5 similar boxes.

17) A machine makes 12 articles in $\frac{1}{2}$ hour. How many articles will it make in (a) 1 hour (b) 2 hours (c) 6 hours?

18) A man earns £6 in two hours. Find how much he earns in (a) $\frac{1}{2}$ hour (b) 5 hours

19) What is the weight of 20 books if 4 similar books weigh 800 g?

20) 100 g of sweets cost 28 p. Find the cost of 400 g.

21) Copy these tables and fill in the spaces

Weight	Cost
50 kg	£1.00
100 kg	
10 kg	
20 kg	

Weight	Cost
8 kg	£9.60
4 kg	
24 kg	
1 kg	

Weight	Cost
7 kg	£8.40
1 kg	
$3\frac{1}{2}$ kg	
14 kg	

Percentage	Cost
15%	£3.00
5%	
1%	
30%	

Weight	Cost
5 kg	£5.15
1 kg	
20 kg	
40 kg	

Percentage	Cost
20%	£1500
10%	
1%	
$\frac{1}{2}$%	

Weight	Cost
15 kg	£3.00
5 kg	
1 kg	
30 kg	

Percentage	Weight
50%	250 g
100%	
10%	
1%	

Weight	Cost
$3\frac{1}{2}$ kg	£7.00
7 kg	
1 kg	
21 kg	

Percentage	Weight
100%	80 g
10%	
5%	
$2\frac{1}{2}$%	

INVERSE PROPORTION

If an increase (or decrease) in one quantity produces a decrease (or increase) in a second quantity in the same ratio, the two quantities are in *inverse proportion*. For example, suppose 3 men working on a job take 6 days. Then 6 men working on a similar job will take *less* time — they will take 3 days if they work at the same rate.

EXAMPLE

5 people building a wall take 20 days. How long will it take 4 people?

  5 people take 20 days

  1 person takes $20 \times 5 = 100$ days

(1 person takes *longer* so *multiply* by 5)

  4 people take $100 \div 4 = 25$ days

(4 people take *less* time so *divide* by 4).

EXERCISE 8

1) 6 men building a wall take 12 days. How long would it take 2 men?

2) A farmer employs 12 men to pick apples. It takes them 2 days. How long would it take 8 men?

3) A bag of sweets is shared between 5 children and each child gets 3 sweets. How many sweets would each get if the same bag were shared between 3 children?

4) 7 men make 21 toys in 1 hour. How long would it take 21 men?

5) 5 men dig up a row of potatoes in 20 minutes. How long will it take 10 men?

6) 3 men dig a hole in the road in 1 hour. How long would it take 12 men?

7) 3 men take 6 hours to complete some work. How long will it take 5 men?

8) A bag of sweets is shared among 6 children and each child gets 8 sweets. How many sweets would each child get if the same bag were shared among 12 children?

9) 8 people take 3 hours to pick a field of strawberries. How long would it take 6 people?

These problems are answered by using direct proportion.

EXAMPLE

Which is the best buy:

  40 g for £2.80   150 g for £9.00

  100 g for £8.00   200 g for £10.00?

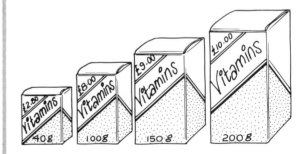

The best way to tackle this is to put the ratios of the weights in their lowest terms:

  40 g : 100 g : 150 g : 200 g

is the same as 4 : 10 : 15 : 20. We have divided through by 10 g.

Now find the price for 10 g of each item:

40 g for £2.80 means that 10 g will cost 70 p
100 g for £8.00 means that 10 g will cost 80 p
150 g for £9.00 means that 10 g will cost 60 p
200 g for £10.00 means that 10 g will cost 50 p

We can see that 200 g for £10.00 is the best buy.

EXERCISE 9

Which is the best buy?

1) 20 g at 60 p or 50 g at £1.00

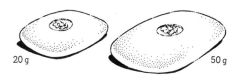

2) 100 mℓ shampoo at 35 p
   150 mℓ shampoo at 42 p or
   200 mℓ shampoo at 70 p

3) 100 g at £1.50
   200 g at £2.20
   300 g at £3.18 or
   400 g at £4.28

4) 850 g of washing powder at £1.02 p or 2500 g at £3.50

5) Is 50 g of cornflour at 35 p better value than 80 g at 64 p?

6) 250 g of cornflakes cost 70 p. A 500 g packet of cornflakes cost £1.20. How much do I save by buying one 500 g packet rather than two 250 g packets?

7) I can buy single nails at 2 p each or 5 for 9 p or 10 for 17 p. I need 36 nails. What is the least I would have to pay? How much do I save over buying them all singly?

— MAPS AND SCALE MODELS —

EXAMPLE

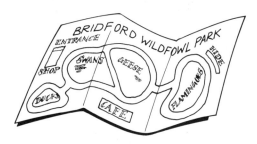

The scale on this map is 1 : 100. 1 cm on the map represents 100 cm = 1 m.

(a) What would 5 cm represent?

(b) What would 0.5 cm represent?

(a) 5 cm would represent
   5×100 cm = 500 cm = 5 m.

(b) 0.5 cm would represent
   0.5×100 cm = 50 cm.

1) A model car is 50 times smaller than a real car.
   (a) 100 cm on the real car would be represented by how much on the model?
   (b) What length on the real car does 3 cm on the model represent?

2) A map has a scale of 1 cm : 1 km. What distance does 2 cm on the map represent? What distance would 5 cm represent? What distance would 2.5 cm represent?

3) A model train is built 72 times smaller than the real train. What is the real measurement represented by (a) 3 cm (b) 0.5 cm (c) 20 cm (d) 25 cm on the model?

4) On a map 5 km is represented by 1 cm. What distance is represented by (a) 2 cm (b) 0.5 cm (c) 5 cm (d) 11 cm?

   What length on the map represents
   (e) 10 km (f) 30 km (g) 2.5 km (h) 7.5 km?

FOREIGN CURRENCY

When people travel abroad they need to change their money into the currency of the country they are visiting.

They might use a calculator to work out the exact amount of money they would get, but it is useful to be able to do a rough check. The methods of direct proportion are used to work out foreign currency exchange. The rate of exchange may vary from day to day, so to find exactly how much foreign money you would get for £1 you would need to look at a *rate of exchange table* such as the one below:

Country	Rate of exchange
France	£1 = 8 French francs
Germany	£1 = 2.2 German marks
Italy	£1 = 2480 lire
Spain	£1 = 186 pesetas
Switzerland	£1 = 1.8 Swiss francs

EXAMPLE

If £1 is roughly 2000 lire, find approximately how many lire you would get for £7.

   £1 = 2000 lire

   £7 = 7 × 2000 = 14 000 lire

EXAMPLE

If £1 is roughly 2 Swiss francs, find how many pounds you would get for 24 Swiss francs.

   2 Swiss francs = £1

   1 Swiss franc = £1 ÷ 2 = 50 p

   24 Swiss francs = 50 p × 24 = £1 × 12 = £12

1) If £1 = 8 French francs find how many francs you would get for £2.

2) If £1 is approximately 2 German marks find how many marks you would get for £3.

3) If £1 is roughly 2000 lire find approximately how many lire you would get for £9.

4) If £1 is roughly 200 pesetas find approximately how many pesetas you would get for £8.

5) If £1 is roughly 2 Swiss francs find approximately how many Swiss francs you would get for £15.

6) If 2 Swiss francs is roughly £1 how much approximately is 20 Swiss francs worth in pounds?

7) If 2000 lire is roughly £1 find how much approximately 1000 lire is worth in pounds and pence.

8) If 8 French francs is roughly £1 find out how much roughly 32 French francs is worth in pounds.

9) If 200 pesetas is roughly £1 find approximately how much 500 pesetas is worth in pounds and pence.

10) If 2000 lire is roughly £1 find how much 10 000 lire is worth in pounds.

11) If 2 German marks is roughly £1 find how much 40 German marks is worth in pounds.

12) If 2000 lire is roughly £1 find approximately how much 50 000 lire is worth in pounds.

13) If 2 Swiss francs is roughly £1 find how much 25 Swiss francs would be worth in pounds and pence.

EXERCISE 12

1) Simplify (a) 5:25 (b) 30:24 (c) 150:200 (d) 42:14.

2) Simplify (a) 3 cm:5 cm (b) 1 m:20 cm (c) 3 mm:1 cm (d) 2 hours:1 hour.

3) Simplify (a) 5 kg:500 g (b) 200 g:1 kg.

4) Simplify (a) $1\frac{1}{8}:\frac{1}{8}$ (b) $\frac{3}{5}:\frac{9}{10}$.

5) (a) Double, (b) treble, (c) halve the quantities in this recipe for meringue

　　4 egg whites
　　200 g caster sugar.

6) Measure these lines and simplify the ratio of their lengths.
　　(a) —————————　　(b) ———————

7) Divide £50 in the ratio 3:2.

8) Divide 16 in the ratio 5:3.

9) Divide 150 g in the ratio 4:1.

10) Divide 2 hours in the ratio 1:3.

11) Divide £40 in the ratio 1:2:1.

12) Divide 52 p in the ratio 6:7.

13) 5 pencils cost 30 p. How much do 7 pencils cost?

14) 6 rubbers cost 42 p. How much do 5 rubbers cost?

15) 9 articles cost 18 p. How much do 12 articles cost?

16) 2 men take 4 hours to do a job. How long will 5 men take?

17) Which is the best value: (a) 35 g for 70 p or (b) 15 g for 45 p?

18) Which is the best value: (a) 12 g for 36 p or (b) 10 g for 20 p?

19) If £1 is roughly 2 German marks how much is £20 worth in marks?

20) If 2 Swiss francs is roughly £1 how much approximately is 48 Swiss francs worth in pounds?

21) If 10 cm on a map represents 40 km how much will 7 cm represent?

22) If the scale on a map is 1:400 how much will 1 cm represent?

10

Averages

The average of a set of numbers is the number typical of that set. The most common sort of average is the *arithmetic mean*.

EXAMPLE

Find the mean of these numbers:

6, 6, 7, 8, 10, 11.

To find the mean we add the numbers and then divide the sum by the number of numbers in the set:

$$\frac{6+6+7+8+10+11}{6}$$

$$= \frac{48}{6}$$

$$= 8$$

So the mean of the set of numbers is 8.

Where the word "average" is used alone, it is the arithmetic mean that is required.

EXAMPLE

Find the average height of five women whose heights are: 4 ft 5 inches, 5 ft 2 inches, 6 ft 1 inch, 5 ft 2 inches and 5 ft exactly.

First we add up their heights:

ft	inches
4	5
5	2
6	1
5	2
5	
25	10

Average height

and then, because there are five women, we divide by 5. Thus:

Average height = (25 ft 10 inches) ÷ 5
              = 5 ft 2 inches

Note carefully that:

$$\text{Average} = \frac{\text{Sum}}{\text{Number in set}}$$

Find the mean of these numbers
1) 2, 3, 5, 6
2) 2, 5, 6, 7, 10
3) 5, 0, 2, 1
4) 2, 6, 7
5) 3, 8, 2, 7
6) 5, 1, 7, 2, 2, 1
7) 8, 1, 1, 2, 3

8) 15, 9, 12
9) 12, 16, 13, 15
10) 2, 5, 1, 3, 4, 5, 1
11) 12, 17, 29, 23, 34
12) 68, 72, 61
13) 0.5, 0.7, 0.3
14) 1.2, 1.8, 1.9, 1.5
15) 0.7, 0.6, 0.8, 0.3
16) 0.9, 0.9, 0.6, 1.7, 1.8, 0.8, 1.7
17) 90, 92, 78, 84
18) 1.58, 2.72, 4.10
19) 7.5, 7.2, 7.2, 7.5, 7.1
20) 8.0, 7.2, 6.1, 3.5, 4.2

Here give your answers as fractions

21) 4, 2, 3, 1
22) 5, 1, 4
23) 8, 1, 0, 5, 2, 1
24) 1, 0, 0, 5, 2
25) $\dfrac{1}{4}, \dfrac{1}{8}, \dfrac{1}{2}, \dfrac{1}{8}$
26) $1\dfrac{1}{2}, 2\dfrac{1}{2}, 3\dfrac{1}{4}, 5\dfrac{3}{4}$
27) $\dfrac{1}{3}, \dfrac{5}{6}, \dfrac{1}{12}$
28) 1, 3, 4, 1, 9, 1, 3

Here give your answers to 3 significant figures

29) 4, 3, 1, 1, 1, 3, 9
30) 2, 5, 0, 3, 4, 0, 0, 1, 1

Here give your answers to three decimal places

31) 5.602, 4.117, 3.281
32) 0.103, 0.105, 0.109

Here give your answers to one decimal place

33) 1.7, 1.3, 1.4, 1.6, 1.2, 1.1
34) 5.2, 5.3, 5.5, 5.6, 6.7, 8.1, 5.5

In Exercises 2, 3 and 4:

Average = Mean

--------- EXERCISE 2 ---------

1) Ten families have 0, 0, 1, 1, 2, 2, 2, 2, 3 and 7 children. Find the average number of children in each family.

2) A batsman scores 25, 30, 110, 80 and 45 runs in five different innings. Find his average score.

3) The weights of six apples are 90 g, 92 g, 93 g, 94 g, 97 g and 92 g. Find their mean weight.

4) The marks given in an examination were 35, 45, 47, 52, 55, 60 and 70. Find the average mark.

5) The hourly wages of seven people are £3.40, £4.20, £2.50, £3.35, £2.10, £4.30 and £5.21. Find their average hourly wage.

6) Four children are aged 11 years, 11 years 1 month, 11 years 3 months and 11 years 4 months. Find their average age.

7) Find the arithmetic mean of 9 cm, 9 cm, 11 cm, 13 cm, 15 cm and 15 cm.

8) Five boxes have weights 120 g, 210 g, 250 g, 300 g and 320 g. Find (a) their total weight (b) their mean weight.

9) Find the mean of 12.3, 12.4, 12.5, 12.5, 12.6, 12.9 and 13.0.

10) Find the mean of three 9s, five 10s and one 13.

Note also that:

Sum = Number in set × Average

--------- EXERCISE 3 ---------

1) A batsman scores an average of 50 runs in 4 innings. What was his total number of runs?

2) Three children have an average age of 11 years 4 months. What is their total age in years?

3) Seven people have a mean hourly wage of £3.10. What is the total wages paid per hour?

4) Six numbers have an average of 13. What is their total?

5) Sixteen families have an average of two children per family. How many children are there altogether?

6) What is the total weight of 11 apples each with average weight of 92 g?

7) Four parcels have an average weight of 200 g. What is their total weight?

8) Five children have an average weekly pocket money of 50 p. What is the total pocket money of these children?

EXAMPLE

A girl gains 10, 15, 20, 30 and 40 marks in five different tests.

(a) What is her average mark?

(b) What mark must she get in her next test to improve her average by 2 marks?

(a) $\dfrac{10 + 15 + 20 + 30 + 40}{5} = \dfrac{115}{5} = 23$

Her average mark is 23.

(b) To improve this by 2 marks her average mark over six tests must be 25, so her total marks over six tests must be:

$6 \times 25 = 150$ marks

Her total marks over five tests were $5 \times 23 = 115$ marks. $150 - 115 = 35$.

She must gain 35 marks in her sixth test.

1) A batsman scores 16, 25, 32 and 63 runs in four innings. What is his average score and what must he score in the next innings to improve his average by 1 run?

2) The average age of five girls is 11 years 2 months. What is their total age in years and months? What must be the age of a sixth girl if their new average age is 11 years 3 months?

3) Five workers earn £80, £95, £95, £100 and £120 a week. What is their average wage? What must a sixth worker earn if the average wage of the six is to be £1 more.

4) The average of five numbers is 11. What must a sixth number be if the average of the six numbers is to be 12?

5) Two parcels weigh 30 g and 45 g. What must a third parcel weigh if the average weight of the three parcels is 40 g?

6) The average length of three gardens is 90 m. What is the length of a fourth garden if the average length of the four gardens is 100 m?

7) In six different tests a boy gains 10, 10, 13, 15, 16 and 14 marks. What is his average mark? What is his total mark for the six tests? What must he get in his seventh test to improve his average by 1 mark?

8) If the average of five numbers is 10 and the average of six numbers is 9, what is the sixth number?

9) Five boxes weigh 500 g in total. One box weighing 140 g is taken away. What is the average weight of the remaining four boxes?

MIXTURES

Grocers often mix two or three types of coffee together and the price will depend on the quantity and price of each type.

EXAMPLE

Coffee at 80 p for 100 g is mixed with coffee at 95 p for 100 g in the ratio 3:2. Find the price of 100 g of the mixture.

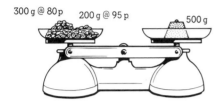

300 g @ 80p   200 g @ 95 p   500 g

Since they are mixed in the ratio 3:2 and 3 + 2 = 5, it will be useful to find the total cost of 500 g of the mixture:

300 g at 80 p per 100 g costs 3 × 80 p = £2.40

200 g at 95 p per 100 g costs 2 × 95 p = £1.90

500 g of the mixture will cost
£2.40 + £1.90 = £4.30

so

100 g of the mixture will cost
£4.30 ÷ 5 = 86 p

──────── EXERCISE 5 ────────

1) 3 kg of pears at 20 p a kg are mixed with 2 kg at 15 p per kg. Find the price of 1 kg of the mixture.

2) 5 kg of apples at 25 p a kg are mixed with 5 kg of apples at 33 p a kg. Find the price of 1 kg of the mixture.

3) 2 litres of petrol at 25 p a litre are mixed with 3 litres at 30 p a litre. Find the price of 1 litre of the mixture.

4) Coffee at 90 p for 100 g is mixed with coffee at 100 p for 100 g in the ratio 2:3. Find the price of 100 g of the mixture.

5) A brass consists of copper and zinc in the ratio 3:1. Find the cost of 1 kg of brass if copper costs 32p per kg and zinc costs 20 p per kg.

6) Three types of wine are mixed in the ratio 5:3:2. Their prices per litre are £1.00, £2.00 and £3.00 respectively. Find the cost per litre of the mixture.

7) Tea at 70 p per 100 g is mixed with tea at 80 p per 100 g in the ratio 3:7. Find the price of 100 g of the mixture.

8) 4 kg of grapes at 60 p a kg are mixed with 6 kg of grapes at 80 p a kg. Find the price of 1 kg of the mixture.

──── WEIGHTED AVERAGES ────

Sometimes some of the numbers are repeated several times.

EXAMPLE

Find the average of:

7, 7, 8, 8, 8, 9, 9, 10.

7 is repeated *twice*, 8 is repeated *three times*, 9 is repeated *twice* and 10 occurs only *once*. We say the "weight" for 7 is 2, for 8 is 3, for 9 is 2 and for 10 is 1. (Sometimes the weights are called "the frequency".)

There are 8 numbers altogether.

The information could have been given in a table like this:

Number	Frequency (or weight)
7	2
8	3
9	2
10	1

To find the total we multiply 7 by 2, 8 by 3, 9 by 2, 10 by 1 and add:

Sum = (7 × 2) + (8 × 3) + (9 × 2) + (10 × 1)

= 14 + 24 + 18 + 10 = 66

Average $\dfrac{66}{8} = 8\dfrac{2}{8} = 8\dfrac{1}{4}$ (or 8.25)

Notice that the number of numbers in the set is $2 + 3 + 2 + 1 = 8$ (the sum of the weights).

The calculations can be recorded in a table:

Number	Frequency (or weight)	Number × frequency
7	2	14
8	3	24
9	2	18
10	1	10
Total	8	66

Average = $66 \div 8 = 8\frac{1}{4}$ or 8.25.

——————— **EXERCISE 6** ———————

Find the average for these numbers

1)

Number	Frequency (or weight)
2	5
3	2
4	1
5	2

2)

Number	Frequency (or weight)
19	2
20	5
25	1
30	3

3)

Number	Frequency (or weight)
1.5	2
3.5	4
5.5	3
6.5	1

4)

Number	Frequency (or weight)
6	4
9	6
11	5
15	5

5)

Number	Frequency (or weight)
33	2
35	2
37	3
38	3

6)

Number	Frequency (or weight)
56	3
62	1
65	1
67	3

EXAMPLE

At 2 football matches no goals were scored, at 5 matches 1 goal was scored, at 9 matches 2 goals were scored and at 4 matches 3 goals were scored. Put this information in a table and find the average number of goals scored per match.

Number of goals	Number of matches
0	2
1	5
2	9
3	4

Total number of goals scored $= (0 \times 2) + (1 \times 5) + (2 \times 9)$
$\qquad + (3 \times 4)$
$= 0 + 5 + 18 + 12$
$= 35$

Number of matches is $2 + 5 + 9 + 4 = 20$

Goal average $= \dfrac{35}{20} = 1\dfrac{15}{20} = 1\dfrac{3}{4}$

EXERCISE 7

1) The wages of ten people are shown by this table:

Wage	Number of people earning this wage
£65	4
£72	5
£80	1

Find:

(a) the total wages paid to all ten people
(b) the average wage.

2) The weights of ten bolts are given by this table:

Weight	Number of bolts with this weight
5 g	2
7 g	3
8 g	3
9 g	2

Find:

(a) the total weight of all ten bolts
(b) the average weight of the bolts.

3) The lengths of five bars of steel are given by this table:

Length	Number of bars of steel with this length
250 cm	3
300 cm	1
400 cm	1

Find the average length of the bars of steel.

4) In 10 hockey matches the goals scored were: no goals in 1 match, 1 goal in 3 matches, 2 goals in 3 matches, 3 goals in 2 matches and 4 goals in 1 match. Put this information in a table. Find (a) the total number of goals scored and (b) the goal average.

5) 3 apples have a weight of 50 g and 7 apples have a weight of 70 g. What is the average weight of an apple?

6) 2 tins are sold at 15 p per tin, 3 tins are sold at 20 p per tin and 5 tins are sold at 30 p per tin. What is the average price per tin?

7) Find the average length using the information in this table:

Length	Number with this length
20 cm	3
24 cm	4
30 cm	2
35 cm	1

8) Find the average weight of a box using the information in this table:

Weight	Number of boxes with this weight
500 g	7
550 g	8
600 g	5

9) Using the information given in this table find the average mark:

Mark	Number with this mark
5	22
6	30
7	18
8	20
9	10

10) Copy this table and fill in the missing spaces. Find the total wage. Find the total number of people and then find the average wage (to the nearest whole pound).

Wage	Number of people with this wage	
£72	15	72 × 15 =
£75	20	75 × 20 =
£80	40	80 × 40 =
£85	35	85 × 35 =
£90	17	90 × 17 =
£92	13	92 × 13 =
£95	60	95 × 60 =

AVERAGE SPEED

There are three formulae to remember:

$$\text{Average speed} = \frac{\text{Distance}}{\text{Time}}$$

(For two-part questions you must find the *total distance* and divide by the *total time*.)

$$\text{Time} = \frac{\text{Distance}}{\text{Average speed}}$$

and

$$\text{Distance} = \text{Average speed} \times \text{Time}$$

As a way of remembering these, try thinking of a simple example such as "If a car travels at 4 miles per hour for 3 hours how far will it travel?" Obviously it will travel 12 miles. The three formulae can be worked out from:

$$4\,\text{miles/hour} = \frac{12\,\text{miles}}{3\,\text{hours}}$$

$$3\,\text{hours} = \frac{12\,\text{miles}}{4\,\text{miles/hour}} \quad \text{and}$$

$$12\,\text{miles} = 4\,\text{miles/hour} \times 3\,\text{hours}$$

where 4 miles/hour (mph) is the speed, 12 miles is the distance and 3 hours is the time.

EXAMPLE

If a car travels for 120 miles at an average speed of 40 miles/hour how long does it take?

$$\text{Use Time} = \frac{\text{Distance}}{\text{Average speed}} = \frac{120\,\text{miles}}{40\,\text{miles/hour}}$$

$$= 3\,\text{hours}$$

EXAMPLE

A car travels at 50 miles/hour for 3 hours; how far has it gone?

$$\text{Use Distance} = \text{Average speed} \times \text{Time}$$

$$= 50\,\text{miles/hour} \times 3\,\text{hours}$$

$$= 150\,\text{miles}$$

EXAMPLE

What is the average speed of a car that travels for 90 miles and takes $2\frac{1}{2}$ hours?

$$\text{Use Average speed} = \frac{\text{Distance}}{\text{Time}} = \frac{90\,\text{miles}}{2\frac{1}{2}\,\text{hours}}$$

$$= 90 \div 2\frac{1}{2}$$

$$= 90 \div \frac{5}{2}$$

(turn second fraction upside down and multiply)

$$= \overset{18}{\cancel{90}} \times \frac{2}{\cancel{5}}$$

$$= 36\,\text{miles/hour}$$

Find the times taken to travel

1) 8 miles at an average speed of 2 mph
2) 12 miles at an average speed of 4 mph
3) 25 miles at an average speed of 5 mph
4) 30 miles at an average speed of 10 mph.

Find the times if the distances and average speeds are as follows

5) Distance 90 miles, average speed 30 mph
6) Distance 120 miles, average speed 40 mph
7) Distance 200 miles, average speed 50 mph.

In some countries, distances are measured in kilometres (km) and speeds in kilometres per hour (km/h or km h^{-1}).

8) Find the time taken to travel 12 km at 3 km/h.
9) How long does it take to travel 15 km at 5 km/h?

Give your answers in hours and minutes.

10) Find the time to travel 10 miles at 4 mph.
11) How long does it take to go 24 miles at 10 mph?
12) 150 miles are travelled at an average speed of 60 mph. How long does this take?
13) How long does it take if you go 2 mph for $4\frac{1}{2}$ miles?

Find the distances travelled if the average speeds and times are as follows

14) Average speed 2 mph, time 4 hours
15) Average speed 5 mph, time 2 hours
16) Average speed 8 mph, time 3 hours.
17) How far do you go if you travel at 20 mph for 3 hours?
18) How far do you go if you travel at 50 mph for $1\frac{1}{2}$ hours?

For the following questions your answers should be in kilometres.

19) Give the distance travelled if you go for $2\frac{1}{2}$ hours at 60 km/h.
20) How many kilometres are travelled in $3\frac{1}{4}$ hours at 40 km/h?

In the following questions you have to find the average speeds.

21) If you travel 2 miles in 1 hour what is your average speed?
22) Find the average speed for a distance of 6 miles in a time of 2 hours.
23) Find the average speed for a distance of 15 miles in a time of 3 hours.

Copy this table and fill in the values for the average speeds.

	Distance	Time	Average speed
24)	6 miles	3 hours	3/mil
25)	8 miles	4 hours	
26)	20 miles	5 hours	
27)	56 miles	8 hours	
28)	60 miles	2 hours	
29)	120 miles	3 hours	
30)	200 miles	4 hours	20
31)	9 km	3 hours	
32)	27 km	3 hours	
33)	250 km	5 hours	
34)	10 km	4 hours	
35)	100 km	8 hours	

1) A car travels for 2 hours at 40 miles/hour. How far has it gone?
2) How long does it take to travel 120 miles at 30 miles/hour?

3) What is the average speed of a car that takes 3 hours to travel 90 miles?

4) A journey lasts for 3 hours at an average speed of 35 miles/hour. How far is it?

5) If you travel at 40 miles/hour for $2\frac{1}{2}$ hours how far do you go?

6) A car travels for 4 hours and goes 160 miles. What is its average speed?

7) What is the average speed if you travel for $2\frac{1}{2}$ hours and go 80 miles?

8) How long does it take to go 60 miles at an average speed of 40 miles/hour?

9) Give the time for a journey of 90 miles at an average speed of 20 miles/hour.

10) After a $3\frac{1}{2}$ hour journey a car has travelled 210 miles. What is its average speed?

EXAMPLE

A car travels 60 miles at 20 miles/hour and 80 miles at 40 miles/hour. What is its overall average speed?

For two-part questions you must work out the *total distance* and the *total time*:

$$\text{Total distance} = 60 \text{ miles} + 80 \text{ miles}$$
$$= 140 \text{ miles}$$

$$\text{Time} = \frac{\text{Distance}}{\text{Average speed}}$$

$$\text{Time for first part} = \frac{60 \text{ miles}}{20 \text{ miles/hour}} = 3 \text{ hours}$$

$$\text{Time for second part} = \frac{80 \text{ miles}}{40 \text{ miles/hour}}$$
$$= 2 \text{ hours}$$

$$\text{Total time} = 3 \text{ hours} + 2 \text{ hours} = 5 \text{ hours}$$

$$\text{Overall average speed} = \frac{\text{Total distance}}{\text{Total time}}$$
$$= \frac{140 \text{ miles}}{5 \text{ hours}}$$
$$= 28 \text{ miles/hour}$$

The results may be clearer if they are put in a table like this:

	Time	Distance	Average speed
1st part	3 hours	60 miles	20 mph
2nd part	2 hours	80 miles	40 mph
Totals	5 hours	140 miles	Overall average speed $= \frac{140}{5}$ $= 28$ mph

(Put your results into a table like the one shown above.)

1) A car travels 120 miles at 40 miles/hour and 120 miles at 60 miles/hour. Find its overall average speed.

2) For 2 hours of a 100 mile journey the average speed is 30 miles/hour and the average speed of the remainder is 40 miles/miles/hour. Find the overall average speed.

3) On an outward journey of 240 miles a motorist takes 6 hours and on the return journey he takes 2 hours less.
   (a) Calculate the average speed on the outward journey.
   (b) Calculate the average speed on the return journey.
   (c) By first calculating the total distance and then the total time find the overall average speed.

4) An aeroplane travels 1000 km in $2\frac{1}{2}$ hours. What is its average speed? It then travels for $5\frac{1}{2}$ hours at an average speed of 800 km/h.
   (a) How far has it travelled altogether?
   (b) What is the total time taken?
   (c) What is the overall average speed?

5) A car travels for 2 hours at 25 miles/hour and 90 miles at 30 miles/hour.
   (a) What is the total time?
   (b) Find the total distance.
   (c) Find the overall average speed.

6) A plane travels for 3 hours at 500 km/hour and for 2 hours at 650 km/hour. Find the total distance travelled and the overall average speed.

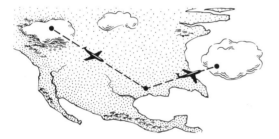

7) Calculate the overall average speed of a journey of 180 miles where the first 120 miles are travelled at an average speed of 40 miles/hour and the remainder at an average speed of 30 miles/hour.

— CHANGING mph TO km/h —

If you are driving abroad you will need your speed in kilometres per hour (sometimes shortened to km/h) rather than in miles per hour (sometimes shortened to mph).

A speed of 40 miles per hour means if you travel at the same speed for one hour you would have gone 40 miles. To change this to km/h we need to know the distance that would have been travelled in one hour in *kilometres*, so we use a ratio method to change 40 miles into kilometres.

EXAMPLE

Change 40 mph to km/h.

   5 miles equals 8 kilometres (approximately)
   1 mile equals (8 ÷ 5) kilometres
        = 1.6 kilometres
   40 miles equals 40 × 1.6 kilometres
        = 64 kilometres

A speed of 40 mph is about 64 km/h.

——— EXERCISE 11 ———

Change the following speeds from mph to km/h.

1)	5 mph	11)	12 mph
2)	10 mph	12)	26 mph
3)	20 mph	13)	32 mph
4)	25 mph	14)	48 mph
5)	40 mph	15)	54 mph
6)	50 mph	16)	23.5 mph
7)	55 mph	17)	41.2 mph
8)	65 mph	18)	54.5 mph
9)	70 mph	19)	120 mph
10)	80 mph	20)	1000 mph

21) Copy the table below and fill in the speeds in km/h.

mph	15	30	45	60	75	90
km/h						

STATISTICAL AVERAGES — — MEAN, MEDIAN AND MODE —

The *mean* is found by adding the numbers and dividing by the number in the set.

The *median* is found by putting the numbers in order and selecting the middle one or the average of the two middle ones.

The *mode* is the number that occurs most frequently.

EXAMPLE

Find the mean, median and mode of these numbers:

6, 6, 6, 7, 7, 8, 8, 10.

The mean is

$$\frac{6+6+6+7+7+8+8+10}{8} = \frac{58}{8} = 7\frac{2}{8}$$
$$= 7\frac{1}{4}$$

The numbers are already in order; the middle ones are both 7, so the median is 7.

The mode is 6 (there are more 6s than any other number).

The mean is the ordinary average that we have already met. Sometimes the median and mode give more suitable information than the mean. The mean can sometimes be distorted by very large or very small numbers.

EXAMPLE

A shop sells 2 pairs of shoes of size 3, 3 pairs of size 4, 5 pairs of size 5 and 2 pairs of size 6 and 1 pair of size 8. Putting these numbers in order:

3, 3, 4, 4, 4, 5, 5, 5, 5, 5, 6, 6, 8

we see that the median is 5. The mode is also 5. More shoes of size 5 are sold than of any other size.

──────────── EXERCISE 12 ────────────

Find the mean, median and mode of these sets of numbers

1) 2, 2, 3, 3, 3, 3, 4, 8
2) 12, 12, 13, 13, 13, 15
3) 1, 1, 1, 1, 5, 10, 10, 11
4) 15, 16, 17, 18, 18
5) 33, 33, 34, 34, 34, 35, 37, 40, 41, 42.
6) A shoe shop sells 3 pairs of shoes of size 4, 5 pairs of size 5, 4 pairs of size 6, 2 pairs of size 7 and 1 pair of size 8. Find the median and the mode.
7) 2 pairs of shoes size 3, 3 pairs size 4, 4 pairs size 5 and 3 pairs size 6 are sold. Find the mean, median and mode.
8) In 8 innings a batsman scores 6, 18, 0, 12, 3, 15, 12 and 10 runs. Find the mean and the median.
9) The numbers of children in 10 neighbouring houses are 2, 3, 3, 5, 0, 1, 2, 6, 0 and 2. Find the mean, median and mode.
10) The weights of 6 apples are 90 g, 90 g, 92 g, 92 g, 92 g and 94 g. Find the mean, median and mode.

The *range* of a set of numbers is the largest minus the smallest.

EXAMPLE

The range of 11, 11, 13, 15, 15, 17 is:

17 − 11 = 6

──────────── EXERCISE 13 ────────────

1) Find the range of 15, 15, 17, 18, 21, 24.
2) Find the range of 17, 17, 19, 24, 26.
3) Find the range of 15, 13, 11, 9, 17.
4) Find the range of 62, 65, 67, 69.
5) Find the range of 51, 48, 47, 54, 51.
6) Find the range of 20, 19, 18, 21, 24.

1) Find the average of 0, 2, 3, 5 and 5.

2) Find the average of 14 cm, 15 cm, 17 cm and 19 cm.

3) If the average of 4 numbers is 7 what is their total?

4) 4 boxes have average weight of 512 g. What is their total weight?

5) A batsman scores a total of 255 runs in 5 innings. What is his average?

6) A motorist travels 80 miles at an average speed of 40 miles/hour and spends 2 hours travelling the remaining 60 miles.
   (a) What is the total distance?
   (b) What is the total time?
   (c) What is the overall average speed?

7) A motorist does the first 50 miles of a 90 mile journey at an average speed of 25 miles/hour. In what time must she do the other 40 miles for her overall average speed to be 30 miles/hour?

8) Find the average price per can of:
   3 cans at 30 p
   5 cans at 36 p
   2 cans at 40 p.

9) 400 g of tea at 90 p per 100 g are blended with 500 g of tea at 99 p per 100 g. Find the price of 100 g of the mixture.

10) The numbers of marks gained in seven tests were

   40, 42, 42, 45, 47, 48 and 51.

   (a) Find the mean.
   (b) Find the mode.
   (c) Find the median.
   (d) Find the range.

11) A girl gains 6, 7, 7, 7 and 8 marks in five different tests. Find her average mark. Her mark in a sixth test meant that her new average was 1 mark lower than it had been before. Find the mark she obtained in her sixth test.

12) A motorist drives for $3\frac{1}{2}$ hours at an average speed of 50 miles/hour. How far has he travelled? He then travels for $2\frac{1}{2}$ hours at an average speed of 60 miles/hour. What is the total distance he has now travelled? What is the total time? What is his overall average speed?

13) Find the mode of these numbers
   2, 3, 0, 0, 1, 2, 0, 0, 4, 0, 5.

14) Find the median of these numbers
   7, 5, 4, 6, 11, 2, 3, 0, 12.

15) Find (a) the mean, and (b) the range, of
   3, 5, 7, 2, 7, 8, 1, 0, 5 and 4.

11

Money in Business

It is important to keep careful records of all money transactions in business. Each business has its own way of keeping records and the forms will all look slightly different but the ideas behind bookkeeping are the same.

A record must be kept of money coming in (from sales, etc.) and of money going out (for expenses, etc.). These records are kept in special books called *ledgers* (or, more likely nowadays, in *computer files*).

Computer files are stored on tapes or disc.

CASH STATEMENTS

Cash statements are usually in two parts: one for money coming in (*income*) and the other for *expenses* (as shown below).

The three items of *essential information* are

1) the date

2) where the money has come from, or where it is going

3) the amount of money.

This is entered into the statement *being careful to put the information on the correct side.*

The left hand side is totalled. This is the total income.

The right hand side is *made to come to the same total* by including "balance carried down" which is worked out by adding the total expenses and taking this away from the total income.

This working out is not done on the statement but will be done on another piece of paper as shown here:

$$\begin{array}{r} 80.00 \\ 15.00 \\ 125.00 + \\ \hline \text{Total expenses } 220.00 \end{array} \qquad \begin{array}{r} 520.00 \\ 220.00 - \\ \hline 300.00 \\ \\ \text{Balance} = £300 \end{array}$$

INCOME			EXPENSES		
Date	Particulars	Receipts	Date	Particulars	Payments
		£			£
1 June	Sale of 100 tickets @ £2	200.00	2 June	Hire of hall	80.00
3 June	Sale of 50 tickets @ £3	150.00	5 June	Printing of tickets	15.00
9 June	Sale of 40 tickets @ £4	160.00	11 June	Cost of entertainment	125.00
11 June	Raffle ticket sale	10.00			
				Balance carried down	300.00
	Total	£520.00			£520.00

Draw up a cash statement (in the same way as just shown) for each of these events

1) Village Fete

1 Aug	Hire of six lorries @ £5 per lorry
3 Aug	Paid to farmer for use of his field £35
3 Aug	Hire of stalls from village hall £15
15 Aug	Donation received from local traders £75
24 Aug	Income from sale of tickets at gate £124
24 Aug	Income from raffle £11
24 Aug	Income from stalls £50

2) Sports Club Gymnastic Display

3 May	Hire of hall £50
7 May	Cost of printing tickets £3.75
11 May	Sale of 20 tickets @ £2 each
15 May	Hire of equipment for event £10
17 May	Sale of 25 tickets @ £2 each
20 May	Income from sale of programmes £14
20 May	Sale of tickets at door £50

3) Annual Works Outing

5 July	Received £2 from each of 70 employees
10 July	Hire of two coaches and drivers £75
10 July	Received £35 subsidy from works
15 July	Cost of 72 teas @ £1 each
15 July	Cost of entertainment £25

4) School Sale of Work

10 Jan	Hire of stalls £15
15 Jan	Cost of decorating stalls £7
17 Jan	Income from raffle tickets £15
17 Jan	Income from tickets at door £12
17 Jan	Good-as-new stall profit £56
17 Jan	Other stalls profit £75

PETTY CASH BOOK

The money that offices keep for small everyday expenses, such as stamps, parcel post, etc., is known as *petty cash*. The record of how the money is spent is kept in a petty cash book which might be set out as shown on the opposite page.

The petty cash book is usually ruled into columns so that it is easy to see how much has been spent on postage, travelling expenses, etc.

The office has some money in the petty cash box at the beginning of the week and this is known as "cash in hand" and is entered as *balance in hand* in the petty cash book. (In our example it is £30.00.)

In this example the total week's expenses are £25.31.

So that the cash in hand will be the *same at the start of the following week*, £25.31 will be put into the petty cash box. This is entered in the petty cash book as *reimbursement*. Since the money *goes into* the box it is entered in the left hand column under "receipts".

The book is then balanced in a similar way to the cash statement on the previous page. The "balance carried down" is calculated on a separate piece of paper by taking the total payments away from the total receipts:

	£
Total receipts	55.31
Total payments	25.31 −
Balance carried down	30.00

This is then put into the petty cash book as shown.

Receipts	Date	Particulars	Payments	Postage	Stationery	Travelling expenses	Office expenses	Cleaning
30.00	3 May	Balance in hand						
	4 May	Postage	1.24	1.24				
	5 May	Envelopes	0.55		0.55			
	5 May	Bus fares	1.70			1.70		
	5 May	String	0.62		0.62			
	6 May	Window cleaner	4.00					4.00
	6 May	Parcel post	0.70	0.70				
	6 May	Pencils	0.40		0.40			
	7 May	Office tea	0.60				0.60	
	7 May	Rail fares	3.50			3.50		
	7 May	Cleaners' wages	12.00					12.00
		Totals	25.31	1.94	1.57	5.20	0.60	16.00
25.31	7 May	Reimbursement						
		Balance carried down	30.00					
55.31			55.31					

--- **EXERCISE 2** ---

1) Rule lines as shown in the example above and enter these items as if you were writing them into a petty cash book

    5 Jan Petty cash balance in hand £40.00
    6 Jan Postage £3.10
    6 Jan Window cleaner £3.50
    6 Jan String 65 p
    7 Jan Ballpoint pens 35 p
    7 Jan Rail fares £4.50
    8 Jan Cleaners' wages £10
    8 Jan Taxi fares £4.50
    9 Jan Office coffee 95 p
    9 Jan Telegram £2.10
    9 Jan Envelopes 65 p
    9 Jan Reimbursement £30.30

2) Draw up a page of a petty cash book with these items

    4 May Cash in hand £30.00
    4 May Envelopes 60 p
    4 May Office tea 70 p
    5 May Pencils 35 p
    5 May Window cleaner £2.50
    5 May Taxi fare £1.75
    5 May Postage £1.35
    6 May Typewriter ribbon £1.10
    6 May Wages for cleaning £15
    6 May Bus fares £1.45
    7 May Parcel post £1.35
    7 May Office coffee 92 p
    8 May Photocopying £1.04
    8 May Ballpoint pens 55 p
    8 May Reimbursement £28.66

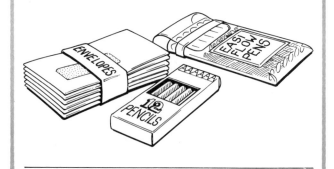

INVOICES

An invoice is a record of goods supplied to a customer, together with the cost. It is *not* a bill although it looks similar.

The invoice has the supplier's name, address and telephone number, the cost, number and type of articles and various totals and reference numbers. Space is also left for discounts and VAT. (By law discounts must be applied before VAT is calculated.)

The column labelled "gross" is calculated by multiplying "qty" (quantity) by "unit cost".

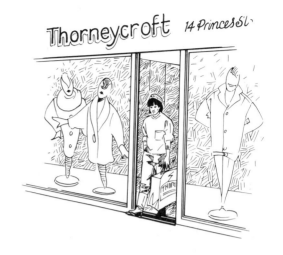

INVOICE

D. THORNEYCROFT
14 PRINCES ST
WILDON
GB6 8LX

125-6301

Ref. No	Qty	Description	Unit Cost	Gross	Discount	VAT
A/501	3	Ladies coats	£45	£135.00	10%	$17\frac{1}{2}$%
A/603	5	Mens jackets	£60	£300.00	10%	$17\frac{1}{2}$%

Less 10% discount	£435
	43.50
	391.50
Plus $17\frac{1}{2}$% VAT	68.51
	£460.01

Despatched
by Harvey &
Peters Deliveries

137

Copy these simplified invoices and complete the last column in each.

1)

	J. SMITH & CO 17 LONDON RD TAXERBY TEL 62-1251		
Qty	**Description**	**Unit Cost**	**Gross**
20	Children's dresses	£2.50	
15	Prs. children's socks	£0.50	
10	Children's cardigans	£1.50	
20	Small woolly hats	£0.50	
Discount 0.00%			
VAT 0.00%			
SALES INVOICE			

2)

	P. WILLIAMS LTD 44 SMITH ST BAXBY TEL 28-1134		
Qty	**Description**	**Unit Cost**	**Gross**
40	Small cakes	6 p	
20	Loaves bread	30 p	
60	Buns	5 p	
25	Pkts. biscuits	4 p	
Discount 0.00%			
VAT 0.00%			
SALES INVOICE			

Copy each of the simplified invoices below and fill in the last column. Calculate the total. Work out 10% discount (by dividing by 10) and take this away from the total. Fill in the final box.

1)

SELBERTON 5428	INVOICE		MR. SINGH 24 ASH AVE. SELBERTON
Qty	**Description**	**Unit cost**	**Gross**
4	Bags apples	£2.00	
5	Boxes oranges	£8.00	
3	Crates pears	£5.00	
TOTAL GROSS			
LESS 10% **DISCOUNT**			
VAT ZERO RATED			

2)

RANDALLS 3 LONG LA TYDFORD	INVOICE		TYDFORD 121
Qty	**Description**	**Unit Cost**	**Gross**
20	Single sheets	£ 5.25	
25	Double sheets	£ 8.00	
10	Pillow cases	£ 4.75	
15	Bedspreads	£24.00	
TOTAL GROSS			
LESS 10% **DISCOUNT**			
VAT 0.00%			

Make up your own invoices for each of the following suppliers

3) From: R. Evans
            101 Dark Street
            Bevington
            Tel. Bevington 3035

Discount 10%
VAT zero rated

33 jars coffee @ 50 p per jar
20 tins soup @ 15 p per tin
10 boxes biscuits @ £2 per box
50 tins fruit @ 20 p per tin
40 pkts sugar @ 30 p per pkt

4) From: J. Saunders
            45 James Street
            Oxminster
            Tel. Oxminster 1139

Discount 10%
VAT zero rated

10 Maths books @ £2.10
30 French books @ £3
15 English books @ £2
10 Geography books @ £4.35
25 History books @ £2.50

12

Wages and Salaries

If you work for an employer you are paid a wage or a salary. Wages are usually paid weekly and salaries monthly.

PAYMENT BY THE HOUR

The amount paid per hour up to a certain limit (called the *basic week*) is known as the hourly or *basic rate*. After that overtime is paid.

EXAMPLE

A man works a basic week of 38 hours at a basic rate of £3 per hour. Calculate his total wage for the week.

38 hours at £3 per hour = 38 × £3 = £114

His total wage is £114.

EXAMPLE

£164 is paid for a basic week of 40 hours. What is the hourly rate?

Hourly rate = $\dfrac{£164}{40}$ = £4.10

1) A woman works a basic week of 40 hours and her basic rate is £6 an hour. Calculate her total wage for the week.

2) £200 is paid for a basic week of 40 hours. Calculate the hourly rate.

3) Calculate the total wage for a 38 hour week if the basic rate is (a) £4 (b) £4.50 (c) £6.15 (d) £7.25

4) Calculate the hourly rate for a 40 hour week if the total wage is (a) £320 (b) £160 (c) £180 (d) £200.

5) The basic week is 37 hours and the basic rate is £4 an hour. Find the total wage for a basic week.

6) What is the total wage for a basic week of $37\frac{1}{2}$ hours at a basic rate of (a) £5 (b) £5.50 (c) £4 (d) £4.50 per hour?

OVERTIME

Work done extra to the basic week is usually paid at "time and a quarter", "time and a third", "time and a half" or "double time".

EXAMPLE

Find the hourly rate of overtime at (a) time and a half (b) time and a third on a basic rate of £4.80.

Overtime at time and a half $= 1\frac{1}{2} \times$ £4.80

$= £4.80 + £2.40$

$= £7.20$

Overtime at time and a third $= 1\frac{1}{3} \times$ £4.80

$= £4.80 + £1.60$

$= £6.40$

EXAMPLE

A basic rate of £5.60 is paid. Find the hourly rate of overtime at (a) double time (b) time and a quarter.

Overtime at double time $= 2 \times$ £5.60

$= £11.20$

Overtime at time and a quarter $= 1\frac{1}{4} \times$ £5.60

$= £5.60$

$+ £1.40$

$= £7.00$

─────── **EXERCISE 2** ───────

1) Find the hourly overtime rate at time and a half on a basic rate of (a) £4 (b) £7 (c) £6.20 (d) £3.60.

2) What is the hourly overtime rate at time and a third on a basic rate of (a) £6 (b) £9 (c) £3.30 (d) £4.20?

3) A basic rate of £4.20 is paid. Find the hourly overtime rate at (a) double time (b) time and a quarter.

4) Time and a quarter is paid for overtime. What is the hourly overtime rate on a basic rate of (a) £4 (b) £8 (c) £2 (d) £3?

5) Find the hourly overtime rate at time and a half on a basic rate of (a) £7.56 (b) £5.98 (c) £3.79 (d) £4.57.

─────── **PIECEWORK** ───────

Sometimes people are paid a fixed amount for each piece of work they make.

Often if they can make more than a certain amount they are paid a *bonus*.

EXAMPLE

A seamstress in a factory is paid 10 p for each handpuppet she makes up to 200 per day. For each handpuppet over 200 she is paid a bonus of 3 p. Calculate how much she earns if she makes 250 handpuppets.

Money earned on first 200 $= 200 \times 10$ p

$= £20.00$

Money earned on next 50 $= 50 \times 13$ p

$= £6.50$

Total earned $= £20 + £6.50 = £26.50$

─────── **EXERCISE 3** ───────

1) A woman is paid 3 p for each lb of sausages that she packs up to a limit of 500. After that she earns a bonus of 1 p. Calculate how much she earns if she packs 600 lb of sausages.

2) A man is paid 4 p for each spot weld he makes up to a limit of 400. After that he earns a bonus of 2 p. Calculate how much he earns if he makes 600 spot welds.

3) 3 p is paid for each handle fixed up to a limit of 300 after which 4 p is paid per handle. Calculate how much is paid if 350 handles are fixed.

4) A pieceworker receives 3p for each buttonhole sewn up to a limit of 80. After that she earns 4p per buttonhole. Find how much she earns if she sews 150 buttonholes.

5) For each article up to a limit of 250 a pieceworker is paid 6 p. After that she is paid 7 p per article. Find how much she earns for completing 300 articles.

6) A man is paid 15 p for each plate he decorates up to a limit of 100 plates. After that he is paid a bonus of 5 p per plate. Find how much he earns if he decorates 120 plates.

COMMISSION

Sales people are usually paid a commission on top of their basic wage. This is calculated as a percentage of the value of the goods they sell.

EXAMPLE

A saleswoman is paid a commission of 2% on goods sold. Calculate the commission if she sells goods worth £500.

$$2\% \text{ of } £500 = \frac{2}{100} \times £500 = £10$$

The commission is £10.

EXAMPLE

A shop assistant is paid a basic wage of £80 and she sells goods to the value of £1000 on which she gets 2% commission. How much does she earn?

$$\text{The commission} = 2\% \text{ of } £1000$$
$$= \frac{2}{100} \times £1000 = £20$$

She earns £80 + £20 = £100.

EXERCISE 4

1) Find the commission at 2% on goods worth
(a) £100 (b) £200 (c) £250 (d) £300 (e) £575.

2) A car is sold for £2000 and the salesman gets 3% commission. How much does the salesman get?

3) 3% commission is paid on goods worth £400. How much is this?

4) Furniture worth £800 is sold by a salesman who gets $2\frac{1}{2}\%$ commission. Find the amount the salesman gets.

5) A woman earns £50 a week and gets 2% commission on goods sold worth £300. Find her earnings for the week.

6) Commission of 2% is paid on a sale of £400. How much is the commission?

SALARIES

Salaried workers, such as teachers or civil servants, earn a fixed amount per year. The money is usually paid monthly. They do not earn commission or get paid for overtime.

EXAMPLE

A civil servant earns £14 400 p.a. How much is this per month?

Monthly salary $= £14\,400 \div 12 = £1200$

EXERCISE 5

These are the salaries of eight people. How much is each paid monthly?

1) £8600
2) £9000
3) £9500
4) £10 000
5) £12 000
6) £16 000
7) £18 000
8) £24 000

Of course these are not the amounts that are actually taken home. To work out someone's *take home pay* certain amounts have to be deducted first. The main items are income tax (see Chapter 13), National Insurance and payments towards a pension fund.

Before deductions the wage (or salary) is known as the GROSS wage (or salary).

After deductions the wage (or salary) is known as the NET wage (or salary)

NATIONAL INSURANCE

Both employers and employees pay National Insurance. Employees pay a certain percentage of their gross wage (2% on the first £58 per week then 10%, in the mid 1990s). Some married women can pay at a reduced rate (3.85% in the mid 1990s). There are upper and lower limits fixed by the Chancellor of the Exchequer. These limits and the rate of National Insurance can vary from year to year. National Insurance pays for such things as sick pay and unemployment benefit.

PENSION FUND

Because the state pension is small, many firms operate their own private pension fund schemes to provide extra pensions. Both employer and employee contribute. The employee's contribution is about 5% or 6% of his or her annual (yearly) salary before deductions. The amount of pension received depends on length of service and earnings at the time of retirement. Some schemes are "transferable" when an employee changes jobs, but many are not.

			NET £1296.21
Payslip	**Totals to date**	**Deductions**	**This payment**
Mr. P. W. Green Assist. Sales Mngr.	Gross £24 000 Tax £4958.75 Pen £1440 NI £2046.72	NI(D) £170.56 Tax £413.23 Pen £120	Basic £2000
Pay Mar 96 by BACS			
Tax month 12 J. Smith Ltd			Total £2000 Dedns £703.79
	NI BK415002B	Code 352L	Net £1296.21
Joined 14 01 93	1 mth @ £24 000 p.a.		

EXAMPLE

A manager earns £24 000 a year. £4958.75 are deducted for income tax, £2046.72 for National Insurance and £1440 for the pension fund. Calculate the take home pay per month.

	£
The deductions are	4958.75
	2046.72
	1440.00
	8445.47

Yearly salary after deductions is

£24 000 − £8445.47 = £15 554.53

Monthly take home pay is

£15 554.53 ÷ 12 = £1296.21

The manager's payslip for the tax last month of the year is shown at the bottom of the previous page.

--- EXERCISE 6 ---

(A calculator will be useful for some of these questions.)

1) A man earns £6000 a year. £450 are deducted for income tax, £360 for National Insurance and £300 for his pension fund. Find his yearly take home pay after deductions. Divide this by 12 to find his monthly take home pay.

2) Calculate the amount paid into a pension fund at a rate of 5% on a yearly salary of £7000.

3) I earn £164 per week. I pay 2% on the first £58 per week and 10% on the rest in National Insurance. How much National Insurance do I pay per week?

4) A woman earns £85 per week. Her deductions are £14 for income tax, £6.50 for National Insurance and £4.25 for the pension fund. Calculate her take home pay for the week.

5) National Insurance is paid at the rate of 2% on the first £3016, then 10% on the rest of an annual salary of £8400. Find how much National Insurance should be paid per year.

6) How much is paid into the pension fund scheme if the rate is 6% on an annual salary of £6500?

7) An employer pays 14% of an employee's wage of £8000 towards National Insurance. How much does the employer pay?

8) A man's weekly wage is £105.75 and his total deductions are £34.73. What is his weekly take home pay?

9) Income tax of £21.25, National Insurance of £21.89 and payment to a pension fund scheme of £21 are deducted from a monthly salary of £420. What is the take home pay?

10) A woman is paid £8 per hour for a 40 hour week. Calculate her total weekly wage. Calculate the amount she pays weekly towards her pension if she pays 5% of her total weekly wage.

--- EXERCISE 7 ---

1) A man works a 40 hour basic week at £5.38 an hour. Calculate his total weekly wage.

2) £156 is paid for a basic week of 39 hours. Calculate the hourly rate.

3) Find the hourly overtime rate at (a) time and a half (b) time and a third on a basic rate of £6.60.

4) A man works a basic week of 40 hours at £6.10 an hour. He then earns overtime at time and a half. Find his weekly wage for 44 hours' work.

5) A woman earns £172. Her basic rate is £4 for a 40 hour week. She works 2 hours overtime. What is her overtime hourly rate?

6) A pieceworker earns 2 p for each article she makes up to a limit of 40 articles. She then earns 3 p per article. How much is she paid for 50 articles?

7) 2% commission is paid on a sale worth £600. How much is this?

8) How much commission at 2% would a salesman get if he sold £350 worth of goods?

9) What is the monthly gross salary of a manager who is paid £19 200 per year?

10) Find the weekly take home pay of a woman who earns £90 gross per week and has £4.36 deducted for National Insurance, £3.52 deducted for income tax and pays £4.50 towards her pension fund.

13

—— Money in the Community ——

—— COUNCIL TAX ——

Council Tax is a local tax collected to help pay for local services such as education, the police and libraries. It is a property tax, so (with a few exceptions) a bill will be sent to each house in the area.

The amount of Council Tax depends on the value of the property, which will fall into a valuation band. This is illustrated by the table below. Of course the band ranges and proportions may change in the future.

Band	Range of values	Proportion of tax payable
A	Up to £40 000	$\frac{6}{9}$
B	£40 000 to £52 000	$\frac{7}{9}$
C	£52 001 to £68 000	$\frac{8}{9}$
D	£68 001 to £88 000	$\frac{9}{9}$
E	£88 001 to £120 000	$\frac{11}{9}$
F	£120 001 to £160 000	$\frac{13}{9}$
G	£160 001 to £320 000	$\frac{15}{9}$
H	£320 001 and over	$\frac{18}{9}$

The council sets the level of Council Tax for Band D properties and properties in other bands pay the portion of the Band D amount as shown in the table.

EXAMPLE

A house is valued as £89 500. In which band is it?

89 500 is more than 88 001 and less than 120 000, so the house is in Band E.

EXAMPLE

A house in Band D is liable for £800 Council Tax per year. How much would a house in Band H be liable for?

Use a ratio method:

$\frac{9}{9}$ = £800

$\frac{1}{9}$ = £(800 ÷ 9)

$\frac{18}{9}$ = £(800 ÷ 9) × 18

  = £1600

The house in Band H would be liable for £1600.

A full Council Tax bill assumes that there are two adults living in the house who both qualify for Council Tax. (Certain people such as school leavers or full-time students aged 18 or 19 for whom Child Benefit is paid, or people resident in hospital, do not qualify.) If only one qualifying adult lives in the house, the Council Tax bill is reduced by 25%.

EXAMPLE

A household consists of one qualifying adult and a school leaver, who does not qualify for Council Tax. The Council Tax for two adults would be £760 a year. How much would their Council Tax bill be?

$$25\% \text{ of } £760 = \frac{25}{100} \times 760$$

$$= £190$$

$$£760 - £190 = £570$$

The Council tax bill would be £570.

EXAMPLE

Mrs. Singh lives alone and pays £525 in Council Tax. She has a 25% reduction. How much would you expect her sister and brother who live together and both qualify for Council Tax to pay if they live in a house in the same valuation band?

Mrs. Singh has a 25% reduction, so she pays 75%.

$$75\% = £525$$

$$1\% = £525 \div 75 = £7$$

$$100\% = 100 \times £7$$

$$= £700$$

You would expect them to pay £700.

EXERCISE 1

Use the table on the previous page to answer these.

1) In which band is a house valued at
   (a) £36 000 (b) £55 000 (c) £750 000?

2) What proportion of Band D tax would be payable on a house in Band C?

3) A house is valued at £50 000. What proportion of Band D tax would be paid?

4) A house in Band D is sent a Council Tax bill for £810. How much would the Council Tax be for a house in (a) Band A (b) Band C (c) Band F?

5) The Council Tax bill for a house worth £70 000 is £783. How much would the Council Tax bill for a £100 000 house be?

6) If a full Council Tax bill is £800, how much would an elderly pensioner living alone be expected to pay if she can claim a 25% reduction?

7) A single person pays £450 after a 25% reduction. What is the full Council Tax bill for a house in the same valuation band?

8) The Council Tax for a house in Band D is £783. How much would the Council Tax bill for a house in Band A be (a) for two qualifying adults (b) for a single person who qualifies for a 25% reduction?

9) The Council Tax bill for a house in Band E is £792 per year. How much is this per calendar month?

10) One Local Authority charges £667 per year for a house in Band D, while another charges £758 per year for a similar house. How much more does the second Local Authority charge per week?

MORTGAGES

If a person buying a house cannot pay outright, he or she arranges a loan called a *mortgage* from a Building Society or a Bank. They generally have to pay a *deposit* of about 5% or 10% although occasionally 100% mortgages are available.

Cost
£42 000

Deposit £2000
Mortgage £40 000
(loan from the Building Society)

Repayments about £299 a month @ 7.5% interest over 25 years

The loan + interest is paid back over perhaps 20 or 25 years. The interest rate sometimes varies according to whether the Building Society has plenty of money available or not. (If more people save with a particular Building Society it will have more money for mortgages.)

EXAMPLE

A woman wishes to buy a house costing £24 000. She pays 10% deposit. How much mortgage does she need?

Deposit = 10% of £24 000 = £2400 (to find 10% divide by 10)

She needs £24 000 − £2400 = £21 600 mortgage

EXAMPLE

A bungalow cost £136 500. The deposit is 10%. How much is the deposit? What mortgage is needed?

Deposit = 10% of £136 500 = £13 650

The mortgage needed is

£136 500 − £13 650 = £122 850.

EXAMPLE

The monthly repayments of a mortgage are £352. What is paid per year?

Yearly repayments = 12 × £352 = £4224

──── **EXERCISE 2** ────

Copy and complete this table

House price	Deposit @ 10%	Mortgage needed	Monthly repayments	Yearly repayments
£37 500			£243	
£42 500			£277	
£53 000			£344	
£55 000			£358	
£60 000			£390	
£75 500			£488	
£103 000			£673	
£125 000			£812	

INSURANCE

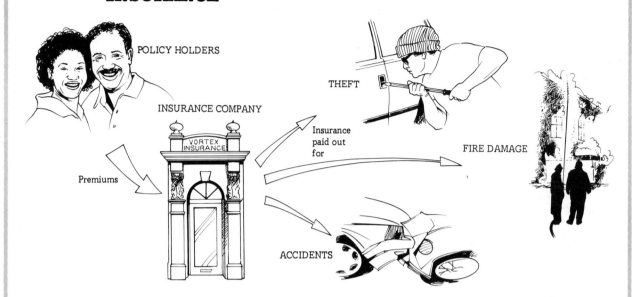

POLICY HOLDERS

INSURANCE COMPANY

VORTEX INSURANCE

Premiums

Insurance paid out for

THEFT

FIRE DAMAGE

ACCIDENTS

The loss from being burgled or suffering fire damage can be so serious that many people take out *insurance policies*. (They are then called *policy holders*.) They pay an insurance *premium* each year. The Insurance Company collects the premiums from thousands of people and invests the money to earn interest. This money is then available to pay out to the people who *claim* for their loss.

People can also insure against injury and car accidents. A car has to be insured by law. Car owners who have had no accidents are usually allowed a *no claims bonus* which means they do not have to pay so much premium.

EXAMPLE

A house and its contents are valued at £60 000. An insurance company charges a premium of £3 per £1000 of insurance. How much premium is charged per year?

Yearly premium $= 60 \times £3 = £180$

EXAMPLE

The insurance on a car is £120 but the owner is allowed $33\frac{1}{3}\%$ no claims bonus. How much does he pay per year?

$$33\frac{1}{3}\% = \frac{1}{3} \text{ so he allowed } \frac{1}{3} \text{ of £120} = £40$$

He has to pay £120 − £40 = £80 per year

EXERCISE 3

1) An insurance company charges £1.50 per £1000 of insurance. Find the yearly premium to insure for £3000.

2) An insurance company charges £1.50 per £1000 of insurance. Find how much has to be paid per year to insure for £8000.

3) The insurance on a car is £100 but 20% no claims bonus is allowed. How much premium is paid per year?

4) 20% no claims bonus is allowed on insurance for a car of £120. What is the premium?

149

5) If the premium on £5000 is £10 how much would the premium be on (a) £1000 (b) £10 000 (c) £4000?

6) If the premium on £6000 is £9, what would it be on (a) £1000 (b) £7000?

———— LIFE ASSURANCE ————

A person may take out life assurance if, when he or she dies, they want their *dependants* (widow/widower, children) to receive a lump sum of money.

The premiums depend on (a) *age* (the older the person, the more money he or she has to pay) and (b) *the amount of money the widow/widower and children are to receive* (the larger the amount the larger the premium).

With *Endowment Assurance*, at the end of a fixed period of time a lump sum is paid anyway.

With *Whole Life Assurance* there is no fixed time.

Some assurance schemes include a *bonus that is based on the profits* that the Insurance Company makes.

———— EXERCISE 4 ————

Look at the table below, which is for a 10 year Endowment Plan with bonus and then answer the questions.

1) What is the minimum guaranteed sum assured if a man pays a net monthly premium of £10 from age (a) 30 (b) 35 (c) 37 (d) 40?

2) What is the possible maturity value if a woman pays a net monthly premium of £50 from age (a) 25 (b) 29 (c) 36 (d) 39?

10 YEAR BONUS ENDOWMENT PLAN					
The *Guaranteed Sum Assured* is the minimum amount your policy is guaranteed to pay. The *Possible Maturity Value* is the possible return with bonus.					
Age today		For net monthly premium of £10		For net monthly premium of £50	
Male	Female	Guaranteed sum assured	Possible maturity value	Guaranteed sum assured	Possible maturity value
Up to 28	Up to 32	£ 1257	£ 2133	£ 6649	£ 11 283
29	33	1257	2133	6648	11 281
30	34	1256	2132	6646	11 278
31	35	1256	2132	6643	11 273
32	36	1255	2129	6641	11 270
33	37	1255	2129	6638	11 264
34	38	1254	2128	6634	11 258
35	39	1253	2126	6630	11 251
36	40	1252	2125	6624	11 241
37	41	1251	2123	6618	11 231
38	42	1249	2119	6610	11 217
39	43	1248	2118	6601	11 202
40	44	1246	2115	6592	11 187

3) What is (a) the minimum guaranteed sum assured and (b) the possible maturity value if a man pays £10 a month from age 32?

4) If a man pays £10 a month over 10 years how much has he paid to the insurance company altogether?

5) What is (a) the minimum guaranteed sum assured and (b) the possible maturity value if a woman pays £50 a month from age 40?

6) If a woman pays £50 a month over 10 years how much has she paid in premiums altogether?

INCOME TAX

Every person who earns money above a certain level pays income tax. The government uses the tax revenue to pay for the National Health Service, roads, the Civil Service, etc.

Tax is not paid on the whole income. Allowances are made for some expenses incurred during work, dependent relatives, etc. A *personal allowance* is made which varies according to whether the tax payer is single or married.

The income left after the allowances have been deducted is known as *taxable income*:

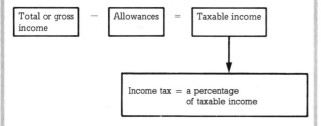

(the Chancellor of the Exchequer may vary the rate of tax in the Budget).

EXAMPLE

A man's salary is £9000. He is allowed a single person's personal allowance of £3000 and income tax relief on his pension fund payments of £540. Calculate his taxable income. He pays tax at the rate of 20%. Find the amount he has to pay in income tax.

Allowances:	£
Personal	3000
Pension fund payments (superannuation)	540
	3540

Taxable income = £9000 − £3540 = £5460

Tax payable = 20% of £5460 = $\frac{20}{100} \times £5460$

= £1092

"PAYE"

People in employment pay income tax by a method known as "Pay As You Earn" or PAYE for short. The tax is taken off their wage or salary before they receive it.

The tax payer and the employer receive a notice of coding which gives the employee's allowances (based on forms he or she has filled in) and gives a code number. The employer then knows from tax tables the amount of tax to deduct from the wages of the employee.

A typical notice of coding is shown overleaf.

EXERCISE 5

1) A man has a taxable income of £2000 and pays tax at the rate of 20%. How much does he pay in income tax?

2) From an income of £6000, allowances of £4200 are deducted. Tax is paid on the rest at a rate of 20%. Find the taxable income and the amount of income tax paid.

3) A man earns £8000. He has a personal allowance of £4000 and is allowed relief on his pension fund payments of £360. Find his total allowable deductions. Work out his taxable income. Find the amount of income tax he pays at 20% on his taxable income.

Notice of your income tax code

Inland Revenue

Your employer or pension payer will use your tax code to ensure you pay the right amount of tax under Pay As You Earn (PAYE).

From: H M Inspector of Taxes

LONDON PROVINCIAL 6
REDE HOUSE
71 CORORATION ROAD
MIDDLESBOROUGH
CLEVELAND

996
340989 007865

MR F JONES
54 KENSINGTON ROAD
DARTFORD

DF9 8WR

You should quote both numbers below if you contact us.

996/K99/Z

| VB | 31 | 66 | 88 | C |

Date 28 02 95

The net allowances figure below shows how much of your pay or occupational pension will not have tax deducted from it. This figure is made up of your total allowances less the total reductions shown below. This notice of tax code replaces any previous notice for the year.

If you think the code is wrong or if your personal circumstances change, you should tell your tax office immediately. The 'See note' column refers to 'Notes in PAYE Understanding Your Tax Code' leaflet.

See note	Allowances	£
17	Personal allowance	3525
	Total allowances	**3525**

See note	Less amounts taken away to cover items shown below	£
25	Medical/health ins.	563
	Less total reductions	**563**

Net allowances – the amount of your pay or occupational pension from which tax will not be deducted **2962**

Your tax code is worked out from your net allowances. Your tax code for the year to 5 April 1996 is 296L. See note A overleaf.

---

4) The total (gross) income of a woman is £5600. Her allowances are

| Personal allowance | £3765 |
| Expenses | £55 |

Calculate her total allowances. Find her taxable income.

She pays income tax at 20%. Find the amount of tax she pays. Find her take home pay when she pays £336 National Insurance.

5) Allowances of £4415 are allowed on an income of £7435. Find the amount of income tax paid at 20% and the take home pay after making National Insurance contributions of £400.

In earlier questions just one rate of income tax has been used. In practice, income tax is charged at different rates, e.g. no tax on £3000 a year.

The tax rate might be 20% for an income of £5000 a year (not on all of the £5000 because there would be some allowances).

For £10 000 a year it might be partly at 20% (lower rate) and partly at 24% (basic rate).

For an income of £40 000 a year on taxable income part could be at 20% (lower rate), part at 24% (basic rate) and part at 40% (higher rate).

EXAMPLE

The rates of income tax proposed for 1996-97 are

		Taxable income
Lower rate	20%	£1–£3900
Basic rate	24%	£3901–£25 500

Work out the tax payable on a taxable income of £7900.

Taxable income is £7900
Lower rate tax on first £3900 =
   20% × £3900 = £780
   £7900 − £3900 = £4000
Basic rate on next £4000 =
   24% × £4000 = £960

Total tax payable is £780 + £960 = £1740

Using the rates in the example above work out
(a) the lower rate tax   (b) the basic rate tax
(c) the total tax payable on a taxable income of

1)	£8900	5)	£12 600
2)	£9900	6)	£13 600
3)	£10 900	7)	£15 600
4)	£11 900	8)	£19 900

1) A Council Tax bill is £612 per year. How much is this per calendar month?

2) Mr Patel pays £590 Council Tax but his brother pays £650. How much more does his brother pay?

3) A woman pays 10% deposit on a house costing £47 500. How much mortgage does she need?

4) The monthly repayments of a mortgage are £320. How much is this per year?

5) How much has to be paid to insure for £7500 if the insurance company quotes £1.50 per £1000?

6) $33\frac{1}{3}$% no claims bonus is allowed on car insurance of £150. What is the premium after the no claims bonus has been deducted?

7) If a man pays £10 a month for life assurance over 15 years how much has he paid altogether?

8) Calculate the income tax to be paid at 20% on a taxable income of £3150.

9) Calculate the total amount of tax to be paid if 20% tax is paid on the first £3900 and 24% tax is paid on the next £20 000 of taxable income.

14

Household Bills and Personal Loans

A typical gas bill is shown opposite. It gives the name and address of the Gas Company, the name and address of the gas user, the present and previous gas meter readings, the amount of gas used in cubic feet, cubic metres and kilowatt hours (shortened to kWh), the cost per kilowatt hour, and the total cost of gas used. There is also a *standing charge* (of 10.10p per day). The total gas bill to be paid is in the "total" box at the bottom of the bill. VAT is added to a gas bill (8% in the mid 1990s). Notice that on the gas bill opposite 8% of 52.12 is *rounded down* to the nearest penny (from £4.1696 to £4.16). Bills are paid *quarterly* so you get *four* bills a year.

─────────── EXERCISE 1 ───────────

1) Look at the gas bill opposite and answer these questions:

   (a) How much does the customer have to pay?

   (b) Write down the present meter reading and subtract from it the previous meter reading. How many hundred cubic feet have been used?

   (c) What are the number of kilowatt hours used?

2) Gas is measured first in hundreds of cubic feet and this is changed into cubic metres and then into kilowatt hours. (The formula to do this is described on the back of the gas bill and involves the calorific value). *To find the cost of gas used we multiply the number of kilowatt hours by the cost per kilowatt hour.* Copy and complete this table.

Cost per kWh	Number of kWh used	Cost of gas used	
		In pence	In pounds
1.400 p	2000		
1.450 p	5000		
1.477 p	3000		
1.496 p	4980		
1.520 p	3356		

3) Some gas meters look like this

which shows a reading of 1720 hundred cubic feet.

154

British Gas

H VAT registration number
232 1770 91

**PO Box 50
Leeds LS1 1LE**

If you have an enquiry please ring

....

Customer reference number
807   333   1054

MR A PATEL
14 SMITH STREET
HUDDERSFIELD   HD2 1JP

Date of bill (tax point)
07 DEC 94

Date of meter reading
06 DEC 94

Calorific value (MJ/m^3)
40.3

Gas Bill

Meter reading _see below for key_		Gas used			Costs
Present	**Previous**	**100s cubic feet**	**Cubic metres**	**Kilowatt hours (kW h)**	
608	516 E	92	260.3	2913	43.03

Standing charge period 07/09/94 to 06/12/94

90 **days at** 10.10 p **per day** — 9.09

£52.12 **at** 8.00% **VAT** — 4.16

Tariff   Standard credit tariff 1.477 **pence per kW h**

Key to types of meter reading	Total **£**	56.28

Key to types of meter reading
E   **Estimated reading**
C   **Customer's reading**
X   **Exchange meter**

155

Copy these sets of dials and put in arrows to show readings of

(a) 3541 hundred cubic feet

(b) 1500 hundred cubic feet

(c) 9513 hundred cubic feet

4) Write down how many hundred cubic feet these meters show

(a)

(b)

(c)

Take care! If the arrows are *between* two numbers then take the smaller number *except* when the arrow is between 9 and 0. In this case take 9 as the reading.

5) Gas meters are read every quarter. If the reading was 2500 at the beginning of the quarter and 3712 at the end of the quarter how many hundred cubic feet were used?

_____

——— ELECTRICITY BILLS ———

Electricity is charged according to the number of units used. There is also a *fixed charge* which is added to the bill and VAT (8% in the mid 1990s) is charged. Bills are paid every quarter so there will be four bills per year.

——— EXERCISE 2 ———

Look at the electricity bill opposite and answer these questions

1) How much does the customer have to pay?

2) How many units were used?

3) What is the cost per unit?

4) What is the fixed charge?

5) Multiply the cost per unit by the number of units used and give your answer in pounds and pence (to the nearest penny).

6) Add the fixed charge and the VAT to the cost of electricity used and check that the bill is correct.

7) Give the total amount payable for this bill.

	£
fixed charge	10.13
1000 units @ 7.22 p per unit	_____
8% VAT	_____
	_____

8) On what date was the meter read?

_____

Your account number	3 191 6132 63

YE **Yorkshire Electricity**

Accounts advice telephone number ....

03/312 3 191 6132 63
MR D JONES
148 OAK LANE
LEEDS LS24 1JP

VAT registration number 545 3681 31

Tariffs and meter readings		Meter reading date	11 MAR 1994	Tax point and date of issue	16 MAR 1994	
Present	Previous	Units supplied		Fixed charges	Amount	VAT % rate
16732 16415 317 Domestic GD 317 @ 7.22p VAT on £33.02 at 8%				10.13	33.02 2.64	8
		AMOUNT NOW DUE AND PAYABLE £			35.66	

D Day units   N Night units   R Customers own reading   E Estimated

---

TELEPHONE BILLS

Telephone bills are paid quarterly so there will be four bills a year. There is a charge for *rental* (under advance charges) and a charge is made per call.

The cost of a call depends on the time of day and whether the call is *local* (within the local call area), *regional* (up to 35miles) or *national* (over 35 miles). There are also different call charges for international calls and for mobile phones. VAT is charged at the rate of 17.5%.

EXERCISE 3

1) Give the total cost of these telephone bills.

	£
(a) rental	24.49
call charges	15.55
VAT @ $17\frac{1}{2}$%	_____
	_____

	£
(b) advance charges	35.49
call charges	108.54
VAT @ $17\frac{1}{2}$%	_____
	_____

2) A telephone bill is £63.92. The call charges were £24.40 and the VAT was £9.52. The rest of the bill was for rental. How much was paid for rental?

3) A family pay £252 a year for their telephone. How much is this per month?

4) Telephone bills for the year were £48.70, £31.90, £46.50 and £35.90. How much was spent on the telephone during the whole year?

5) A family expects to pay about £156 a year on their telephone. How much would they expect to pay per quarter?

Note: *There is an exercise on direct dialled phone charges in Chapter 15, p. 171.*

This is a typical telephone bill:

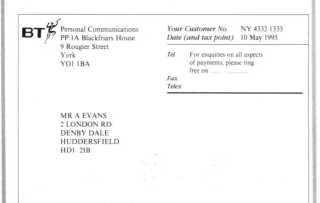

MOBILE PHONES

The tables that follow show the price per minute of calls to mobile phones and Personal Communication Network phones (shortened to PCNs) in the mid 1990s.

All calls are subject to a minimum charge of 5p. **Daytime (Mon. to Fri. 8 a.m.–6 p.m.)**		
	Mobile phones	*PCN phones*
pence per minute	41.05 p	16.71 p
Evenings, night-time & weekends (all other times)		
pence per minute	28.32 p	9.85 p

Note: *On the telephone bill each call charge is rounded down to the nearest tenth of a penny. The call charges are then added together and the total call bill is rounded down to the nearest penny before VAT is added.*

So the charge on a telephone bill for a 1 minute call in the evening to a PCN phone would be rounded down from 9.85 to 9.8 p (9 whole pence plus 8 tenths of a penny).

A total call bill of say £62.879 would be rounded down to £62.87 before VAT was added.

EXERCISE 4

Use the charts above. Give the answers to a tenth of a penny (rounded down if necessary) before VAT is added.

1) What is the cost of a 1 minute daytime call to a mobile phone?

2) Give the charge of a 5 minute evening call to a mobile phone.

3) How much does a 10 minute call to a PCN in the daytime cost?

4) Give the cost of a 2 minute call to a mobile phone at the weekend.

5) What is the cost of a 7 minute PCN call at noon on a Tuesday?

6) How much more would it cost to make a 10 minute call to a mobile phone in the daytime on a Monday rather than at the weekend?

7) A business call in the daytime to a mobile phone must not cost more than £5. What should the maximum time be, to the nearest minute?

8) How long was a PCN call on Saturday if it cost 98.5 p?

9) A girl makes a call to a mobile phone for 30 minutes each weekend. How much would she save per week if she halved the length of time she was on the phone?

10) Find the cost of (a) a 3 minute evening call to a PCN (b) a 2 minute daytime call to a mobile phone and (c) a 7 minute weekend call to a mobile phone. (d) Give the total cost of the three calls rounded down to the nearest penny before VAT is added.

EXERCISE 5

Shown below is part of an advertisement for a mobile phone.

What it costs you (VAT is included)

Mobile phone initial purchase	£4.99
Connection, normally £29.50	FREE!
Monthly rental	£14.99
Itemised billing (per month)	99 p
FIRST THREE MONTHS FREE!	
Peak rate calls (per min)	50 p
Off-peak calls (per min)	20 p

Use the above information to answer these questions.

1) What does a 4 minute peak rate call from a mobile phone cost?

2) How much does a 5 minute off-peak rate call cost?

3) Calculate the rental per year.

4) What would the itemised billing cost in the first year if the first three months are free?

5) Work out the cost of making 3 peak calls each lasting 4 minutes. (Use your answer to question 1.)

6) Jan makes, on average, 3 peak calls each lasting 4 minutes per week. How much would this cost per year?

7) Calculate the cost of making 7 off-peak calls each lasting 5 minutes. (Use your answer to question 2.)

8) Jan also makes, on average, 7 off-peak calls each lasting 5 minutes per week. What would this cost per year?

9) Estimate Jan's total cost during the first year. Include the cost of the initial purchase, the rental, the itemised billing charge and the cost of 3 four minute peak calls per week and 7 five minute off-peak calls per week.

— COST OF RUNNING A CAR —

To find how much it costs to run a car, *tax, insurance, petrol, repairs and maintenance*, and *depreciation* must all be taken into account. "Depreciation" means by how much the car has gone down in value during the year.

EXAMPLE

A car is bought for £2500 and after 1 year is valued at £2000. During the year it did 7500 miles at an average of 25 miles to the gallon. Petrol cost £2.00 a gallon on average through the year. Insurance cost £320, tax £140 and repairs and maintenance £500. Find the total cost of the year's motoring.

	£
Depreciation	500
Insurance	320
Tax	140
Repairs and maintenance	500
Cost of petrol	
$\dfrac{\overset{300}{\cancel{7500}}}{\underset{1}{\cancel{25}}} \times £2$	600
Total cost of year's motoring	£2060

EXERCISE 6

1) A man estimates the cost of his motoring as follows

Depreciation	£850
Insurance	£270
Tax	£140
Cost of petrol	£650
Repairs	£242

Find the total cost of the year's motoring.

2) A car is bought for £3000 and sold the next year for £2050. It travelled 5200 miles at 26 miles to the gallon and petrol cost £2.40 per gallon on average. Insurance was £220. Tax was £140 and repairs and maintenance £170. Find (a) the depreciation and (b) the cost of petrol for 5200 miles. (c) Calculate the total cost of the year's motoring.

3) How much does it cost to travel 150 miles at the average cost of 6p per mile?

4) If a man travels 2600 miles at a total cost of £130 find the average cost per mile.

5) A car is bought for £3500 and sold 4 years later for £1500. Find by how much the car has depreciated in value per year.

6) A woman travels 320 miles a week. She uses 8 gallons of petrol. How many miles per gallon is this?

7) A litre of petrol costs 52 p. How much do 7 gallons cost? (4.5 l = 1 gall).

8) If £12 a week are spent on petrol how much is spent per year?

9) A car averages 25 miles to the gallon. How far will it travel on 8 gallons? How many gallons are needed to travel 300 miles? If petrol costs £2.30 per gallon how much does it cost to travel 300 miles? The car is changed for a new one that does 30 miles to the gallon. How far will the new car travel on 8 gallons? How many gallons will be needed to travel 300 miles? How much will it cost to travel 300 miles in the new car at £2.30 per gallon for petrol? How much money is saved over 300 miles by having the new car?

VALUE ADDED TAX

Value Added Tax (VAT for short) is a tax which is added to some goods and services such as furniture and telephone bills. Some goods and services are exempt (bear no tax). VAT (at 8% in the mid 1990s) is added to electricity and gas bills. VAT on other goods was raised from 15% to $17\frac{1}{2}$% in the early 1990s.

EXAMPLE

8% VAT is charged on a fuel bill of £560. How much VAT will be charged?

$$\text{VAT} = 8\% \text{ of } £560$$
$$= \frac{8}{100} \times £560$$
$$= £44.80$$

press keys

$\boxed{8} \boxed{\div} \boxed{1} \boxed{0} \boxed{0} \boxed{\times} \boxed{5} \boxed{6} \boxed{0} \boxed{=}$

to get the result $\boxed{44.8}$.

EXAMPLE

A table costs £40. VAT at $17\frac{1}{2}\%$ is charged. Calculate the cost of the table plus VAT.

$$\text{VAT} = 17\frac{1}{2}\% \text{ of } £40$$

$$= \frac{17\frac{1}{2}}{100} \times £40$$

press keys

$\boxed{1} \boxed{7} \boxed{\cdot} \boxed{5} \boxed{\div} \boxed{1} \boxed{0} \boxed{0} \boxed{\times} \boxed{4} \boxed{0} \boxed{=}$

to get the result $\boxed{7}$.

Without a calculator, VAT at $17\frac{1}{2}\%$ could be worked out this way:

10% of £40 = £4 (divide by ten)

5% of £40 = £2 (half of ten)

$2\frac{1}{2}\%$ of £40 = £1 (half of five)

$17\frac{1}{2}\%$ of £40 = £7 $(17\frac{1}{2}\% = 10\% + 5\% + 2\frac{1}{2}\%)$

──────── **EXERCISE 7** ────────

1) Find 8% of (a) £60 (b) £400 (c) £612 (d) £1.20.

2) A woman's gas bill is £120 but VAT at 8% is charged on top of this. Find how much she pays altogether.

3) Find the VAT payable on an electricity bill of £20 if the rate of VAT is 8%.

4) A telephone bill excluding VAT is £50. Find the cost when VAT is added on. The rate of VAT is $17\frac{1}{2}\%$.

5) A man buys some tools for £4.60 but VAT at $17\frac{1}{2}\%$ has to be added to the bill. How much did the tools cost with VAT?

6) VAT at $17\frac{1}{2}\%$ is added to a bill for a meal which came to £3.60. How much was paid altogether?

7) Calculate the VAT payable on a garage bill of £240 if the rate of VAT is $17\frac{1}{2}\%$.

8) How much VAT is there on a bill of (a) £30 (b) £300 (c) £3000 if the rate of VAT is $17\frac{1}{2}\%$?

9) A lawnmower is priced at £80 plus VAT. How much is the lawnmower with VAT if the rate of VAT is $17\frac{1}{2}\%$?

10) A washing machine costs £220 exclusive of VAT. VAT is charged at $17\frac{1}{2}\%$. How much is actually paid for the washing machine?

───────────────────────

If you know the price *including* VAT and wish to know the cost without VAT then you have to use the following method:

EXAMPLE

The price of a coat is £23.50 including VAT. The rate of VAT is $17\frac{1}{2}\%$. How much does the coat cost without VAT?

Let 100% be the price without VAT so $117\frac{1}{2}\%$ is the price with VAT.
Use a ratio method.

$$117\frac{1}{2}\% = £23.50$$

$$1\% = £23.50 \div 117\frac{1}{2}$$

$$= £23.50 \div 117.5$$

$$100\% = £100 \times (23.50 \div 117.5)$$

This is best worked out using a calculator

$$= £20$$

So the coat costs £20 without VAT.

161

Take the rate of VAT to be $17\frac{1}{2}$%.

1) If a table costs £70.50 with VAT how much does it cost without VAT?

2) A radio costs £47 inclusive of VAT. What is the cost without VAT?

3) The cost of a box of chocolates is £2.35 with VAT. Find the cost of the chocolates exclusive of VAT.

4) A restaurant bill is £117.50 with VAT included. How much is it without VAT?

5) With VAT included a telephone bill came to £94. How much was the bill with VAT excluded?

6) A carpet cost £235 with VAT. How much was it without VAT?

——— **HIRE PURCHASE** ———

Hire purchase (or HP for short) is used when we buy things *now* but pay for them over the next few months. Some money is usually paid straight away. This is called a *deposit*. The remaining money to be paid is called the *balance*. Interest is charged on the balance.

EXAMPLE

A woman buys a washing machine for £240 and pays a deposit of 20%. 10% interest is charged on the balance. The balance + interest is paid in 12 equal amounts per month for 1 year.

Find (a) the deposit, (b) the balance, (c) the interest, (d) the total to be repaid and (e) the amount of each instalment.

Deposit = 20% of £240

$$= \frac{\overset{2}{\cancel{20}}}{\underset{10}{\cancel{100}}} \times \cancel{£240}^{24} = £48$$

Balance = £240 − £48 = £192

Interest = £10% of £192 = £19.20

Total to be repaid = £192 + £19.20

$= £211.20$

Monthly instalments = £211.20 ÷ 12

$= £17.60$

——— **BANK LOANS** ———

Many people take out personal loans from a Bank. The Bank charges *interest*. The *loan + interest* is usually paid back in equal monthly instalments.

EXAMPLE

A man borrows £300 from his Bank. The Bank charges 20% interest. He pays back the loan + interest in 12 equal amounts. Find each amount.

Loan = £300

Interest = 20% of £300 (or say 10% = £30)

$$= \frac{20}{100} \times £300 \quad \text{(so that 20\% = £60)}$$

$$= £60$$

Total amount to be repaid = £300 + £60

$$= £360$$

Amount of each instalment = £360 ÷ 12

$$= £30$$

──────── EXERCISE 9 ────────

1) If the HP payments on a record player are £4 per month, how much will be paid at the end of 3 years?

2) A woman buys £120 worth of kitchen equipment. She pays a deposit of £20. She is charged 10% interest on the balance.
   (a) How much is the balance?
   (b) How much is the interest?
   (c) If the balance + interest is paid back in 10 equal instalments how much is each instalment?
   (d) How much has she paid altogether?

3) The cash price of some furniture is £200. To buy on HP a deposit of 25% is made followed by 10 equal payments of £21. Find
   (a) the total paid for the furniture on HP
   (b) the difference between the cash price and the HP price.

4) A man borrows £100 from a Bank. 20% interest is charged. If he repays the loan plus interest in 12 equal amounts find how much he pays each time.

5) A Bank lends a woman £600. They charge interest at 20%. She pays the loan + interest back in 12 equal amounts. Find how much she pays back each time.

──────── EXERCISE 10 ────────

1) What is the cost of a gas bill of 3000 kW h at 1.5 p per kW h?

2) Find the amount to be paid on these gas bills

	£
(a) standing charge, 100 days at 10.13 p per day 1000 kW h at 1.5 p per kW h	_____
VAT at 8%	_____
	_____
(b) standing charge, 90 days at 10.30 p per day 3000 kW h at 1.6 p per kW h	_____
VAT at 8%	_____
	_____

3) How many hundred cubic feet do these meters show?

   (a)

   (b)

4) Give the total amount before VAT to be paid on this electricity bill

	£
fixed charge	11.20
1000 units @ 6.4 p per unit	_____
	_____

5) If a family pays about £52 per year on their telephone bill find how much they would expect to pay per quarter.

163

6) Give the total cost of this telephone bill

	£
Charge for rental of line	24.79
Call charges	58.26
VAT at $17\frac{1}{2}$%	_____
	_____

7) The cost of a year's motoring is

	£
depreciation	£750
insurance	£290
tax	£140
cost of petrol	£750
repairs	£190

Find the total cost of the year's motoring.

8) A car is bought for £2000 and sold 2 years later for £1000. Find the depreciation per year.

9) How much does it cost to buy 22 gallons of petrol at 55p per litre?

10) VAT at 8% is added to a bill of £500. How much VAT is paid? How much is the total bill?

11) With VAT at $17\frac{1}{2}$% included, electrical goods cost £2350. How much do they cost without VAT?

12) A woman borrows £500 from her Bank. The Bank charges 20% interest. She pays back the loan + interest in 10 equal amounts. Find (a) the interest (b) the total she pays back (c) the value of each of the 10 equal amounts.

15

Graphs and Charts

AXES, SCALES AND COORDINATES

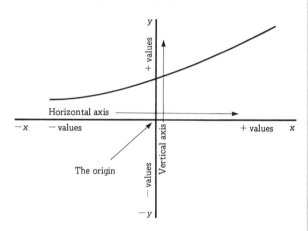

A graph is a picture of numerical information.

There are two axes: the horizontal axis and the vertical axis. The point where the axes cross is called the *origin*.

If all the numbers on the graph are positive (greater than zero) then the graph will look like this:

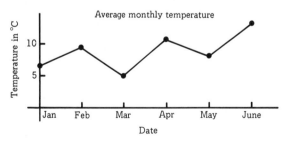

The horizontal and vertical axes should both be labelled.

There should be a *heading* at the top of the graph.

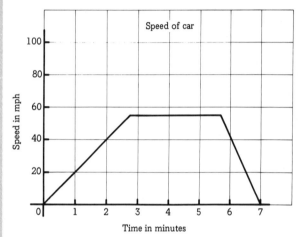

It is important when drawing a graph to choose a sensible *scale* so the graph fills the whole page (without going over the edge or being squashed into one corner!).

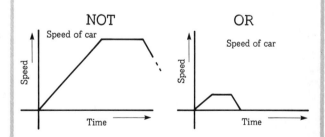

The *scale* on a graph is the *spacing* between the labels on the axes. Number labels on the **axes** should always be spaced *evenly*.

The scale of a graph can be used to deceive!

Wages in L.B. Davis Ltd.

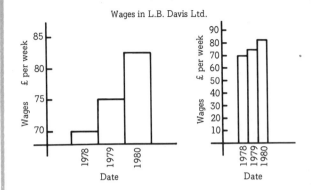

On the graph on the left it looks as if wages have risen a great deal from 1978 to 1980 but on the graph on the right they don't seem to have risen by very much.

The *coordinates* say where the points are on a graph.

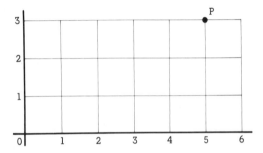

The point P has coordinates (5, 3). The first number, 5, says how far *along* and the second number, 3, says how far *up*.

──────── EXERCISE 1 ────────

1) Give the coordinates of the points A, B, C, D, E and F.

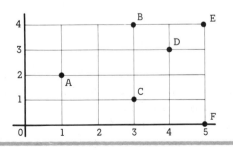

2) Give the coordinates of the points A, B, C, D, E, F, G, H, I and J. Copy the graph on to squared paper and draw a smooth curve through the points.

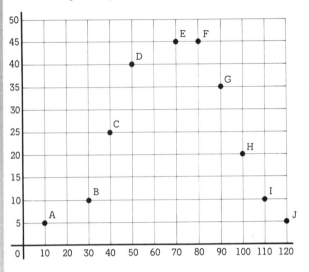

──────── **DRAWING A GRAPH** ────────

Graphs are drawn on squared paper. They show a relationship between two sets of numbers.

EXAMPLE

This table shows the average wage in J. Brown & Co. over 16 years:

Year	1950	1952	1954	1956	1958
Wage	£10	£12	£14	£16	£21

Year	1960	1962	1964	1966
Wage	£24	£29	£33	£40

Plot these points on a graph and from it estimate (a) the average wage in 1955 (b) the probable wage in 1968.

First draw two axes and then choose a suitable scale. (Our scale is 1 cm for 2 years and 1 cm for £10.)

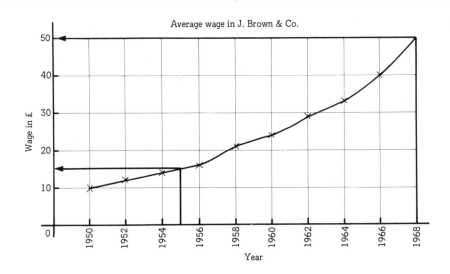

Average wage in J. Brown & Co.

Plot the points with a sharp pencil and mark them neatly with a small cross. For this graph you can draw a smooth curve through the points.

A graph may be used to find values not given in the table.

Reading from the graph: the average wage in 1955 was £15, the probable average wage in 1968 was £50.

Of course it is not always possible to draw a smooth curve through the points – some graphs are best drawn with straight lines joining the points. This must be decided by using common sense.

─────────── **EXERCISE 2** ───────────

1) Plot these points on a graph choosing suitable scales for the axes.

Age of child in years	2	4	6	8	10
Average height in cm	85	101	115	129	140

Draw a smooth curve through the points.

From your graph estimate (a) the height of a 5 year old child (b) the age of a child whose height is 135 cm.

2) A car was tested for petrol consumption at various speeds.

Speed in mph	10	30	50	70	80
Consumption in miles/gallon	20	35	40	30	23

Plot the points and draw a smooth curve through them.
(a) Give the petrol consumption at 40 mph.
(b) At what speeds is the consumption 25 miles to the gallon?
(c) What is the most efficient speed at which to drive the car?

3) Temperature can be measured in degrees centigrade (°C) or degrees Fahrenheit (°F). This diagram can be used to change °C to °F.

Temperature in °C

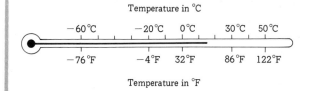

Temperature in °F

Plot a graph with this information. Join the points with a straight line. From your graph
(a) what is 20 °C in °F?
(b) what is 98 °F in °C?

167

4) This is part of a table to change cm to inches.

Cm	60	70	80	90	100
Inches	23.6	27.6	31.5	35.4	39.4

Plot these points on a graph and join them with a straight line. Use your graph to convert
   (a) 65 cm to inches (correct to one decimal place)
   (b) 37.5 inches to the nearest cm.

Look at the following sketches of graphs and then answer the questions.

5) (a) How many ounces in 250 grams (approximately)?
   (b) How many grams in 14 ounces (approximately)?

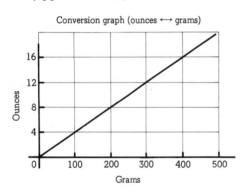

Conversion graph (ounces ⟷ grams)

6) (a) What is the fastest speed?
   (b) How long does it take to achieve the fastest speed?

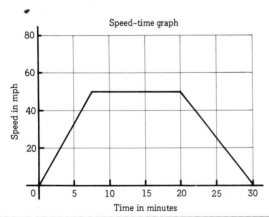

Speed-time graph

(c) For how long is it travelling at this speed?
(d) How long does it take to come to rest?

7) (a) How old was the tree when it measured 50 cm round its trunk?
   (b) When the tree was 50 years old what was its measurement round its trunk?

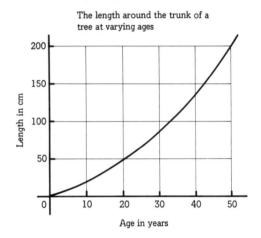

The length around the trunk of a tree at varying ages

8) (a) How far has the motorist driven?
   (b) For how long did the motorist stop driving?
   (c) What was the average speed between 6 and 7 p.m.?
   (d) What was the average speed between 3 and 5 p.m.?

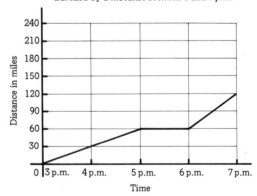

Distance-time graph showing the distance travelled by a motorist between 3 and 7 p.m.

The *rate of increase* or *slope* of a graph is found by drawing two dashed lines as shown and working out the value of:

Distance up ÷ Distance along

using the units of the graph.

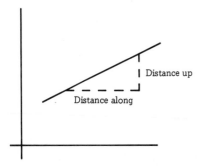

For a straight line graph this value should be the same no matter where the triangle is drawn.

This value is sometimes known as the *gradient* of the graph.

EXAMPLE

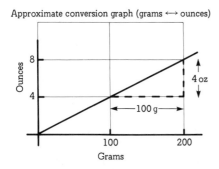

Approximate conversion graph (grams ⟷ ounces)

The rate of increase of ounces to grams is:

$$\frac{4\,oz}{100\,g} = 0.04\,oz/g$$

EXAMPLE

The slope of the next graph is $\dfrac{30\text{ miles}}{1\text{ hour}}$

$= 30$ mph

For a distance–time graph the slope gives the speed.

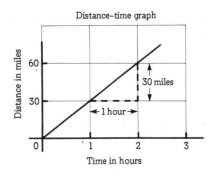

Distance–time graph

──────── **EXERCISE 3** ────────

1) Find the rate of increase of wages with respect to time in J. Smith Ltd.

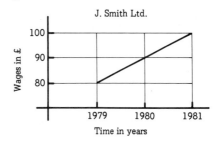

J. Smith Ltd.

2) Find the slope of this graph during the first 3 seconds.

(The slope of a speed–time graph gives the acceleration.)

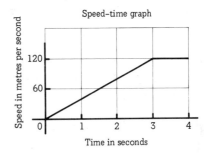

Speed–time graph

3) Find the gradient of this graph.

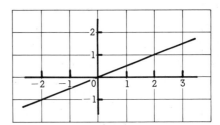

4) The slope of this graph gives a car's average speed. What is the average speed of the car?

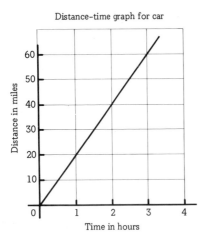

Distance–time graph for car

CHARTS

Charts should present information in a clear and easily understood form.

Bus fare charts are set out in this way:

The Grange

25	Royal Parade						
40	25	Queen St					
60	40	25	Fir Tree Drive				
64	60	40	25	West Road			
72	64	60	40	25	Mitford St		
76	72	65	60	40	25	The Croft	
80	76	72	64	60	40	25	Southdene St
80	80	76	72	64	60	40	25	Bus Station

To find the cost of travelling from Queen St to The Croft look down the column under the label Queen St until the row containing The Croft is reached. Then read off the fare: 65 p.

Look at the chart and answer these questions.

EXERCISE 4

How much does it cost to travel from
1) West Road to Southdene St?
2) The Grange to the Bus Station?
3) Royal Parade to West Road?
4) West Road to the Bus Station?
5) Queen St to Mitford St?
6) The Croft to Southdene St?
7) A girl has to go by bus from Fir Tree Drive to Mitford St to visit a friend and then go on from there to Southdene St for a meeting. How much more does it cost to break her journey than to go direct from Fir Tree Drive to Southdene St?

The chart opposite shows telephone charges for local calls, calls up to 35 miles away and calls over 35 miles away at daytime, evenings and night-time, and weekend rates. Prices are in pence and include VAT.

The next chart shows when the rates apply. For example, it is cheapest to make a call on Saturday and Sunday (weekend rate). The most expensive time to make a call is between 8 a.m. and 6 p.m. on Mondays to Fridays.

Telephone calls are priced according to *when* the call is made (daytime, evenings and night-time, or at the weekend) the *length* of the call (measured to the nearest second) and the *destination* of the call, i.e. whether it is *local, regional* (up to 35 miles away) or *national* (over 35 miles away).

On the bill, each call charge is rounded *down* to the nearest tenth of a penny. So if the charge of a 1 minute daytime national call was 9.78 p, it would be rounded down to 9.7 p.

The total call bill is rounded *down* to the nearest whole penny before VAT is added, so if the total call bill was £52.189 it would be rounded down to £52.18 before VAT was added.

The minimum charge for a call is 5 p.

Use the telephone charges charts given below to answer these questions. All answers should be given rounded down to a whole number of pence.

Telephone Charges

		Local	Regional (up to 35 miles)	National (over 35 miles)
Daytime	per minute	3.95 p	8.23 p	9.78 p
	5 min	19.7 p	41.1 p	48.9 p
	10 min	39.5 p	82.3 p	97.8 p
Evenings and night-time	per minute	1.65 p	3.95 p	5.81 p
	5 min	8.2 p	19.7 p	29.0 p
	10 min	16.5 p	39.5 p	58.1 p
Weekends	per minute	1.0 p	3.29 p	3.29 p
	5 min	5.0 p	16.4 p	16.4 p
	10 min	10.0 p	32.9 p	32.9 p

All calls are subject to a minimum charge of 5p.

	Mon Tue Wed Thu Fri	Sat Sun
Midnight–8 a.m.	Evenings and night-time rate	Weekend rate
8 a.m.–6 p.m.	Daytime rate	Weekend rate
6 p.m.–midnight	Evenings and night-time rate	Weekend rate

Different prices apply to mobile and pay phones.

1) A 5 minute local call is made at the weekend. What would this cost?

2) Give the cost of a 10 minute local call in the daytime.

3) What does a regional call lasting 5 minutes in the evening cost?

4) A man is on the telephone for 15 minutes in the evening to his aunt who lives 74 miles away. What would the call cost?

5) What would you expect a 20 minute local call made on a Sunday to cost?

6) What rate (daytime, evenings and night-time or weekend) is charged on a call made on a Friday at 11 p.m.?

7) Give the cost of a 4 minute regional call made at 2 p.m. on a Tuesday.

8) A woman is telephoned by her grandson who lives 25 miles away. He phones at the weekend and the call lasts 20 minutes. Work out the cost of the call. (It is not a local call).

9) I want to make a 10 minute phone call to someone who lives over 50 miles away. How much do I save by phoning in the evening rather than in the daytime?

10) How much time is allowed for a weekend local call costing 35 p?

11) A daytime call to a destination 80 miles away costs 97 p. How long was the call?

12) A national call lasting 10 minutes costs £0.97. Which rate was charged, daytime, evenings and night-time or weekend?

13) How much does a 2 minute weekend local call cost?

14) What is the total cost of 100 local daytime calls of 5 minutes each?

PIE CHARTS

A pie chart is another way of showing numerical information in picture form.

EXAMPLE

24 pupils were asked what sort of career they would like:

  8 chose work with animals

  3 chose office work

  2 chose teaching

  5 said they would like to work outdoors

  the rest said "don't know"

Hints for drawing pie charts:

(a) use a sharp pencil

(b) draw a reasonable size circle with a pair of compasses

(c) calculate the angles at the centre and put them in a table as shown opposite

(d) use a protractor to draw the angles

(e) label each sector carefully

(f) label the angles.

How to calculate the angles at the centre of the pie chart:

There are $360°$ in the whole circle. *Divide 360 by the total number of pupils.* This gives the angle for *one* pupil. (In our example it works out exactly but usually you have to *round to the nearest whole degree.*)

To find the angle of each section *multiply* by the number of pupils in this section. (It is often not possible to draw a pie chart completely accurately.)

Set the work out like this:

$360 \div 24 = 15$ There are $15°$ per pupil		
Sector label	*Number of pupils*	*Angle*
Work with animals	8	$8 \times 15° = 120°$
Office work	3	$3 \times 15° = 45°$
Teaching	2	$2 \times 15° = 30°$
Work outdoors	5	$5 \times 15° = 75°$
Don't know	6	$6 \times 15° = 90°$

EXERCISE 6

1) This pie chart shows the proportion of money spent by men, women and children on magazines in a shop.

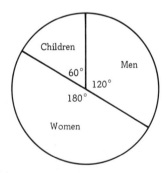

(a) Who spends the most?

(b) If the shop takes about £360 a day on magazines how much is this from
(i) women (ii) men (iii) children?

172

Hint: The best way to solve this problem is to think in terms of fractions of the circle (e.g. women buy $\frac{1}{2}$ the magazines, men $\frac{1}{3}$ and children the rest).

2) How a boy spends his £1.28 pocket money is shown by the pie chart below. How much does he spend on sweets? How much on comics? How much does he save? How much does he spend at the swimming pool?

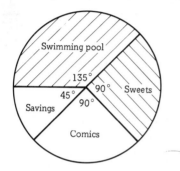

3) Draw a pie chart showing the proportion of trees in Radley Wood, using this table

Oak trees	45
Ash trees	90
Beech trees	120
Fir trees	45
Other	60

4) A family spends their week's money in this way

Food	£48
Clothes	£24
Electricity and gas	£30
Rent and Community Charge	£28
Other	£14

Show their expenditure on a pie chart.

BAR CHARTS

There are three types of bar chart: a single bar chart, a vertical bar chart and a horizontal bar chart.

EXAMPLE

A family spends a monthly wage of £600 in this way:

Food	£200
Clothes	£100
Heating	£80
Mortgage and Community Charge	£180
Other	£40
Total	£600

This information can be shown on:

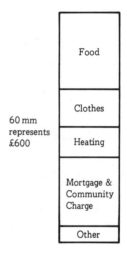

a *single bar chart* (1 mm height represents £10) or

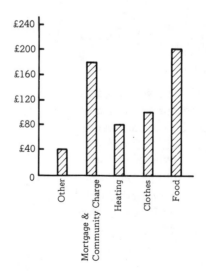

a *vertical bar chart* or

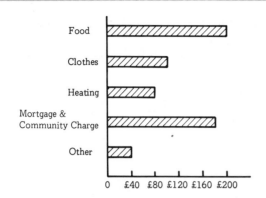

Food

Clothes

Heating

Mortgage &
Community Charge

Other

0   £40 £80 £120 £160 £200

a horizontal bar chart

On a single bar chart the *heights* of the various sections show how much is spent on each item in relation to the other items. The *lengths of the bars* on a vertical or horizontal bar chart show how much is spent.

─────────── **EXERCISE 7** ───────────

1) 100 people were asked how they travelled to work. 50 said by car, 30 said bus, 5 said train and 15 said they walked. Present this information on a single bar chart.

2) A sample survey was done to market a new range of soups. 500 people were questioned. 260 said they preferred Tomato. 140 said they liked Chicken best. 40 preferred Oxtail. 15 preferred Vegetable and 25 liked Mushroom best. The rest didn't know. Show this information on a vertical bar chart.

3) This horizontal bar chart shows how the sales of two types of washing machine varied from 1984 to 1990

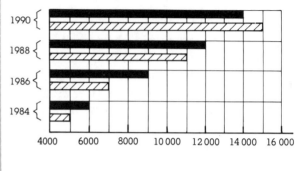

 Type A
Type B

(a) Was type A or type B more popular in 1984?
(b) Which was the more popular type in 1990?
(c) How many washing machines of type B were sold in 1988?
(d) How many more washing machines of type A were sold in 1990 than 1984?
(e) Would the manufacturer do better to make more washing machines of type B than type A in 1993?

─────────── **HISTOGRAMS** ───────────

The histogram is a special kind of graph with the area representing the frequency. The simplest type of histogram is a bar chart with all the bars the same width.

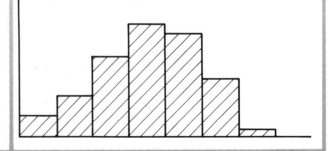

EXAMPLE

The diameters of 50 ball bearings were measured and the readings (in mm) were:

15.01	15.02	14.99	15.00	15.01
15.00	15.01	15.00	15.00	14.99
14.99	15.00	15.00	15.02	14.98
15.01	15.00	15.01	15.00	15.00
15.00	15.00	15.00	15.01	14.99
14.99	15.00	14.99	15.00	15.00
15.00	15.01	15.00	15.00	15.00
15.00	14.99	15.00	14.99	15.01
15.00	15.00	14.99	15.00	15.00
14.98	15.00	15.00	15.02	14.99

These numbers as they stand don't mean very much but if the measurements are grouped and the results put into a table then it is easier to see a pattern emerging.

Measurement	Tally marks	Frequency
14.98	I I	2
14.99	LHT LHT	10
15.00	LHT LHT LHT / LHT LHT I I	27
15.01	LHT I I I	8
15.02	I I I	3

Tally marks are simply an aid to counting. Every time a measurement of, say, 15.01, is counted a mark is put by 15.01 in the frequency table. The fifth mark is drawn across a group of four. The *frequency* of 15.01 is simply the number of times 15.01 appears in the above readings.

Once the results are displayed on a histogram it can be seen quite clearly that the most frequent measurement is 15.00 mm with a sharp falling off on either side (so the manufacturing process must be fairly accurate).

A histogram showing the diameters of a sample of 50 ball bearings:

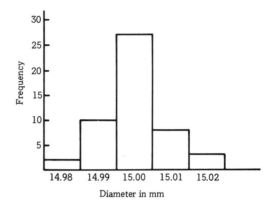

If the shape of the histogram had been more like this:

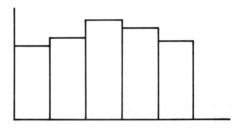

then it would become obvious that the manufacturing process was not so accurate, and the manufacturer would have either to take steps to remedy this or to decide whether the limits of accuracy were acceptable.

—————— **EXERCISE 8** ——————

1) The lengths of 36 steel bars were measured (in centimetres) with these results

35.95	35.98	35.95	35.94	35.96	35.94
35.93	35.96	35.94	35.95	35.92	35.94
35.96	35.97	35.95	35.93	35.93	35.95
35.94	35.95	35.93	35.92	35.95	35.93
35.93	35.97	35.94	35.94	35.94	35.94
35.94	35.92	35.95	35.96	35.93	35.94

Draw up a frequency table and from that draw a histogram.

2) These were the results of a "speed of reaction" test given to 72 people. The scores are out of 10.

```
5  3  7  6  1  4   6  3  9  4  9   4
4  5  4  3  4  5   5  6  5  7  8   5
2  3  2  7  6  3   9  5  4  6  4   7
3  4  3  2  5  6   2  8  2  3  2   1
9  3  4  5  7  2   3  7  6  4  8  10
5  6  8  3  5  7  10  8  1  5  4   4
```

Draw a histogram to show this information.

3) This histogram shows the population of six towns:

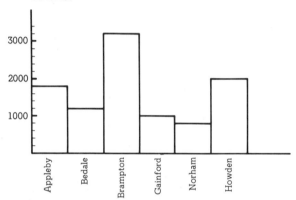

(a) Which town has the most people?
(b) Which town has the fewest?
(c) Approximately how many people are there in the six towns altogether?

PICTOGRAMS

Suppose a manufacturer sold 3500 cans of soup to a particular supermarket in 1 week. They might want to show this in picture form:

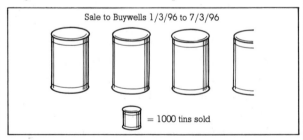

Such a diagram is called a *pictogram*.

---

1) This is a pictogram showing how many new houses were built in a certain city

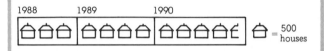

How many new houses were built in 1988? in 1989? in 1990?

2) Using the pictogram below say how many days of sunshine (a) Selsea (b) Fulton and (c) Exford received in 1989.

1989		
Selsea	☼☼☼☼☼☼☼☼	
Fulton	☼☼☼☼☼☼☼	☼ = 15 days' sunshine
Exford	☼☼☼☼☼	

3) This pictogram shows car sales for the years 1985 to 1989.

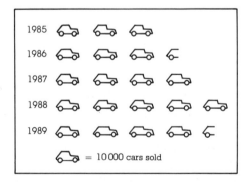

(a) In which year were the most cars sold?
(b) How many cars were sold in 1986?
(c) How many more cars were sold in 1989 than 1985?
(d) Give the increase in sales from 1987 to 1988 and express this as a percentage.

176

EXERCISE 10

1) Plot the following points on a graph choosing a suitable scale for the axes. Draw a smooth curve through the points.

Age of baby in months	Average weight of baby in kg
birth	3.5
3	5.8
6	7.7
9	9.0
12	10.0
15	10.7
18	11.2

From your graph estimate (a) the average weight of a baby aged 2 months (b) the probable age of a baby weighing 10.9 kg.

2) This table is used for changing miles to kilometres

Miles	Km
5	8
10	16
15	24
20	32
25	40
30	48

Plot these points on a graph, putting miles horizontally and kilometres vertically. Draw a straight line through the points. Use the graph to change
(a) 12 miles to km
(b) 30 km to miles.

What is the rate of increase of km with respect to miles?

3) What is the slope of this graph?

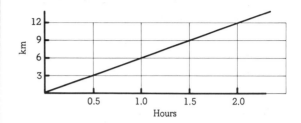

4) Use this bus fare chart to find the fare from
(a) Oak Lane to Maple St
(b) Elm Avenue to Ash Close
(c) Beech Drive to the Bus Station.

Oak Lane

30p	Beech Drive				
39p	30p	Elm Avenue			
45p	39p	30p	Ash Close		
51p	45p	39p	30p	Maple St	
60p	51p	45p	39p	30p	Bus Station

5) Draw a pie chart (and label the angles) to show how a girl spends her time on her hobbies

Music	2 hours
Reading	3 hours
Dancing	1 hour
Skating	5 hours
Judo	1 hour

6) Draw a single bar chart to show the proportion of skilled, unskilled, clerical and managerial workers in a factory if the proportions are: skilled 50%, unskilled 25%, clerical 10% and managerial 15%. Make sure that the height of each section is in the correct proportion.

7) The Local Authority made a survey of how a Leisure Centre was used. 1000 people were asked which facility they thought they used most. The results were

Swimming pool	450
Badminton	150
Table tennis	125
Squash	180
Gymnastics	95

Draw a vertical bar chart to show this information.

8) A tea company made a comparison of tea sold loose and in tea bags during the years 1980 to 1992.

In 1980 £600 000 worth of loose tea was sold and £200 000 worth of tea bags
In 1984 £750 000 worth of loose tea was sold and £350 000 worth of tea bags
In 1988 £500 000 worth of loose tea was sold and £550 000 worth of tea bags
In 1992 £450 000 worth of loose tea was sold and £950 000 worth of tea bags

Put this information into a horizontal bar chart (distinguishing between the bars of each type — loose tea and tea bags — by contrasting shading).

9) The marks obtained in a test were as follows

19	15	14	18	17
15	14	18	17	19
20	13	14	19	18
15	15	17	16	17
17	17	13	14	16
12	17	13	15	16
17	17	16	15	13
16	18	16	16	17
17	16	15	18	14

Group these results in a frequency table. Draw a histogram. Which is the most frequent mark?

These are the marks obtained by a second group of people

Mark	Frequency
11	1
12	3
13	5
14	9
15	16
16	11
17	7
18	5
19	3
20	2

Draw a second histogram showing these results and say how they differ from the first set. (The frequency should be plotted vertically and the mark horizontally.)

10)

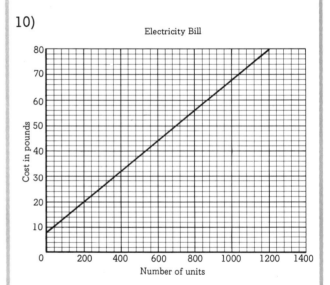

Electricity Bill

An electricity bill consists of a fixed charge and the cost per unit used.

From the graph what is

(a) the fixed charge?

(b) the cost per unit?

(Hint: The fixed charge is the cost of the bill when *no* units have been used. The cost per unit is the *slope* of the graph.)

16

Maps and Timetables

ROAD MAPS

EXERCISE 1

1) Smith St is in square C2. Which squares are these streets in?
   (a) John William St
   (b) West Road
   (c) Finchley Avenue
   (d) Charles St
   (e) Orchard Close

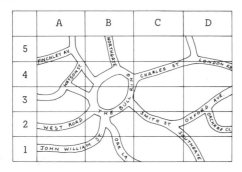

2) All distances on the map opposite are in miles.
   (a) Which is shorter: Leeds to Huddersfield via Brighouse or Leeds to Huddersfield via Dewsbury?
   (b) Which is shorter: Huddersfield to Halifax via Brighouse or Huddersfield to Halifax via Sowerby Bridge.
   (c) What is the length of the shortest route from Morley to Sowerby Bridge?

(d) How far is it by road from Dewsbury to Brighouse?
(e) Give the distance by road from Ripponden to Leeds (i) via Brighouse (ii) via Halifax and Bradford (iii) via Huddersfield and Morley.

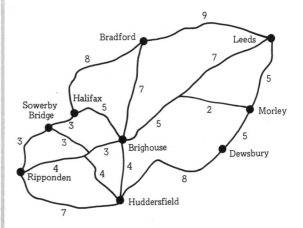

3) The scale on the map overleaf is 1 cm to 0.5 km.

Measure the direct distances and give the number of kilometres between

(a) Cheltenham and Leckhampton
(b) Cheltenham and Charlton Kings
(c) Charlton Kings and Pilley
(d) Andoversford and Charlton Kings
(e) Pilley and Leckhampton
(f) Andoversford and Pilley.

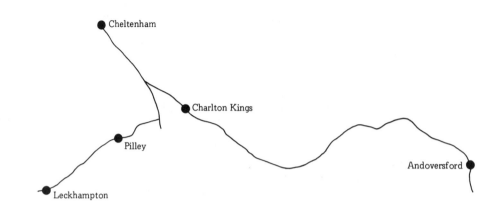

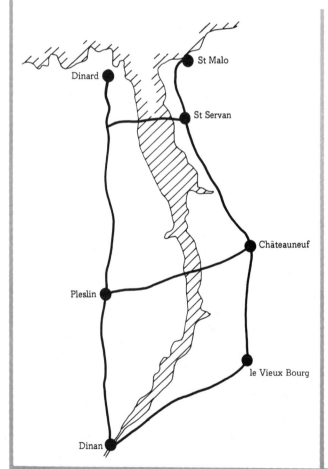

4) 1 cm on the map on the left represents 2 km on the ground.

Estimate how far it is from
(a) Dinard to Pleslin
(b) Châteauneuf to le Vieux Bourg
(c) St Malo to Dinard
(d) St Servan to Pleslin by the shortest route
(e) Châteauneuf to Dinan by the shortest route.

BUS TIMETABLES AND ROUTE MAPS

Use the bus timetable and bus route map shown on the opposite page to answer the following questions.

EXERCISE 2

1) How long does the bus take to travel from Stroud to Cirencester?

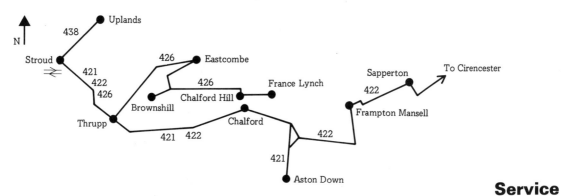

STROUD · CIRENCESTER via Chalford

Service 422

Service 422 via Bowbridge, Thrupp, Brimscombe, Chalford, Aston Down Turn, White Horse Inn, Frampton Mansell, Sapperton, Park Corner, Duntisbourne Rouse Turn, Daglingworth, Stratton.

Mondays to Fridays*

STROUD, Bus Station	0855	1055	1255	CIRENCESTER, Bus Station	0950	1150	1450
Thrupp, Brewery Lane	0903	1103	1303	Daglingworth	1000	1200	1500
Bourne Bridge	0907	1107	1307	Sapperton	1010	1210	1510
Chalford, Marle Hill	0912	1112	1312	White Horse Inn	1017	1217	1517
White Horse Inn	0917	1117	1317	Chalford, Marle Hill	1022	1222	1522
Sapperton	0924	1124	1324	Bourne Bridge	1027	1227	1527
Daglingworth	0934	1134	1334	Thrupp, Brewery Lane	1031	1231	1531
CIRENCESTER, Bus Station	0944	1144	1344	STROUD, Bus Station	1039	1239	1539

CODE

* — Not Bank Holiday Mondays, Good Friday or Boxing Day.

---

2) What time must I arrive at Sapperton bus stop in order to catch the bus to Stroud that arrives at 1039?

3) Give the time for the bus to travel between Daglingworth and Cirencester.

4) What time must I arrive at Aston Down Turn in order to make sure of catching the first bus in the morning to Cirencester?

5) What is the number of the bus service from Stroud right into Aston Down?

6) What do you think this sign means ⇌ ?

7) Give the number of the bus service to (a) Uplands (b) France Lynch.

8) Is there a bus from Cirencester to Stroud at 10 minutes to 3 in the afternoon?

9) Between which stops marked on the timetable is Duntisbourne Rouse Turn?

10) What time is the first bus after 11 a.m. from the White Horse Inn to Stroud?

RAILWAY TIMETABLES

EXERCISE 3

Use the railway timetable overleaf to answer the following questions

1) What is the first train after 9 a.m. on Saturday from Worcester to London?

2) How long does it take the 1046 from Swindon to reach London?

3) What time must I arrive on Kemble Station in order to catch the 0712 from Cheltenham to London on Saturday? How many minutes is this before 8 a.m.?

4) Give the length of the journey in minutes from Worcester to Gloucester on the 1818 on a Sunday.

5) What time does the first train after 11 a.m. on a Friday leave Worcester for London?

WORCESTER · CHELTENHAM · GLOUCESTER · SWINDON · LONDON

Mondays to Fridays

WORCESTER,																	
Shrub Hill	...	0653	0805	0915	...	1107	...	1317	...	1508	1554	...	1757	2022	...	...	...
CHELTENHAM.	0620	0722	0828	0943	1041	1135	1230	1344	1437	1536	1624	1718	1826	2051	2138	...	...
Gloucester	0633	0741	0847	0956	1055	1148	1244	1358	1451	1550	1638	1732	1843	2105	2152	...	...
Stonehouse . . .	0645	0755	0902	1008	1107	1200	1256	1410	1503	1602	1650	1744	1855	2117	2204	...	...
Stroud	0650	0802	0909	1013	1112	1206	1302	1415	1508	1607	1656	1749	1900	2123	2209	...	...
Kemble.	0706	0819	0926	1029	1128	1222	1318	1431	1524	1623	1713	1805	1916	2140	2225	...	...
SWINDON. . . .	0723	0836	0943	1046	1145	1239	1335	1448	1541	1640	1730	1824	1933	2157	2242	...	...
Reading	*0803*	*0915*	1017	*1143*	*1245*	*1353*	*1443*	*1545*	*1643*	*1718*	*1853*	*1908*	*2010*	2245	...	...	...
LONDON,																	
Paddington. . .	0834	0938	1053	*1216*	*1308*	*1426*	*1517*	*1618*	*1717*	*1747*	*1920*	*1937*	*2039*	2314	...	...	...

Saturdays

WORCESTER,																	
Shrub Hill	...	...	...	0915	...	1114	...	1317	...	1508	1554	...	1757	...	...	...	...
CHELTENHAM.	0620	0712	0828	0943	1047	1142	1235	1344	1428	1536	1624	1723	1826	2051	2138	...	...
Gloucester	0633	0724	0847	0959	1100	1155	1248	1358	1442	1550	1638	1738	1843	2105	2152	...	...
Stonehouse . . .	0645	0737	0902	1012	1112	1208	1300	1410	1454	1602	1650	1750	1855	2117	2204	...	...
Stroud	0650	0742	0909	1017	1117	1213	1305	1415	1459	1607	1656	1756	1900	2123	2209	...	...
Kemble.	0706	0758	0926	1033	1133	1229	1321	1431	1515	1623	1713	1813	1916	2140	2225	...	...
SWINDON. . . .	0723	0815	0943	1050	1150	1246	1338	1448	1532	1640	1730	1830	1933	2157	2242	...	...
Reading	*0803*	*0848*	1017	*1143*	*1245*	*1353*	*1443*	*1545*	*1643*	*1718*	*1853*	*1908*	*2010*	2245	...	...	...
LONDON,																	
Paddington. . .	0834	0917	1053	*1216*	*1308*	*1426*	*1517*	*1618*	*1717*	*1747*	*1920*	*1937*	*2039*	2314	...	...	...

Sundays

	Up to 1.10.90			9.10.90 to 2.1.91			10.1.91 to 15.5.91			All Sundays				
WORCESTER,														
Shrub Hill	...	...	1325	...	...	...	...	...	...	1525	...	1818	...	...
CHELTENHAM.	0850	1120	1355	0840	1040	1430	0840	1045	1430	1555	1716	1845	1930	2215
Gloucester	0910	1140	1415	0900	1100	1448	0910	1112	1448	1615	1735	1900	1949	2230
Stonehouse . . .	0930	1200	1427	0920	1120	1500	0922	1124	1500	1627	1749	1912	...	2242
Stroud	0940	1210	1432	0930	1130	1505	0927	1129	1505	1632	1756	1917	2007	2247
Kemble.	1005	1235	1448	0955	1155	1521	0943	1145	1521	1648	1812	1933	2023	2303
SWINDON. . . .	1035	1305	1507	1025	1225	1540	1000	1202	1540	1705	1830	1950	2041	2320
Reading	*1124*	*1351*	*1551*	*1201*	*1401*	*1620*	*1051*	*1251*	*1620*	*1738*	*1926*	*2032*	*2120*	...
LONDON,														
Paddington. . .	*1250*	*1428*	*1628*	*1242*	*1438*	*1657*	*1128*	*1333*	*1657*	*1808*	1933	*2108*	2144	...

CODE
Connecting services in italics.

---

6) What train must I catch from Kemble on a Monday if I want to be sure of arriving in Reading by 4 p.m.?

7) If I am travelling on Sunday 14.10.90 (14 October 1990) what time is the first train from Cheltenham to London? How long does it take to get there?

8) Give the time of the last train from Worcester to London on (a) Monday to Friday, (b) Saturday, (c) Sunday, and the length in hours and minutes of each of these journeys.

THE CALENDAR

EXERCISE 4

Use the calendar opposite to answer these.

1) How many days are there altogether in April, May and June?

2) On which day does 5 June fall in 1996?

3) How many Sundays are there in August 1996?

1996 Calendar

JANUARY
Mon	1	8	15	22	29	.
Tue	2	9	16	23	30	.
Wed	3	10	17	24	31	.
Thu	4	11	18	25	.	.
Fri	5	12	19	26	.	.
Sat	6	13	20	27	.	.
Sun	7	14	21	28	.	.

FEBRUARY
Mon	.	5	12	19	26	.
Tue	.	6	13	20	27	.
Wed	.	7	14	21	28	.
Thu	1	8	15	22	29	.
Fri	2	9	16	23	.	.
Sat	3	10	17	24	.	.
Sun	4	11	18	25	.	.

MARCH
Mon	.	4	11	18	25	.
Tue	.	5	12	19	26	.
Wed	.	6	13	20	27	.
Thu	.	7	14	21	28	.
Fri	1	8	15	22	29	.
Sat	2	9	16	23	30	.
Sun	3	10	17	24	31	.

APRIL
Mon	1	8	15	22	29	.
Tue	2	9	16	23	30	.
Wed	3	10	17	24	.	.
Thu	4	11	18	25	.	.
Fri	5	12	19	26	.	.
Sat	6	13	20	27	.	.
Sun	7	14	21	28	.	.

MAY
Mon	.	6	13	20	27	.
Tue	.	7	14	21	28	.
Wed	1	8	15	22	29	.
Thu	2	9	16	23	30	.
Fri	3	10	17	24	31	.
Sat	4	11	18	25	.	.
Sun	5	12	19	26	.	.

JUNE
Mon	.	3	10	17	24	.
Tue	.	4	11	18	25	.
Wed	.	5	12	19	26	.
Thu	.	6	13	20	27	.
Fri	.	7	14	21	28	.
Sat	1	8	15	22	29	.
Sun	2	9	16	23	30	.

JULY
Mon	1	8	15	22	29	.
Tue	2	9	16	23	30	.
Wed	3	10	17	24	31	.
Thu	4	11	18	25	.	.
Fri	5	12	19	26	.	.
Sat	6	13	20	27	.	.
Sun	7	14	21	28	.	.

AUGUST
Mon	.	5	12	19	26	.
Tue	.	6	13	20	27	.
Wed	.	7	14	21	28	.
Thu	1	8	15	22	29	.
Fri	2	9	16	23	30	.
Sat	3	10	17	24	31	.
Sun	4	11	18	25	.	.

SEPTEMBER
Mon	.	2	9	16	23	30
Tue	.	3	10	17	24	.
Wed	.	4	11	18	25	.
Thu	.	5	12	19	26	.
Fri	.	6	13	20	27	.
Sat	.	7	14	21	28	.
Sun	1	8	15	22	29	.

OCTOBER
Mon	.	7	14	21	28	.
Tue	1	8	15	22	29	.
Wed	2	9	16	23	30	.
Thu	3	10	17	24	31	.
Fri	4	11	18	25	.	.
Sat	5	12	19	26	.	.
Sun	6	13	20	27	.	.

NOVEMBER
Mon	.	4	11	18	25	.
Tue	.	5	12	19	26	.
Wed	.	6	13	20	27	.
Thu	.	7	14	21	28	.
Fri	1	8	15	22	29	.
Sat	2	9	16	23	30	.
Sun	3	10	17	24	.	.

DECEMBER
Mon	.	2	9	16	23	30
Tue	.	3	10	17	24	31
Wed	.	4	11	18	25	.
Thu	.	5	12	19	26	.
Fri	.	6	13	20	27	.
Sat	.	7	14	21	28	.
Sun	1	8	15	22	29	.

4) Shrove Tuesday is 40 days before Palm Sunday. If Palm Sunday is 31 March, what day is Shrove Tuesday?

5) How many months are 30 days long?

6) A school holiday last 6 weeks. School breaks up on a Friday and reopens on a Monday. If school breaks up on July 19th, on what day does it reopen?

7) How many days are there between but not including 1 February and 8 March in 1996?

8) How many weekdays (Monday to Friday but not Saturday or Sunday) fall between 6 July and 11 August?

9) Give the date of the first Sunday in October in 1996.

10) If 31 December 1996 falls on a Tuesday what day will 5 January 1997 be?

THE 24 HOUR CLOCK

The time on this clock could be read as showing 0300 hours or 1500 hours in 24 hour clock notation.

The first two figures show the hours and the second two the minutes.

0300 = 3 a.m.
  (three o'clock in the morning)

1500 = 3 p.m.
  (three o'clock in the afternoon)

EXERCISE 5

Give these times in 24 hour clock notation
1) 2 p.m.
2) 7 p.m.
3) noon
4) 6.30 a.m.
5) midnight

6) 4.25 p.m.
7) eleven o'clock in the morning
8) quarter past two in the afternoon
9) five past six in the evening
10) quarter to eight in the morning
11) twenty-five past eleven at night
12) two o'clock in the morning

How many minutes are there between
13) 1300 and 1325
14) 0820 and 0950
15) 1715 and 1755
16) 1917 and 2108?

Give these 24 hour clock times in terms of a.m. or p.m. times
17) 0715
18) 1935
19) 2314
20) 0648

Reminder There are 60 minutes in 1 hour.
There are 24 hours in 1 day.

EXAMPLE

What is the time 17 minutes after 0255?

After 5 minutes the time will be 0300.
There are 12 minutes still to go.
The time will be 0312.

EXAMPLE

What is the time 12 h 25 min before 0845?

25 minutes before 0845 the time is 0820.
8 h 25 min before 0845 the time is 0020.
A further 4 hours earlier the time is 2020.

——————————— **EXERCISE 6** ———————————

What is the time

1) 5 minutes after 0455?
2) 16 minutes after 0830?
3) 25 minutes after 0355?
4) 1 h 20 min after 1645?
5) 2 h 15 min after 1845?
6) 13 min before 0825?
7) 28 min before 1824?
8) 1 h 10 min before 1750?
9) 2 h 45 min before 0840?
10) 4 h 10 min before 0215?

17

Basic Algebra

SIMPLE EQUATIONS

Letters are used in place of numbers in algebra, for example:

$x + 4 = 6$

This means some number plus 4 equals 6. Since 2 is the only number which, when added to 4 gives 6, then x must equal 2.

In algebra we would write:

if $x + 4 = 6$

then $x = 2$

$x + 4 = 6$ is an example of an *equation*
$x = 2$ is the *solution* to this equation.

EXAMPLE

Solve the equation $x + 5 = 8$.

If $x + 5 = 8$

then $x = 3$

EXAMPLE

Solve the equation $x - 3 = 7$.

The *only* number that gives 7 when 3 is taken from it is 10.

If $x - 3 = 7$

then $x = 10$

EXERCISE 1

By using common sense solve these equations in the same way as above

1) $x + 6 = 8$ 11) $x + 4 = 8$
2) $x + 2 = 5$ 12) $x + 9 = 12$
3) $x + 3 = 10$ 13) $x + 7 = 8$
4) $x + 4 = 11$ 14) $x - 4 = 8$
5) $x + 1 = 9$ 15) $x - 3 = 6$
6) $x + 8 = 15$ 16) $x - 5 = 9$
7) $x + 7 = 14$ 17) $x - 1 = 7$
8) $x + 7 = 10$ 18) $x + 3 = 5$
9) $x - 2 = 5$ 19) $x - 7 = 9$
10) $x - 1 = 9$ 20) $x - 3 = 10$

$x + x + x + x + x$ is written as 5x in algebra. It means 5 times x.

$y + y + y$ would be written as 3y.

EXERCISE 2

In the same way write these numbers algebraically

1) $z + z + z$ 7) $n + n + n + n$
2) $y + y$ 8) $t + t + t + t + t$
3) $t + t$ 9) $m + m$
4) $p + p + p + p + p + p$ 10) $s + s + s$
5) $q + q + q + q$ 11) $p + p + p$
6) $r + r + r$ 12) $w + w + w$

Algebra is used to solve problems. The following problem is easy to solve in your head but it shows how letters can be used for lengths or quantities whose values are not known directly.

EXAMPLE

A square has a perimeter of length 36 cm. How long is each side?

(The perimeter is the distance all the way round the edge.)

First we draw a sketch:

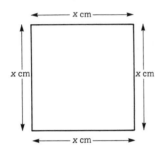

and then say:

let the length of each side be x cm.

Then the perimeter $= x + x + x + x$ cm

$$= 4x \text{ cm}$$

$$= 36 \text{ cm}$$

Since 9 is the only number which, when multiplied by 4 gives 36 then:

$x = 9$

$$\frac{4x}{4} = \frac{36 \text{ cm}}{4}$$
$$= 9$$

The length of each side is 9 cm.

——— EXERCISE 3 ———

Solve these equations using common sense

1) $5x = 20$		7) $6z = 42$
2) $9p = 27$		8) $3t = 24$
3) $3w = 12$		9) $7z = 21$
4) $5h = 25$		10) $2t = 24$
5) $3y = 18$		11) $5y = 60$
6) $3m = 12$		12) $5s = 35$

COLLECTING LIKE AND UNLIKE ——— TERMS ———

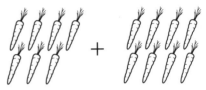

Put 7 carrots together with 8 carrots and you get 15 carrots.

In the same way in algebra:

$7c + 8c = 15c$

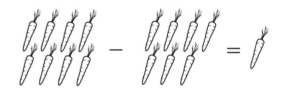

If you take 7 carrots away from 8 carrots only 1 carrot is left.

In algebra:

$8c - 7c = c$

(Notice that $1 \times c$ or $1c$ is written c in algebra.)

The sum of 3 donkeys and 7 carrots cannot be made any simpler! In the same way $3d + 7c$ cannot be simplified. If, however, there are 3 donkeys and 7 carrots already in a field and another 2 donkeys and 8 carrots were put into the field as well, then we could say there are now 5 donkeys plus 15 carrots, and in the same way, in algebra:

$3d + 7c + 2d + 8c = 5d + 15c$

the ds have been collected together and the cs have been collected together.

(7c and 8c are known as *like terms* and 3d and 7c are known as *unlike terms*.)

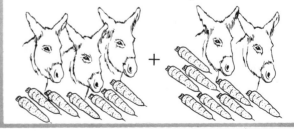

186

EXAMPLE

Simplify $4c - 5c + 3c$.

($4c$ and $3c$ are positive terms whereas $-5c$ is a negative term. Group the positive terms first and then group the negative terms.)

$$4c - 5c + 3c = 4c + 3c - 5c$$
$$= 7c - 5c$$
$$= 2c$$

EXAMPLE

Simplify $7t - 3 - 3t + 9$.

$$7t - 3 - 3t + 9 = 7t - 3t + 9 - 3$$
$$= 4t + 6$$

EXAMPLE

Simplify $a - 3b - 2a + 3c + 5a + 6b$.

$$a - 3b - 2a + 3c + 5a + 6b$$
$$= a + 5a - 2a + 6b - 3b + 3c$$
$$= 6a - 2a + 3b + 3c$$
$$= 4a + 3b + 3c$$

──────── **EXERCISE 4** ────────

Simplify if possible
1) $3x + 6x = 9x$.
2) $6d + 2d =$
3) $5c - c = 4c$.
4) $5y + 11y$
5) $3h + 9h$
6) $c + 6c = 7c$.
7) $11z - 5z$
8) $3s + 2s + s$

9) $3t - 2t + 6t$
10) $14j - 6j$
11) $4c + 3d$
12) $8n - 5n$
13) $4p - p + 3p$
14) $5z - 11z + 6z$
15) $9e + 5e - 6e$
16) $3c + 5d + 2c + d$
17) $8m + 3n + m + 2n$
18) $13w + 2v + 2w + 3v$
19) $15z - 11z + 4q + 3q$
20) $a - 2b + 4a + 5b$
21) $7a + 5b - 2a - 3b$
22) $5x + 3 + 2x + 6$
23) $17y + 11 - 12y - 2$
24) $8p - 2q + 3p + 8q$
25) $2a + 3b + 3c + a + b + c$
26) $7a + 5 + 3b - 4 + a$
27) $3a + 4b + 5c$
28) $13c + 5d - 11c - 5d$
29) $2x - 2x + 3y - 3y + 6$

──────────

──────── **SUBSTITUTION** ────────

$V = l \times w \times h$ is an example of an algebraic formula. It is, in fact, the formula for the volume (or space), V, taken up by a box with length l, width w and height h.

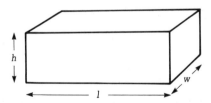

If the length is 5 cm, the width 4 cm and the height 2 cm then the volume will be $5 \times 4 \times 2$ cm^3 (volume is measured in cubic centimetres, written cm^3)

$$= 20 \times 2 \text{ cm}^3$$
$$= 40 \text{ cm}^3$$

187

In algebra we would write:

Find V if

$V = l \times w \times h$ and

$l = 5$, $w = 4$ and $h = 2$

$V = 5 \times 4 \times 2$

$= 20 \times 2$

$= 40$

substituting values for the letters. (Substituting means putting numbers in place of the letters.)

EXAMPLE

Find $x + 3$ if $x = 4$.

$x + 3 = 4 + 3$

$= 7$

EXAMPLE

Find $5y$ if $y = 3$.

$5y = 5 \times 3$ (5y means 5 times y and the

$= 15$ $\times$ sign reappears when
numbers are put in
place of the letters.)

EXAMPLE

Find $5a + 3$ if $a = 4$.

$5a + 3 = 5 \times 4 + 3$ (since 5a means 5 times a)

$= 20 + 3$ (multiplication must be
done before addition)

$= 23$

EXAMPLE

Find $3p + 4q$ if $p = 8$ and $q = 2$.

$3p + 4q = 3 \times 8 + 4 \times 2$

$= 24 + 8$

$= 32$

EXAMPLE

Find $\dfrac{s + t}{v}$ if $s = 6$, $t = 8$ and $v = 2$.

$$\frac{s + t}{v} = \frac{6 + 8}{2} = \frac{14}{2} = 7$$

──────────── **EXERCISE 5** ────────────

Find

1) $x + 3$ if $x = 11$
2) $5z$ if $z = 3$
3) $3w + 2$ if $w = 4$
4) $5s - 3$ if $s = 6$
5) $c - 2$ if $c = 8$
6) $6 - x$ if $x = 1$
7) $4p + 7q$ if $p = 2$ and $q = 3$
8) $3a - 2b$ if $a = 7$ and $b = 4$
9) $7a + 4$ if $a = 10$
10) $7t - 5$ if $t = 6$
11) $\dfrac{p + q}{3}$ if $p = 9$ and $q = 6$
12) $\dfrac{6s}{5}$ if $s = 10$
13) $\dfrac{4 + 3p}{2}$ if $p = 8$
14) $\dfrac{3m + 2n}{5}$ if $m = 6$ and $n = 6$
15) $\dfrac{f + g}{h}$ if $f = 15$, $g = 11$ and $h = 2$
16) $\dfrac{9p + 11q}{r}$ if $p = 2$, $q = 10$ and $r = 8$
17) $\dfrac{12s + 16t}{8}$ if $s = 2$ and $t = 3$
18) $\dfrac{15p + 30q}{10}$ if $p = 4$ and $q = 3$
19) $28a + 20b$ if $a = 2$ and $b = 4$
20) $12x - 11$ if $x = 3$

188

In algebra $a \times a \times a$ is written a^3 (we say "a cubed") and $r \times r$ is written r^2 (we say "r squared").

Notice carefully that r^2 is not the same as 2r.

$$\boxed{r^2 = r \times r} \quad \text{whereas} \quad \boxed{2r = 2 \times r}$$

If $r = 10$ then $r^2 = r \times r = 10 \times 10 = 100$ but $2r = 2 \times r = 2 \times 10 = 20$.

Writing $r \times r$ as r^2 is known as using *index form*.

EXAMPLE

Write $b \times b \times b \times b \times b$ in index form.

$b \times b \times b \times b \times b = b^5$ in index form

(we say "b to the fifth")

--- **EXERCISE 6** ---

Write the following using index form

1) $b \times b \times b$
2) $m \times m \times m \times m$
3) $n \times n$
4) $p \times p \times p$
5) $s \times s$
6) $q \times q \times q \times q \times q \times q$
7) $r \times r \times r$
8) $w \times w \times w \times w \times w$
9) $z \times z \times z$

If $p = 2$, $q = 4$, $r = 10$, $s = 9$ and $t = 11$ find the following. (In questions 16 to 19 square before adding or subtracting.)

10) p^2
11) q^2
12) s^2
13) t^2
14) r^3
15) q^3
16) $p^2 + q^2$
17) $t^2 - r^2$
18) $s^2 + t^2$
19) $s^2 - p^2$

The multiplication sign is often left out when using algebra, for example:

$u \times v \times w$ is written uvw and

$3 \times a \times b$ is written $3ab$

--- **EXERCISE 7** ---

Write these leaving out the multiplication signs

1) $p \times q \times r$
2) $5 \times s \times t$
3) $m \times n$
4) $a \times b \times c$
5) $8 \times x \times y \times z$
6) $8 \times p \times q \times r$
7) $14 \times a \times b$
8) $12 \times p \times q \times r$
9) $3 \times p \times q \times r$
10) $11 \times a \times b$
11) $3 \times a \times b \times c$
12) $2 \times p \times q$

If $a = 3$, $b = 5$, and $c = 8$ find the values of

13) ab
14) bc
15) ac
16) $3ab$
17) $5bc$
18) $7ca$
19) abc
20) bac
21) bca
22) $4abc$
23) $4bca$
24) $4cab$

If $p = 4$, $q = 5$ and $r = 6$ find the values of these expressions, remembering that multiplication and division must always be done before addition and subtraction

25) $pq + r$
26) $6pr + 2pq$
27) $pqr + 4$
28) $p^2 + qr$
29) $2pr^2$
30) $4q^3$
31) $pr + 2rqp$
32) $qr + pq$
33) $5q + 3p + 2r$
34) $pq - r$
35) $6pq \div r$
36) $2pq - 3r$
37) $(3pq \div r) - (2r \div 4)$
38) $\dfrac{3pq}{r}$
39) $\dfrac{8q}{p} + \dfrac{qr}{10}$

HARDER EQUATIONS

Suppose we wanted to solve the equation:

$$7x + 5 = 26$$

we could, by trial and error, find the solution:

$$x = 3$$

but it is much better to use a proper method to find the answer. First we show the method and then the explanation.

Method: $7x + 5 = 26$

$7x = 26 - 5$ (taking 5 from both sides)

$7x = 21$

$x = 21 \div 7$ (dividing both sides by 7)

$x = 3$

An equation is like a balance.

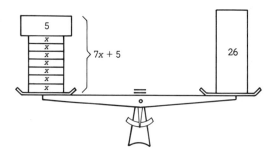

If we take 5 units away from both sides of the scales the scales will still balance.

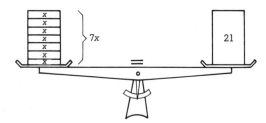

If we divide each side into 7 equal parts and only take 1 part the scales will still balance.

The scales will still balance (and the equation will still be correct) if we:

1) add or subtract the same quantity from both sides

2) multiply or divide both sides by the same quantity.

EXAMPLE

Solve $2x - 3 = 17$.

When we add 3 to both sides we get:

$$2x - 3 + 3 = 17 + 3$$

and the -3 and the $+3$ "cancel each other out" leaving $2x$ on the left hand side.

So:

$2x - 3 = 17$

$2x = 17 + 3$ (adding 3 to both sides)

$2x = 20$

$x = 20 \div 2$ (dividing both sides by 2)

$x = 10$

EXERCISE 8

Solve

1) $4x - 1 = 3$	7) $4x - 5 = 11$
2) $7x - 5 = 16$	8) $5x - 5 = 35$
3) $8x - 3 = 21$	9) $2x - 4 = 20$
4) $9x - 4 = 23$	10) $3x - 7 = 17$
5) $6x - 1 = 35$	11) $9x - 7 = 20$
6) $8x - 11 = 5$	12) $6x - 2 = 40$

EXAMPLE

Solve $3x + 5 = 11$.

$3x = 11 - 5$ (taking 5 from both sides)

$3x = 6$

$x = 6 \div 3$ (dividing both sides by 3)

$x = 2$

Solve

1) $3x + 2 = 8$
2) $5x + 7 = 17$
3) $6x + 3 = 51$
4) $2x + 1 = 19$
5) $7x + 5 = 40$
6) $4x + 1 = 13$
7) $7x + 8 = 36$
8) $9x + 2 = 20$
9) $11x + 2 = 35$
10) $6x + 9 = 39$
11) $5x + 4 = 39$
12) $2x + 11 = 17$

We may also:

1) change sides completely
2) square the quantities on both sides.

EXAMPLE

Solve $5 = 3 + x$.

$5 - 3 = x$ (taking 3 from both sides)

$2 = x$

$x = 2$ (changing sides)

It is possible to change sides at the beginning (or indeed at any stage in the working).

EXAMPLE

Solve $12 = 4x$.

$4x = 12$ (changing sides)

$x = 12 \div 4$ (dividing both sides by 4)

$x = 3$

EXAMPLE

Solve $\sqrt{x} = 8$.

$\sqrt{x} \times \sqrt{x} = 8 \times 8$ (squaring both sides)

$x = 64$

Solve

1) $8 = 5 + x$
2) $10 = 4 + x$
3) $14 = 8 + x$
4) $21 = 7 + x$
5) $11 = 2 + x$
6) $20 = 5x$
7) $25 = 5x$
8) $14 = 2x$
9) $15 = 5x$
10) $\sqrt{x} = 5$
11) $\sqrt{x} = 11$
12) $\sqrt{x} = \dfrac{1}{2}$

EXAMPLE

Solve $5x + 3 = 3x + 7$.

We need to get the xs on one side and the numbers on the other.

Subtract $3x$ from both sides:

$2x + 3 = 7$

Subtract 3 from both sides:

$2x = 4$

Divide both sides by 2:

$x = 2$

EXAMPLE

Solve $4x - 5 = 2x + 3$.

Subtract $2x$ from both sides:

$2x - 5 = 3$

Add 5 to both sides:

$2x = 8$

Divide both sides by 2:

$x = 4$

Solve

1) $7x + 5 = 5x + 7$
2) $8x + 2 = 3x + 12$

3) $5x + 4 = 2x + 13$
4) $11x + 2 = 9x + 4$
5) $8x + 3 = 6x + 13$
6) $8x + 1 = 2x + 7$
7) $7x + 5 = 5x + 15$
8) $10x + 2 = 7x + 14$
9) $11x + 3 = 2x + 39$
10) $5x + 1 = x + 49$
11) $3x - 2 = x + 8$
12) $7x - 3 = 4x + 12$
13) $9x - 7 = 5x + 9$
14) $11x - 9 = 5x + 27$
15) $17x - 11 = 10x + 3$
16) $6x - 1 = x + 14$
17) $20x - 9 = 14x + 3$
18) $9x - 9 = 3x + 9$
19) $5x - 10 = 2x + 20$
20) $6x - 1 = 2x + 3$

FORMULAE

In physics and mathematics, formulae are often used when one quantity is related to other measurements in a definite way.

EXAMPLE

Find v if $v = u + at$ and $u = 2$, $a = 10$ and $t = 4$.

This is one of the most frequently used formulae in physics relating final velocity of a moving object to the initial velocity, constant acceleration and time.

$\quad v = u + at$

$\therefore v = 2 + 10 \times 4$ (multiplication must be done before addition)

$\quad\quad = 2 + 40$

$\quad\quad = 42$

($\therefore$ means "therefore".)

The Greek symbol π is often used in connection with measurements of the circle as will be explained in the next chapter. It is usually given the approximate value of 3.14 or $3\frac{1}{7}$ which as a top-heavy fraction is $\frac{22}{7}$. The formulae in the next exercise are all in common use.

EXERCISE 12

1) Find A if $A = l \times b$ and $l = 7$ and $b = 9$.

2) Find C if $C = 2\pi r$ and $\pi = 3.14$ and $r = 10$.

3) Find A if $A = \pi r^2$ and $\pi = 3.14$ and $r = 10$.

4) Find v if $v = \sqrt{2gh}$ and $g = 10$ and $h = 20$.

5) Find f if $\dfrac{1}{f} = \dfrac{1}{v} + \dfrac{1}{u}$ and $v = 12$ and $u = 6$.

6) Find s if $s = ut + \dfrac{1}{2}at^2$ and $u = 2$, $t = 4$ and $a = 10$.

7) Find s if $s = \dfrac{v^2 - u^2}{2a}$ and $v = 5$, $u = 1$, and $a = 4$.

8) Find V if $V = \dfrac{1}{3}\pi r^2 h$ and $\pi = 3.14$, $r = 10$ and $h = 6$.

9) Using the formula $I = \dfrac{PRT}{100}$, find I if $P = 500$, $R = 5$ and $T = 4$.

10) If $A = P\left(1 + \dfrac{R}{100}\right)^n$ find A when $P = 200$, $R = 10$ and $n = 2$.

PLOTTING EQUATIONS ON A GRAPH

The equation $y = 5x + 1$ can be plotted on a graph.

A table of values is constructed by substituting different values of x into the equation. It is easier to choose simple values such as 0, 1, 2, 3, etc.

$y = 5x + 1$ which means y is 5 times the value of x plus 1

when $x = 0$ $y = 1$
   (since $5 \times 0 + 1 = 0 + 1 = 1$)

when $x = 1$ $y = 6$
   (since $5 \times 1 + 1 = 5 + 1 = 6$, etc.)

when $x = 2$ $y = 11$

when $x = 3$ $y = 16$

when $x = 4$ $y = 21$

The table of values is thus:

x	0	1	2	3	4
y	1	6	11	16	21

These values can be plotted on a graph:

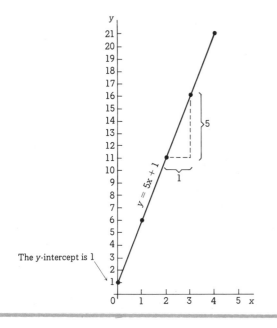

The y-intercept is 1

Gradient = Distance up ÷ Distance along
   $= 5 \div 1$
   $= 5$ (see Chapter 15 on graphs and charts)

Notice that:

The equation of the graph is $y = 5x + 1$.

The gradient (slope) of the graph is 5.

The line cuts the y-axis at $y = 1$ (this is called the y-intercept).

The equation of a straight line graph is

$$y = (\text{gradient})x + (y\text{-intercept})$$

EXAMPLE

Plot the equation $y = 4x + 1$ on a graph.
(a) What is the gradient?
(b) What is the value of the y-intercept?
(c) What is the value of y when $x = -1$?

The table of values will be:

x	0	1	2	3	4
y	1	5	9	13	17

Because of question (c) we will have to include negative values on the x and y-axes:

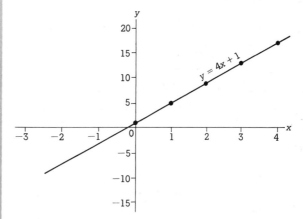

(a) the gradient is 4
(b) the y-intercept is 1
(c) when $x = -1$ then $y = -3$.

1) Plot the equation $y = 2x + 7$ on a graph. Label the x and y-axes and label the straight line with the equation.
   (a) What is the gradient of the straight line?
   (b) What is the value of the y-intercept?

2) Plot the equation $y = 3x + 5$ on a graph.
   (a) State the gradient (slope) of the graph.
   (b) Where does the line cut the y-axis?

3) Plot these values on a graph, choosing a suitable scale and draw a straight line through them:

x	0	1	2	3	4	5
y	3	5	7	9	11	13

   (a) What is the gradient of the straight line?
   (b) What is the y-intercept?
   (c) What is the equation of the straight line?

4) $y = 3x - 7$.

x	0	1	2	3	4	5
y	−7	−4	−1	2	5	8

   Plot these points on a graph and write down
   (a) the value of the gradient
   (b) the value of y when $x = 0$. Draw a straight line through the points.

5) Plot these values on a graph and draw a straight line through them:

x	−3	−2	−1	0	1	2	3
y	3	4	5	6	7	8	9

   (a) What is the gradient of the straight line?
   (b) Where does the line cut the y-axis?
   (c) What is the equation of the straight line?

BASIC

BASIC is a computer programming language. BASIC instructions are numbered (usually in tens) and are executed in sequence.

In BASIC the instruction $X = 5$ means "Let X take the value 5" and the instruction $Y = X + 1$ means "Let Y take the value of X plus 1".

Running the program

```
10 X = 5
20 Y = X + 1
30 PRINT X, Y
```

would result in 5 and 6 being printed.

In BASIC * means "multiplied by" and / means "divided by".

To repeat a set of instructions a FOR ... NEXT loop is used. The program below will print out the first fifteen square numbers.

```
10 FOR NUMBER = 1 to 15
20 PRINT NUMBER * NUMBER
30 NEXT
```

To print out the first 20 square numbers the number 15 would be changed to 20.

To print out the first ten multiples of 7 one of the words "NUMBER" in line 20 would be exchanged with a figure 7.

What values would be printed if these programs were run?

1)
```
10 N = 20
20 M = N + 10
30 PRINT N, M
```

2)
```
10 V = 4
20 W = V + V + V
30 PRINT V, W
```

3)
```
10 S = 4
20 T = 2 * S
30 U = S/2
40 PRINT S, T, U
```

4)
```
10 X = 12
20 Y = 8
30 PRINT X − Y, X + Y, X * Y, X/Y
```

```
5)  10  FOR NUMBER = 1 to 5
    20  PRINT NUMBER
    30  NEXT

6)  10  FOR NUMBER = 1 to 10
    20  PRINT 3 * NUMBER
    30  NEXT

7)  10  FOR NUMBER = 1 to 8
    20  PRINT NUMBER * NUMBER
        * NUMBER
    30  NEXT

8)  10  FOR NUMBER = 1 to 6
    20  PRINT "HELLO"
    30  NEXT
```

```
9)  10  FOR NUMBER = 1 to 5
    20  PRINT 100 − NUMBER
    30  NEXT

10) 10  FOR NUMBER = 1 to 10
    20  PRINT 2 * NUMBER + 1
    30  NEXT

11) 10  FOR NUMBER = 1 to 8
    20  PRINT 3 * NUMBER − 1
    30  NEXT

12) 10  FOR NUMBER = 1 to 10
    20  PRINT NUMBER/2
    30  NEXT
```

18

── Plane Shapes and Solid Figures ──

─── PLANE SHAPES ───

A "plane" shape is made from lines, known as the *sides* of the shape. It is flat. Here are some of the more familiar plane shapes:

The Triangle (Three Sides)

An *angle* is formed where the sides meet. It is a measure of the amount of opening between the lines.

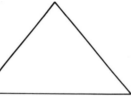

A triangle with three sides of equal length and with three equal angles is known as an *equilateral* triangle.

The Square (Four Sides)

All the angles are equal. The sides are of equal length.

The Quadrilateral (Four Sides)

Any plane shape with four sides is called a "quadrilateral". The square and the rectangle are quadrilaterals.

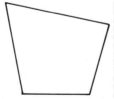

The Rectangle (Four Sides)

All the angles are equal. Two pairs of opposite sides are of equal length.

The Circle

The length across the centre of a circle is called the *diameter*. The length round the edge is known as the *circumference*.

A shape is *regular* if all the sides are of equal length and all the angles are equal. The equilateral triangle and the square are regular. The pentagon (a five-sided figure), hexagon (a six-sided figure) and the octagon (an eight-sided figure) as drawn below are all regular:

Regular Pentagon (Five Sides)

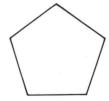

Regular Hexagon (Six Sides)

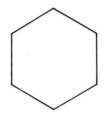

Regular Octagon (Eight Sides)

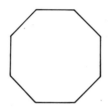

The length all the way round the edge of a plane shape is called the *perimeter*.

——— SOLID FIGURES ———

Sphere

A ball bearing and a cricket ball are examples of spheres.

Cube

A lump of sugar is sometimes called a sugar cube.

It has six faces, eight corners (or *vertices*) and twelve edges.

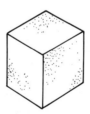

Prisms

A prism has the same shape all the way along its length.

Cube

All the sides are of equal length and each face is a square.

Cuboid

Each face is a rectangle. The cuboid is a box shape.

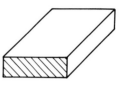

Cylinder

Most tins are cylindrical.

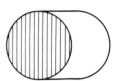

The cuboid could be called a rectangular prism and the cylinder could be called a circular prism.

Triangular prism

The front face is a triangle.

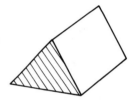

Hexagonal prism

The front face is a hexagon.

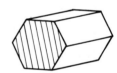

An *irregular prism* has a front face with no particularly well-known shape. The front face is sometimes called the "cross-section".

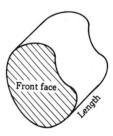

Pyramids

Pyramids sit on a flat base and come up to a point. The shape of the base gives the name to the type of pyramid:

Triangular based pyramid

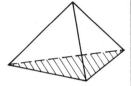

Square based pyramid

Circular pyramid (cone)

A triangular pyramid with all its faces equilateral triangles (triangles with three equal sides) is known as a *tetrahedron*.

A tetrahedron has all its sides of equal length.

A *net* is the shape you would cut from a piece of cardboard in order to make a model of a solid figure. This is the net of a cuboid:

You could fold this

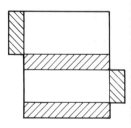

to make this

Different nets can make the same shape:

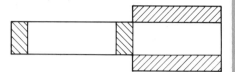

This net would also make the same cuboid shape:

──────── **EXERCISE 1** ────────

1) Write down the number of sides of these plane shapes
   (a) triangle (b) hexagon (c) quadrilateral (d) pentagon (e) rectangle (f) octagon (g) square.

2) Work out the perimeter (length all the way round the edge) of these shapes

   (a)
   3 cm
   3 cm

   (b)
   3.5 cm 4.6 cm
   7.7 cm

   (c)
   7.5 mm
   3.5 mm

   (d)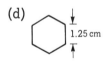
   1.25 cm

3) Write down the number of faces of these solid figures
   (a) cuboid (b) triangular prism (c) square based pyramid (d) hexagonal prism (e) tetrahedron.

4) How many edges are there on (a) a square based pyramid (b) a cuboid (c) a triangular prism (d) a tetrahedron?

5) How many vertices do these solid figures have
   (a) a cuboid (b) a square based pyramid (c) a triangular prism (d) an octagonal prism?

6) Draw a suitable net for (a) a cube (b) a triangular prism (c) a square based pyramid.

7) Which of these nets would make a cube and which would not?

(a)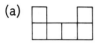

(b)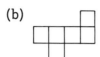

(c)

(d)

(e)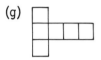

(f)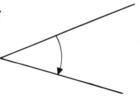

(g)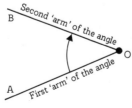

8) Draw three different types of prism and three different types of pyramid and label each carefully.

— MEASUREMENT OF ANGLES —

An angle is the amount of opening between two lines. It does not depend on the lengths of the lines.

An angle can also be regarded as a measure of *turning*.

AO and OB are called the arms of the angle. If the line AO is turned about O until it is in the same position as OB, then it has "*turned through the angle AÔB*" (compare the turning of the hands of a clock).

If you face in a certain direction and then turn completely round so that you face the same way again, we can say you have turned through 360 degrees, written 360° with the small symbol ° standing for the word "degrees".

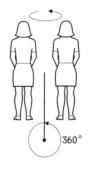

Half a complete turn is 180°

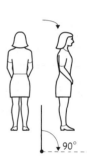

and a quarter turn is 90°, called a *right angle*.

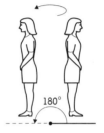

A right angle is often shown like this:

or

There are *two right angles in 180°* and *four right angles in 360°*.

Angles less than 90° are called *acute*, e.g. 15°, 60°, 76°.

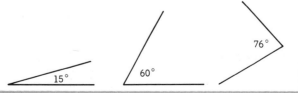

199

Angles more than 90° but less than 180° are called *obtuse*, e.g. 110°, 150°, 138°.

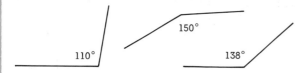

Reflex angles are those angles which are bigger than 180°, e.g. 286°, 250°, 300°.

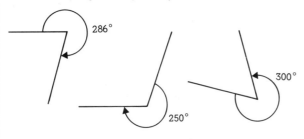

──────── **EXERCISE 2** ────────

Draw up a table with three headings: acute, obtuse and reflex, and put these angles in the correct columns

17°, 190°, 240°, 173°, 65°, 230°, 185°,
54°, 135°, 270°, 155°, 72°, 85°, 169°,
315°, 198°, 32°, 88°, 350°, 120°, 12°,
110°, 51°, 95°, 103°, 114°, 63°, 210°

Angles are measured using an instrument called a *protractor*.

There are 180° in a semi-circle (half a circle).

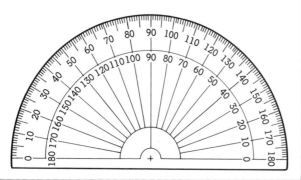

How to Measure an Angle

Place the protractor so that its centre lies on the point where the arms of the angle meet and the bottom line of the protractor lies along one of the arms as shown.

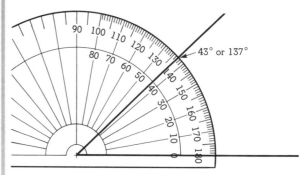

Then read off the measurement of the angles as shown by the arrow. Care must be taken to choose the correct number. This angle is obviously less than 90° so we choose 43° rather than 137°.

There are 60 minutes in a degree but these are too small to be shown on a protractor. We write 60 minutes as 60'.

How to Draw a Given Angle

Draw a straight line with a pencil and ruler leaving enough space for the angle to be drawn and mark a small fine line across it to make a small cross. Place the protractor so that the bottom line lies along the line you have drawn and the centre of the protractor lies on the cross. Mark off the required angle with a sharp pencil and join the cross and marked line with a pencil and ruler.

EXAMPLE

Draw an angle of 60°.

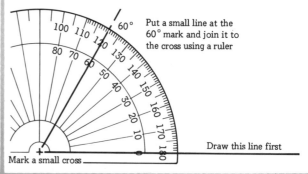

Put a small line at the 60° mark and join it to the cross using a ruler

Draw this line first

Mark a small cross

EXERCISE 3

(a) Measure these angles using a protractor.
(b) On a separate piece of paper draw them (with a protractor) as accurately as possible.

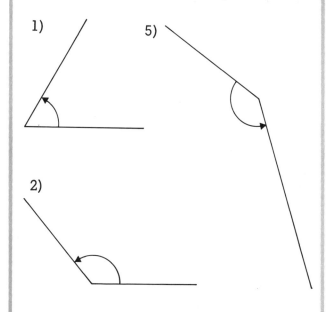

1)

2)

3)

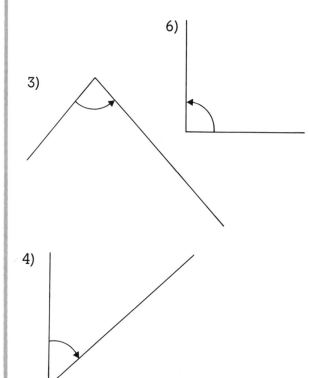

4)

5)

6)

NAMING OF ANGLES

Angles are named either with one small letter, for example, *a*

or with three capital letters, letters, for example, BÂC

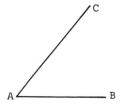

A sign like this ⌢ is put over the middle letter to show that an angle is being denoted.

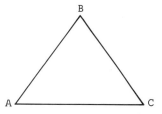

If the angles are in a triangle, to find which angle is meant by AĈB, say, the angle can be traced out in the order of the letters.

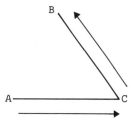

The angles AĈB and BĈA are considered to be the same, the ordering of the "outside" letters making no difference.

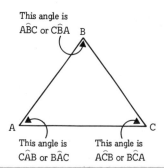

This angle is
AB̂C or CB̂A

This angle is
CÂB or BÂC

This angle is
AĈB or BĈA

If small letters are used to label angles the usual labelling for triangle ABC is as shown here:

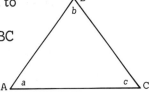

2)

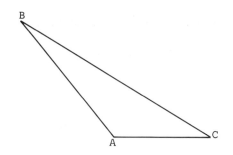

──────── **EXERCISE 4** ────────

Rename the angles $a, b, c, d, e, f, g, p, q$ and r using capital letters.

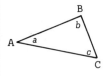

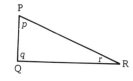

3)

4)

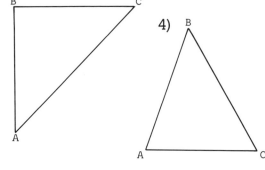

──────── **EXERCISE 5** ────────

(a) Measure these angles using a protractor.

(b) For each triangle copy a table similar to this one:

$C\hat{A}B$	
$A\hat{B}C$	
$B\hat{C}A$	
Sum	

Put your results into the table.

(c) In each case find the sum of the angles.

5)

6)

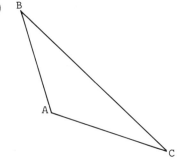

1)

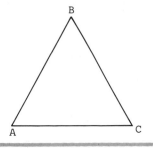

7)

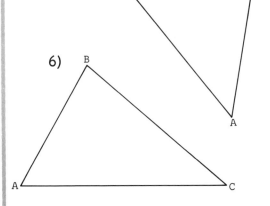

202

– SIMPLE ANGLE PROPERTIES –

This is a pair of *parallel lines*:

Parallel lines never meet. They are always the same distance apart and are in the same direction. Railway lines are parallel.

In geometry a pair of parallel lines is shown by a pair of single or double arrows:

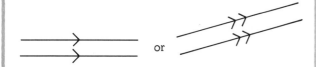

or

A four-sided shape with one pair of parallel sides is called a *trapezium*:

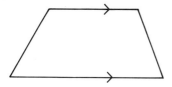

This shape formed by two sets of parallel lines is called a *parallelogram*:

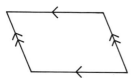

A parallelogram with equal sides is called a *rhombus*.

A line drawn across a pair of parallel lines is called a *transverse line* and it forms eight angles. Some of these angles are equal in size to each other and have special names.

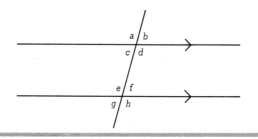

Alternate Angles

Angles *c* and *f* are equal. They are called alternate angles. The two angles which are formed by the letter Z are alternate angles.

A special geometrical property of angles is that:

   Alternate angles are equal

Angles *d* and *e* are also a pair of equal alternate angles.

Corresponding Angles

Angles *c* and *g* are equal. They are called corresponding angles.

   Corresponding angles
   are equal

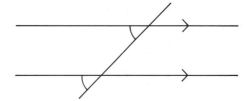

d and *h* are another pair of corresponding angles, as are *a* and *e*. Angles *f* and *b* also form a pair of corresponding angles.

Angles on a Straight Line

Angles *a* and *b* lie on a straight line. Therefore the sum of these two angles must be 180°.

   Angles on a straight
   line add up to 180°

Vertically Opposite Angles

Angles b and c lie opposite to each other and are equal. Such angles as b and c are called vertically opposite angles.

> Vertically opposite angles are equal

Angles f and g are also vertically opposite and thus equal to each other. The obtuse angles a and d are vertically opposite to each other as are the obtuse angles e and h.

In the diagram below angles q and s are equal (alternate angles) and angles p and t are equal (corresponding angles)

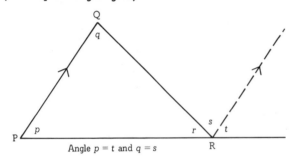

Angle $p = t$ and $q = s$

$$\therefore p + q + r = t + s + r = 180°$$

(since r, s and t lie on a straight line).

Angle Sum of a Triangle

It can be shown (by a method similar to that indicated in the last example) that:

> The angle sum of a triangle is $180°$

EXAMPLE

Find angle x.

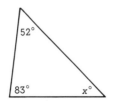

The angle sum of a triangle is $180°$

$$\therefore\ x + 52 + 83 = 180$$
$$x + 135 = 180$$
$$x = 180 - 135$$
$$x = 45$$

The unknown angle is $45°$.

Similarly, using the angle properties just described, other unknown angles can be worked out.

EXAMPLE

Find angles a, b, c, d and e in this diagram:

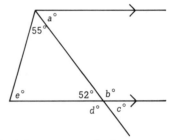

$a = 52$ (alternate angles)

$b = 180 - 52 = 128$ (angles on a straight line add up to $180°$)

$c = 52$ (vertically opposite angles)

$d = 128$ (vertically opposite to b)

$e = 180 - (55 + 52) = 73$ (angle sum of a triangle is $180°$)

Note that $b + c + d + 52 = 360$.
Angles at a point add up to $360°$.

────────── **EXERCISE 6** ──────────

Give reasons for your answers in this exercise.

1) Find x.

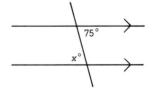

204

2) Find a and b.

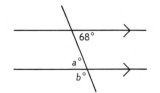

3) Find x, y and z.

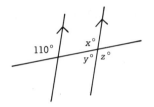

4) Find x.

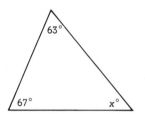

5) Find x and y.

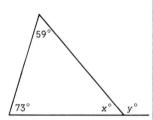

6) Find a, b and c.

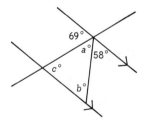

7) Find x.

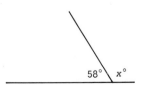

8) Find x and y.

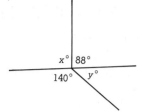

9) Find x and y.

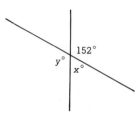

10) Find x, y, z, w and t.

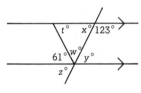

11) Find x, w and t.

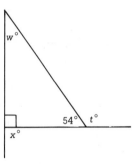

12) Find y.

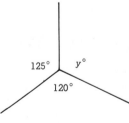

13) Find w, x, y and z.

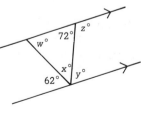

TRIANGLES

A triangle which has every angle less than 90° is called *acute-angled* or *scalene*.

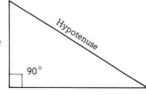

Every angle is less than 90°

A scalene triangle

A right-angled triangle has one of its angles equal to 90°.

The longest side is opposite the right angle and is called the *hypotenuse*.

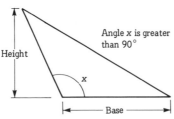

Hypotenuse

90°

A right-angled triangle

A triangle with one angle greater than 90° is called an obtuse-angled triangle.

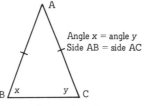

Height

Angle *x* is greater than 90°

x

Base

An obtuse-angled triangle

An isosceles triangle has two equal sides and two angles equal. The equal angles are opposite the equal sides.

A

Angle *x* = angle *y*
Side AB = side AC

B *x*   *y* C

An isosceles triangle

An equilateral triangle has all its sides and all its angles equal. Each angle is 60°.

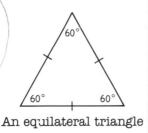

60°

60°   60°

An equilateral triangle

A small line drawn across the sides of a triangle shows that the lengths of those sides are equal. In this triangle sides AC and BC are equal.

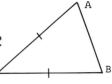

A

C   B

Sometimes a pair of lines is drawn to show equal sides:

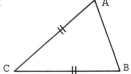

A

C   B

The Right-angled Triangle

In a right-angled triangle the square of the length of the hypotenuse is equal to the sum of the squares of the lengths of the other two sides. This is known as Pythagoras' theorem:

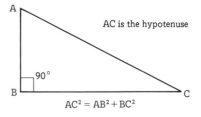

A

AC is the hypotenuse

90°

B   C

$$AC^2 = AB^2 + BC^2$$

We think the Egyptians may have used Pythagoras' theorem to build tools to help them construct the Pyramids. There is actually no known record of Pythagoras having written anything about the rule which bears his name!

EXAMPLE

Triangle ABC has a right angle at B. The length of AB is 6 cm and the length of BC is 8 cm. Find the length of AC.

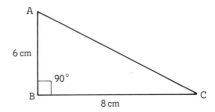

A

6 cm

90°

B   8 cm   C

206

$$AC^2 = AB^2 + BC^2$$
$$= 6^2 + 8^2$$
$$= 36 + 64$$
$$= 100$$
$$AC = \sqrt{100} = 10$$

The length of side AC is 10 cm.

EXAMPLE

A rectangle is 9 cm long and 5 cm wide. What is the length of the diagonal of the rectangle?

The *diagonal* of the rectangle is the distance from one corner to the opposite corner, so it can be seen that it is the hypotenuse of the right-angled triangle.

If we label the rectangle ABCD as shown then we have to find the length of AC:

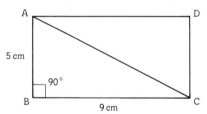

$$AC^2 = AB^2 + BC^2 = 5^2 + 9^2 = 25 + 81 = 106$$
$$AC = \sqrt{106} = 10.3 \text{ cm (using a calculator}$$
pressing $\boxed{106}\ \boxed{\sqrt{\ }}$).

When the length of the hypotenuse (the longest side) and the length of one other side are given, then we rearrange the equation before finding the third side.

EXAMPLE

In a right-angled triangle the length of the hypotenuse, AC, is 10 mm and the length of side BC is 5 mm. Find the length of the third side.

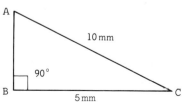

$$AC^2 = AB^2 + BC^2$$

Rearranging this equation:

$$AB^2 = AC^2 - BC^2 = 10^2 - 5^2 = 100 - 25$$
$$= 75$$
$$AB = \sqrt{75} = 8.66 \text{ mm (using a calculator}$$
pressing $\boxed{75}\ \boxed{\sqrt{\ }}$).

You should use a calculator to find the square roots.

1) Find the length of side AC.

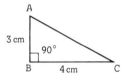

2) Find the length of the hypotenuse.

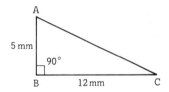

3) Find the lengths of the sides marked x.

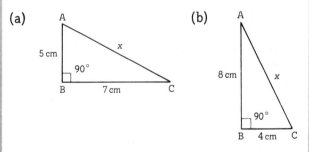

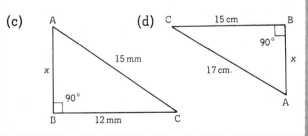

207

nd the length of the diagonal of a
:tangle with width 10 cm and length
24 cm.

5) Find the length of the diagonal of a square of side 5 cm.

6) In a right-angled triangle the lengths of the two shorter sides are 2 cm and 1 cm. Find the length of the longest side.

The Isosceles Triangle

Triangle ABC is isosceles with length AB = length AC. D is the middle point of the line BC.

Angle $B\hat{A}D$ = Angle $D\hat{A}C$

Length BD = Length DC

Angle $B\hat{D}A$ = Angle $A\hat{D}C$ = $90°$

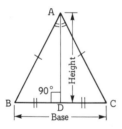

A line from A to D (the middle point of BC) forms two right-angled triangles. The triangles are BDA and CDA and both have a right angle at D. This fact is frequently used to find the height of an isosceles triangle.

EXAMPLE

Find the height of this isosceles triangle:

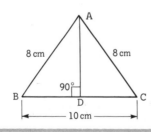

The height is the length AD

$$BC = 10\,cm \quad so \quad BD = \frac{1}{2} \times 10 = 5\,cm$$

Using Pythagoras' theorem

$$AB^2 = AD^2 + BD^2 \quad so$$
$$AD^2 = AB^2 - BD^2 = 8^2 - 5^2$$
$$= 64 - 25 = 39$$
$$AD = \sqrt{39} = 6.24\,cm \text{ (using a calculator}$$

pressing $\boxed{39}$ $\boxed{\sqrt{}}$).

Square root

Find the heights of these isosceles triangles

1)
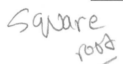

4)
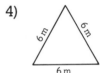

2)

5)
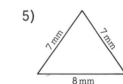

3)

6)
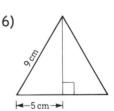

7) Find the height of an isosceles triangle with two equal sides of length 11 cm and base 12 cm.

8) Find the height of an equilateral triangle with sides of length 8 cm.

208

CIRCLES

The *radius* of a circle is the distance from the centre to the edge of the circle.

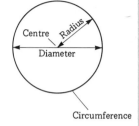

The *diameter* of a circle is the distance across the centre from edge to edge.

The diameter = 2 × The radius

The distance round the edge of a circle is called the *circumference* and, for *any* circle you care to draw, the *circumference is just over three times the length of the diameter*. (It is difficult to measure round the edge of a circle but you could try with a piece of cotton and then measure the length of the cotton to verify that this is so.)

A better approximation is to say that:

Circumference = 3.14 × Diameter

The exact value of circumference ÷ diameter can never be worked out but it is so important that it has been given the special symbol π (the Greek letter "pi"). We generally take π as being 3.14 or $\frac{22}{7}$ when working in fractions.

Using an algebraic formula:

$$C = \pi \times d$$

where C is the circumference, π is 3.14 or $\frac{22}{7}$ and d is the diameter.

Since the diameter is twice the length of the radius, an alternative form is:

$$C = 2 \times \pi \times r \quad \text{or} \quad 2\pi r$$

where C is the circumference, π is 3.14 or $\frac{22}{7}$ and r is the radius.

EXAMPLE

The diameter of a circle is 10 cm. Find the length of the circumference of the circle.

$$C = \pi \times d$$
$$= 3.14 \times 10$$
$$= 31.4 \text{ cm}$$

The circumference is 31.4 cm.

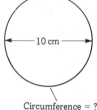

Circumference = ?

EXAMPLE

The radius of a circle is 14 cm. What is its circumference?

$$C = 2 \times \pi \times r$$
$$= 2 \times \frac{22}{7} \times 14$$
$$= 88 \text{ cm}$$

Circumference = ?

(it is easier to use $= \frac{22}{7}$ when the radius is a multiple of 7 since the numbers cancel)

If your calculator has a π button, then pressing this puts the value of π into the display:

| π | gives | 3.141592654 |

Notice that this is still only an approximate value.

EXERCISE 9

Find the circumferences of the following circles

1) Diameter 2 cm
2) Diameter 5 cm
3) Diameter 7 mm
4) Diameter 21 cm
5) Radius 3 cm
6) Radius 7 m
7) Radius 10 mm
8) Radius $3\frac{1}{2}$ cm.

9) A wheel has a diameter of 70 cm. What is the length around the rim?

10) A circular flowerbed measures 4 m across the centre. How far is it round the edge?

SCALE DRAWINGS

Architects use scale drawings when submitting plans of a house or building to the Local Authority for planning permission.

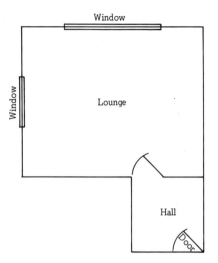

On a scale drawing every measurement is in proportion to the actual or real measurement. Above is a scale drawing of a room with 1 cm on the drawing representing 1 m in the actual room.

From this drawing we could tell that the lounge is 4 m wide by 5 m long and that the hall is 2 m square.

EXERCISE 10

1) This is a sketch of the plan of a boat. Make an accurate scale drawing with 1 cm representing 1 m. You will need a ruler and a pair of compasses.

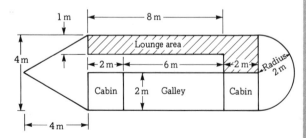

2) Draw quarter size an equilateral triangle which has sides 24 cm long, using a protractor and a ruler. Measure the height on your drawing. What would be the height of the original triangle?

3) Using a pair of compasses and a ruler make a scale drawing of this clock tower.

Use a scale of 1 cm to 1 metre. Measure the perimeter of the scale drawing.

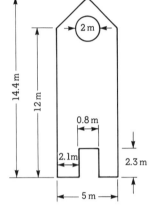

4) Make a full size drawing of these nets, cut them out and decide which will fold to make the net of a cuboid.

(a)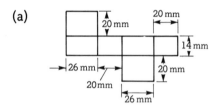

(b)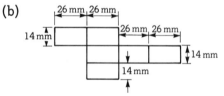

(c)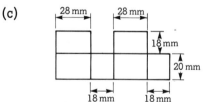

(d)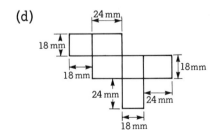

210

5) Using a protractor, make a full size or half size scale drawing of the net of this tetrahedron and fold it to make the solid figure as shown.

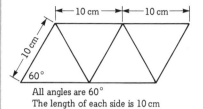

All angles are 60°
The length of each side is 10 cm

Tetrahedron

BEARINGS

The more familiar points of the compass are shown on this diagram.

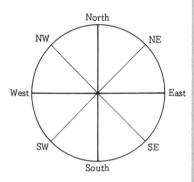

The four main points are north, south, west and east with the intermediate points, north-east (NE), north-west (NW), south-east (SE), and south-west (SW) as shown.

Directions can also be expressed in degrees measured from north. Usually three figures are used, so 45° would be written 045°.

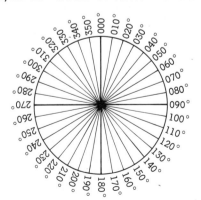

This is a sketch of a ship steaming in a direction of 045° (which is the same as NE).

North would be 000°, east 090°, south 180° and west 270°.

Bearings are always measured clockwise:

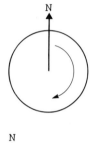

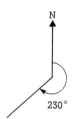

EXERCISE 11

1) Give three-figure bearings for the directions (a) south-west (b) north-west.

2) Draw four separate diagrams to show the directions of these bearings
(a) 030° (b) 125° (c) 240° (d) 300°.

3) Using a protractor and *measuring the angle clockwise* express these directions using a three-figure bearing

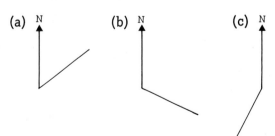

211

4) An aircraft leaves an aerodrome A and flies on a bearing of 080° for a distance of 90 km to a point P. It then changes direction and flies due west for a distance of 60 km until it reaches a point Q. Draw a scale diagram to represent this flight using a scale of 1 cm to 10 km. Hence find the distance from A to Q and write down the bearing of Q from A.

5) As shown in the diagram below, a boat sails from a port P to a point X. It then sails to a second point Y. The diagram has been drawn to scale. Use it to find

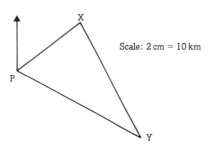

Scale: 2 cm = 10 km

(a) the distance of X from P, (b) the bearing of X from P, (c) the distance of Y from P, (d) the bearing of Y from P.

LOGO

LOGO is another computer programming language. It can be used to draw shapes. An imaginary turtle walks across the screen with a pen in his hand. It either has the pen up (does not draw) or has the pen down (draws a straight line).

The simplest instructions are

CS	clear the screen
PU	pen up
PD	pen down
FD 100	forward one hundred units
RT 90	turn right by 90 degrees
LT 90	turn left by 90 degrees

The turtle starts facing the top of the screen with its pen down.

The LOGO commands to draw a rectangle 100 units long by 60 units wide are

FD 100   RT 90   FD 60   RT 90
FD 100   RT 90   FD 60

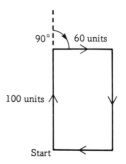

The 100 is a measure of how far to go. 200 would take you twice as far and 50 only half as far. RT 90 means turn right through 90° and RT 60 would mean turn right by 60°.

EXERCISE 12

What shapes would be drawn by using these LOGO instructions? You will need drawing paper, a ruler and a protractor. Take 100 units to mean a distance of 5 cm.

1) CS   FD 100   RT 90   FD 100   RT 90
   FD 100   RT 90   FD 100

2) CS   FD 100   RT 120   FD 100
   RT 120   FD 100

3) CS   FD 50   PU   FD 50   PD   FD 50

4) CS   FD 100   RT 72   FD 100   RT 72
   FD 100   RT 72   FD 100   RT 72
   FD 100

5) CS   FD 100   RT 90   FD 50   LT 90
   FD 50   LT 90   FD 150   LT 90
   FD 50   LT 90   FD 50   RT 90
   FD 100   LT 90   FD 50

6) (a) (A hard one!) CS   PD   RT 30
FD 100   RT 300   FD 100   RT 240
FD 100   RT 300   FD 100

(b) Rewrite (a) using the LT instruction instead of RT.

(c) Rewrite (a) using RT and LT so that no angle is greater than 180°.

It is tedious to keep repeating instructions, so fixed sets of instructions called PROCEDURES are used. Try to find a book to learn some more about LOGO and PROCEDURES. Turtle geometry is great fun!

_____

19

Area

Area is the space taken up by a flat surface such as a table top:

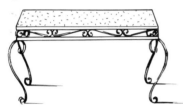

or a curved surface such as the outside of a tin:

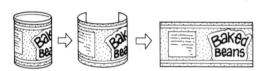

It is measured in *square units*.

This is a unit square of side 1 cm:

It is called a "1 cm square" and it has an *area* of 1 square centimetre (1 cm²).

UNITS OF AREA

The *metric* units of area are:

square millimetres	(mm²)
square centimetres	(cm²)
square metres	(m²)
or square kilometres	(km²)

A square metre is the amount of space taken up by a square of side 1 metre.

The *imperial* units of area are:

square inches	(sq. in or in²)
square feet	(sq. ft or ft²)
square yards	(sq. yd or yd²)
or square miles	(sq. miles or miles²)

A square inch is the amount of space taken up by a square of side 1 inch.

Land is measured in *hectares* (1 hectare is 10 000 m²) or *acres* (1 acre is 4840 sq. yd):

1 hectare is about $2\frac{1}{2}$ acres

Carpets and vinyl flooring are generally sold by the square metre (metric) or square yard (imperial).

AREAS OF RECTANGLES AND — SHAPES MADE FROM THEM —

Twenty 1 cm squares would fit inside this rectangle:

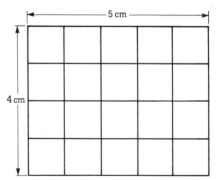

We say its area is 20 cm².

214

*The area of a rectangle is defined as length ×
width* (even when the length and width are not
whole numbers).

EXAMPLE

Find the areas of these rectangles

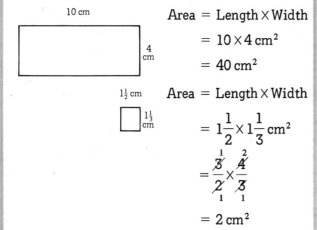

Area = Length × Width

$= 10 \times 4 \, cm^2$

$= 40 \, cm^2$

Area = Length × Width

$= 1\frac{1}{2} \times 1\frac{1}{3} \, cm^2$

$= \frac{\overset{1}{\cancel{3}}}{\underset{1}{\cancel{2}}} \times \frac{\overset{2}{\cancel{4}}}{\underset{1}{\cancel{3}}}$

$= 2 \, cm^2$

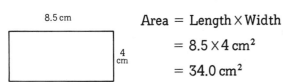

Area = Length × Width

$= 8.5 \times 4 \, cm^2$

$= 34.0 \, cm^2$

EXAMPLE

Find the area of this rectangle (a) exactly (b) to
the nearest square metre.

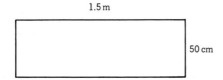

It is essential to make both measurements of
the *same type*.

Change centimetres to metres:

50 cm = 0.5 m

Area = Length × Width

$= 1.5 \times 0.5 \, m^2$

$= 0.75 \, m^2$ exactly

$= 1 \, m^2$ (to the nearest square metre)

Working:

$$\begin{array}{r} 1.5 \\ 0.5\times \\ \hline 0.75 \end{array}$$

1 d.p. plus 1 d.p. gives 2 d.p. in
answer.

Find the areas of these rectangles

1)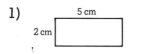

2)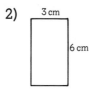

3)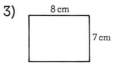

4)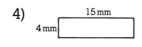

5)

6)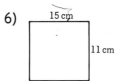

7)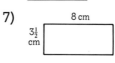

8)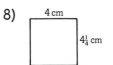

9)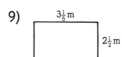

10)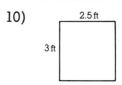

11)

12)

13)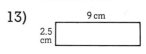

14)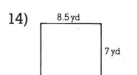

15)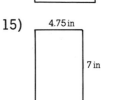

Find the areas of these rectangles

16) Length 10 mm width 8 mm
17) Length 3 m width 50 cm
18) Length 5 cm width 28 mm
19) Length 4.5 cm width 2.3 cm
20) Length 13.5 cm width 8.6 cm.
21) Find the area of a carpet which is 2 m wide and 350 cm long.
22) What is the area of a 15 cm square tile?
23) Calculate the area in square feet of a pane of glass which is 3'6" by 8'.
24) A piece of vinyl flooring is 2 m wide and 2.8 m long. What is its area to the nearest square metre? If it costs £2.25 per square metre how much will it cost? (Assume that you are charged for a whole number of square metres.)
25) Sheet metal is sold at £4.50 per square yard. What is the cost of a piece 6 feet long and 3 feet wide?
26) What is the area of a wall consisting of 150 tiles which are each 10 cm by 10 cm?
27) How many 6 inch square tiles will be needed to tile two walls of a shower unit, each wall being 7 ft 6 in high and 2 ft 6 in wide? What will be the cost if tiles are £6 a square yard?

EXAMPLE

Find the area of the side of the L-shaped piece of wood shown:

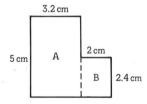

This shape can be divided into two rectangles. Call one rectangle A and the other B.

The area of A = Length × Width

$$= 5 \times 3.2 = 16.0 \, \text{cm}^2$$

The area of B = Length × Width

$$= 2.4 \times 2 = 4.8 \, \text{cm}^2$$

Total area $= 20.8 \, \text{cm}^2$

--- **EXERCISE 2** ---

By dividing these shapes into rectangles find their total areas. (All measurements are in cm.)

1)
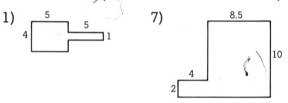

7)
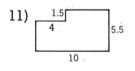

2)
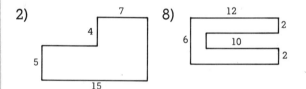

8)
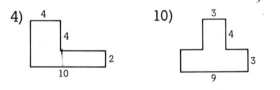

3)
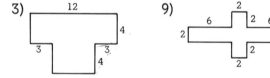

9)

4)

10)

5)

11)

6)
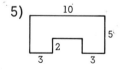

12)
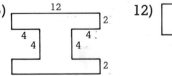

216

AREAS OF BORDERS

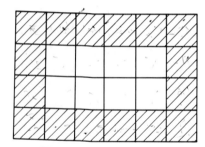

There are $6 \times 4 = 24$ squares altogether in this rectangle.

There are $4 \times 2 = 8$ white squares.

There are 16 shaded squares (count them!)

The number of shaded squares = total number of squares − white squares.

The shaded squares form a *border* round the edge of the white squares.

To find the area of a border:

Calculate the total area (by multiplying length × width of the outside rectangle)

and *take away* the area of the central rectangle.

EXAMPLE

A lawn is 6 m by 3 m and has a 1 m wide path all the way round the edge. Find the area of the path.

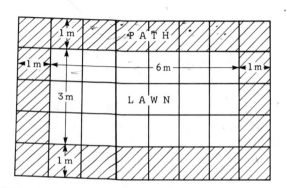

Total area = length × width of outer rectangle.

The length of the outer rectangle is 8 m and the width 5 m.

Total area = $8 \times 5 = 40 \, m^2$

Area of central rectangle = $6 \times 3 = 18 \, m^2$

Area of border = $40 - 18 \, m^2 = 22 \, m^2$
(check by counting the shaded squares)

EXAMPLE

Find the area of the shaded border of the rectangle below which is 0.5 cm wide all the way round.

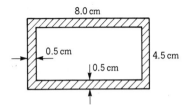

First we must find the length and width of the inner rectangle.

Length of inner rectangle = $8.0 - (2 \times 0.5)$

$= 8.0 - 1.0$

$= 7.0 \, cm$

Width of inner rectangle = $4.5 - (2 \times 0.5)$

$= 4.5 - 1.0$

$= 3.5 \, cm$

Total area = Length × Width of outer rectangle

$= 8.0 \times 4.5 \, cm^2$

$= 36.0 \, cm^2$

Area of inner rectangle = $7.0 \times 3.5 \, cm^2$

$= 24.5 \, cm^2$

Area of border = Total area − Area of central rectangle

$= 36.0 - 24.5 \, cm^2 = 11.5 \, cm^2$

EXERCISE 3

Find the areas of the borders. (It might be helpful to draw these on squared paper and to check your answers by counting the squares.)

1)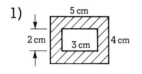

2)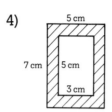

3)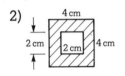

4)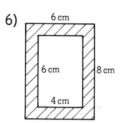

5)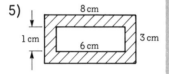

6)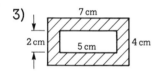

7) This is a sketch of the top of an iron pillar. Find the area of the iron.

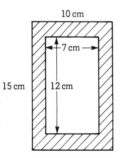

8) A lawn which is 7 m by 10 m has a path which is 1 m wide all round it. Draw a sketch of the lawn and path and mark the length and width of (i) the lawn and (ii) the lawn plus the path on your diagram. Find (a) the total area of lawn plus path (b) the area of the lawn (c) the area of the path (d) the cost of putting weedkiller on the path if it costs 2 p per square metre.

9) A picture measures 1.5 m by 1 m. It has a wooden frame around it which is 0.25 m wide. Draw a diagram of the picture and mark the length and width on it. Draw the frame round the picture and work out the length and width of the frame and mark them on your sketch. Calculate (a) the total area of picture plus frame (b) the area of the picture alone (c) the area of the frame (d) the cost of buying the wood for the frame if the wood was sold at £2 per square metre.

AREAS OF CIRCLES

It can be shown that the area of a circle = $\pi \times$ radius $\times$ radius where π is about 3.14 or $\frac{22}{7}$.

The *radius* is the distance from the centre to the edge. The *area* is the space taken up by the inside of the circle.

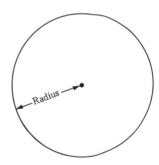

EXAMPLE

Find the area of a circle of radius 2 cm.

$$\text{Area} = \pi \times \text{Radius} \times \text{Radius}$$
$$= 3.14 \times 2 \times 2 \text{ cm}^2$$
$$= 6.28 \times 2 \text{ cm}^2$$
$$= 12.56 \text{ cm}^2$$

It is easier when using a calculator to take π as 3.14. However, if the radius is 7 or 14 or another multiple of 7 it is sometimes easier to take π as $\frac{22}{7}$. When working with *fractions* it is often best to take $\pi = \frac{22}{7}$.

EXAMPLE

Find the area of a circle of radius 7 cm.

Area = $\pi \times$ Radius $\times$ Radius

$= \dfrac{22}{7} \times 7 \times 7$ cm² (make the whole numbers into top-heavy fractions and *cancel* by 7)

$= \dfrac{22}{\cancel{7}} \times \dfrac{\cancel{7}}{1} \times \dfrac{7}{1}$ cm²

$= 22 \times 7$ cm²

$= 154$ cm²

You may *choose* whichever value of π is most convenient. Some people prefer always to take the value of π as about 3.14 because they find working with decimals easier than working with fractions. In this case all fractions must be changed to decimals first.

--- **EXERCISE 4** ---

Find the areas of these circles. The answers will vary slightly according to whether you use $\pi = 3.14$ or $\pi = \frac{22}{7}$.

1) Radius 3 cm
2) Radius 5 cm
3) Radius 10 cm
4) Radius 14 cm
5) Radius $3\frac{1}{2}$ cm
6) Radius $1\frac{3}{4}$ cm

Sometimes the diameter is given instead of the radius:

EXAMPLE

Find the area of a circle of diameter 8 cm.

Diameter = $2 \times$ Radius

$\therefore$ Radius = 4 cm

Area of the circle = $\pi \times$ Radius $\times$ Radius

$= 3.14 \times 4 \times 4$ cm²

$= 12.56 \times 4$ cm²

$= 50.24$ cm²

A *composite* figure is made up of several simpler shapes. The shape in the example below is made up of a square and half a circle (a semi-circle). To find the total area we find the areas of the separate shapes and *add them together*.

EXAMPLE

Find the area of this composite figure:

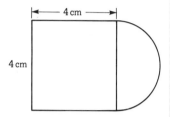

The area of the square is 4×4 cm² = 16 cm²

The circle has diameter 4 cm.

Diameter = $2 \times$ Radius

$\therefore$ Radius is 2 cm

The area of a whole circle with radius 2 cm is given by

$\pi \times$ Radius $\times$ Radius = $3.14 \times 2 \times 2$ cm²

$= 6.28 \times 2$ cm²

$= 12.56$ cm²

The area of half such a circle = $12.56 \div 2$

$= 6.28$ cm²

The total area of the shape = The area of the square + The area of the semi-circle

$= 16 + 6.28$ cm²

$= 22.28$ cm²

--- **EXERCISE 5** ---

These are harder questions. Take care with the units. Give your answers correct to 3 significant figures.

1) Find the area of a circle with diameter 12 cm.

2) Find the area of a circle with diameter 16 mm.

3) Find the area of a circle with diameter 40 m.

4) Find the area of a circle with diameter 1.4 m.

Find the areas of these shapes (they are all made from rectangles or squares and circles or semi-circles).

5)

8)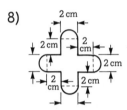

6)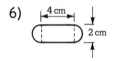

9)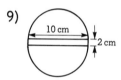

7)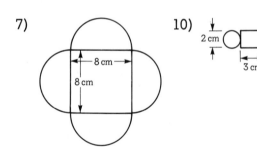

10)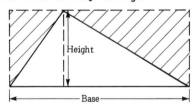

In the next six questions find the area of (a) the square, (b) the circle or part of a circle and (c) the shaded area by *taking away the area of the circle, or part of a circle, from the area of the square.*

11)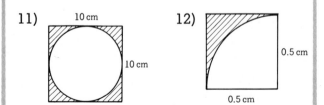

12)

13)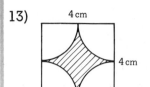

14)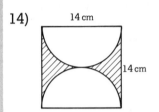

15)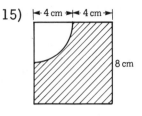

16)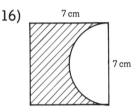

It is easy to see that the area of a right-angled triangle is *half* the area of a rectangle with the same height and base. (Imagine a piece of paper folded in half.)

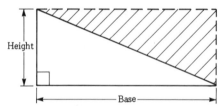

This is also true for *any* triangle:

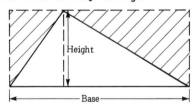

Area of a triangle $= \dfrac{1}{2}$ Base $\times$ Height

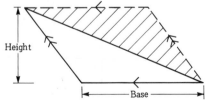

220

It is not so easy to see that this is true for a triangle that "leans over". However, such a triangle can be shown to have an area half that of a parallelogram (which looks like a pushed over rectangle) with the same base and height. In turn, the parallelogram has the same area as a rectangle with the same base and height as shown by this diagram (by transferring the triangle on the end to the shaded area):

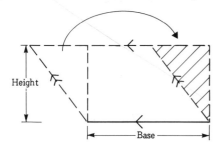

Reminder: Area is the space taken up by a flat (or curved) surface.

The *area of a triangle* is the amount of space inside the triangle.

EXAMPLE

Find the areas of these triangles:

(a)

(b)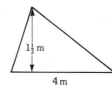

(c)

$$\text{Area of a triangle} = \frac{1}{2} \text{Base} \times \text{Height}$$

(Care must be taken with the units.)

(a) $\text{Area} = \frac{1}{2} \times 8 \times 10 \text{ cm}^2$

$= 4 \times 10 \text{ cm}^2$

$= 40 \text{ cm}^2$

(b) $\text{Area} = \frac{1}{2} \times 1\frac{1}{2} \times 4 \text{ m}^2$

$= \frac{1}{2} \times \frac{3}{2} \times \frac{4}{1} \text{ m}^2$

$= 3 \text{ m}^2$

(c) $\text{Area} = \frac{1}{2} \times 5.6 \times 3 \text{ mm}^2$

$= 2.8 \times 3 \text{ mm}^2$

$= 8.4 \text{ mm}^2$

EXERCISE 6

Find the areas of these triangles

1)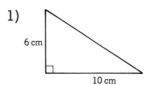

2)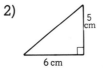

3)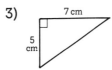

4)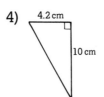

5)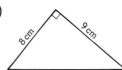

6)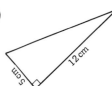

7)

8)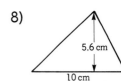

221

9)

11)

10)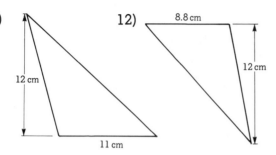

12)

Find the areas of these triangles

13) Base 16 mm height 7 mm
14) Base 3.2 m height 4 m
15) Base 26 mm height 8 mm
16) Base 12.5 cm height 6 cm
17) Base 24 m height 5 m
18) Base $4\frac{1}{2}$ cm height $1\frac{1}{3}$ cm
19) Base 50 mm height 9 mm
20) Base 46 mm height 8 mm

— THE AREA OF A TRAPEZIUM —

The area of a trapezium may be found by dividing the shape into two triangles.

EXAMPLE

Find the area of this shape:

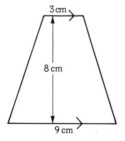

First divide the shape into two triangles, labelling them A and B:

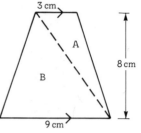

Area of triangle A $= \frac{1}{2} \times 3 \times 8 = 12\,\text{cm}^2$

Area of triangle B $= \frac{1}{2} \times 9 \times 8 = 36\,\text{cm}^2$

Total area of trapezium $= 12 + 36 = 48\,\text{cm}^2$

── **EXERCISE 7** ──

Find the areas of these shapes

1)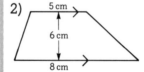

5)

2)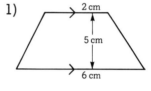

6)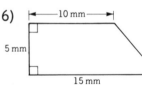

3)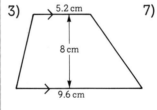

7)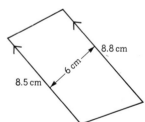

4)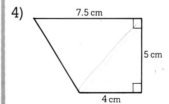

222

8) Find the area of a trapezium with two parallel sides of length 5 cm and 7 cm and height 8 cm.

9) The distance between the two parallel sides of a trapezium is 10 cm. The lengths of the parallel sides are 8 cm and 9 cm. What is the area of the trapezium?

10) Calculate the area of a trapezium with parallel sides 3 m and 5 m and distance between them 12 m.

— REARRANGING FORMULAE —

If we are given the *area* of a shape and need to find the *height* or *length* it is often useful to use the algebraic formula for the area and *rearrange* it to find the unknown height or length.

Area of a rectangle $=$ Length $\times$ Width

$$A = l \times w$$

$$\therefore l = A \div w \quad \text{and} \quad w = A \div l$$

EXAMPLE

Find the width of this rectangle:

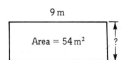

$$w = A \div l$$
$$\text{width} = 132 \div 12$$
$$= 11 \text{ cm}$$

Area of a triangle $= \dfrac{1}{2}$ Base $\times$ Height

$$A = \dfrac{1}{2} \times b \times h$$

$$\therefore b = 2 \times A \div h \quad \text{and} \quad h = 2 \times A \div b$$

EXAMPLE

Find the height of this triangle:

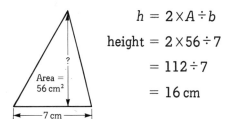

$$h = 2 \times A \div b$$
$$\text{height} = 2 \times 56 \div 7$$
$$= 112 \div 7$$
$$= 16 \text{ cm}$$

Area of a circle $= \pi \times$ Radius $\times$ Radius

$$A = \pi \times r \times r \quad (\text{or } A = \pi r^2)$$

$$\therefore r \times r = A \div \pi$$

$$\text{and} \quad r = \sqrt{A \div \pi}$$

EXAMPLE

Find the radius of this circle:

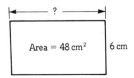

$$r = \sqrt{A \div \pi}$$
$$\text{Radius} = \sqrt{314 \div 3.14} \text{ cm}$$
$$= \sqrt{100} \text{ cm}$$
$$= 10 \text{ cm}$$

(For more difficult numbers the use of a calculator is recommended.)

EXERCISE 8

1) Find the length of this rectangle

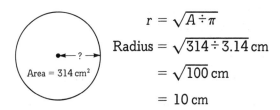

2) Find the width of this rectangle

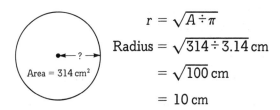

3) (a) Find the height of this triangle

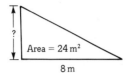

Area = 24 m²
8 m

(b) Find the height of this triangle

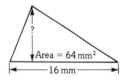

?
Area = 64 mm²
16 mm

4) Find the length of the base of this triangle

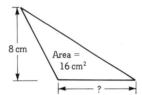

8 cm
Area = 16 cm²
?

5) Find the radius of this circle

?
Area = 12.56 cm²

6) Find the length of a rectangle with area 56 cm² and width 4 cm.

7) Find the width of a rectangle with length 9 mm and area 45 mm².

8) Find the height of a triangle with base 100 cm and area 300 cm².

9) Find the height of a triangle with area 60 mm² and base 5 mm.

10) Find the base of a triangle with area 84 cm² and height 12 cm.

1) Find the areas of these rectangles

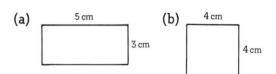

(a) 5 cm   3 cm
(b) 4 cm   4 cm

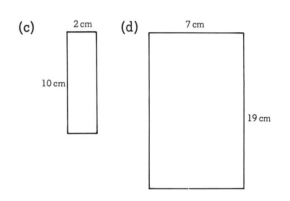

(c) 2 cm   10 cm
(d) 7 cm   19 cm

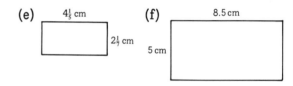

(e) $4\frac{1}{5}$ cm $2\frac{1}{7}$ cm
(f) 8.5 cm   5 cm

2) Find the area of a rectangle with length 26 cm and width 5 cm.

3) By dividing these shapes into rectangles find their total areas

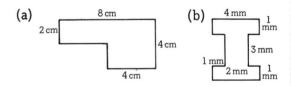

(a) 8 cm   2 cm   4 cm   4 cm
(b) 4 mm   1 mm   3 mm   1 mm   1 mm   2 mm

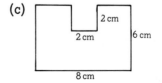

(c) 2 cm   2 cm   6 cm   8 cm

4)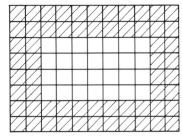

(a) Write down the number of white squares.
(b) Write down the number of shaded squares.
(c) Write down the total number of squares.

5) Find the area of a rectangular lawn measuring 2.5 m by 3 m. There is a path 0.5 m wide all round such a lawn. Sketch the lawn plus path and put on the measurements of the larger rectangle. What is the total area of lawn plus path?

Find the area of the path and calculate the cost of putting on weedkiller at 4p per square metre.

6) What is the area of a circle of radius (a) 2 cm (b) 35 cm?

7) Give the areas of these triangles
(a) base 16 cm height 9 cm
(b) base 1.8 m height 10 m.

8) Find the areas of these shapes
(a)

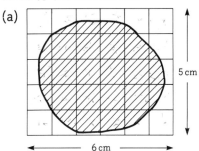

3 cm

6 cm

(b)

10 cm

6 cm

6 cm

9) Find the shaded areas of these shapes

(a)
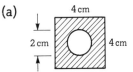
4 cm
2 cm
4 cm

(b)

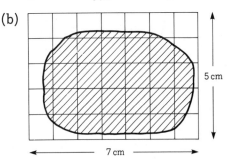
10 cm
8 cm
5 cm
5 cm

10) Find (a) the length of a rectangle area 81 cm² and width 8.1 cm (b) the height of a triangle area 20 mm² and base 10 mm.

11) For an irregular shape the area can be estimated by counting squares. (Try to match two half squares to make a whole one.)

By counting squares estimate these shaded areas to the nearest cm².

(a)

5 cm

6 cm

(b)

5 cm

7 cm

20

———— Volume ————

Volume is the space taken up by a solid object such as a cardboard carton or a football:

It is measured in *cubic units*.

This is a unit cube of side 1 cm:

It is called a "1 cm cube" and has a *volume* of 1 cubic centimetre (1 cm³).

———— UNITS OF VOLUME ————

The *metric* units of volume are:

	cubic millimetres	(mm³)
	cubic centimetres	(cm³)
	cubic metres	(m³)
or	cubic kilometres	(km³)

The *imperial* units of volume are:

	cubic inches	(cu. in or in³)
	cubic feet	(cu. ft or ft³)
	cubic yards	(cu. yd or yd³)
or	cubic miles	(cu. miles or miles³)

A cubic metre is the amount of space contained in a cube whose edge is 1 metre. Similarly a cubic foot is the amount of space contained in a cube whose edge is 1 foot long.

The volume of liquids is sometimes measured in litres and

$$1 \text{ litre} = 1000 \text{ cm}^3$$

In imperial measure, pints, quarts and gallons are used as was described in Chapter 5.

VOLUMES OF CUBOIDS AND — SHAPES MADE FROM THEM —

Thirty 1 cm cubes would fit inside this shape:

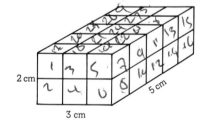

We say its volume is 30 cm³.

The volume of a cuboid is defined as length × width × height (even when the length, width and height are not whole numbers).

EXAMPLE

Find the volumes of these cuboids

(a)

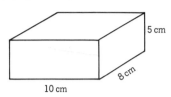

$$\text{Volume} = \text{Length} \times \text{Width} \times \text{Height}$$
$$= 10 \times 8 \times 5 \text{ cm}^3$$
$$= 80 \times 5 \text{ cm}^3$$
$$= 400 \text{ cm}^3$$

(b)

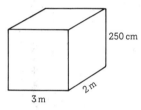

$$\text{Volume} = \text{Length} \times \text{Width} \times \text{Height}$$

All measurements must be in the *same units* so changing 250 cm to metres

$$250 \text{ cm} = 2.5 \text{ m}$$
$$\text{Volume} = 3 \times 2 \times 2.5 \text{ m}^3$$
$$= 6 \times 2.5 \text{ m}^3$$
$$= 15.0 \text{ m}^3$$

──────── **EXERCISE 1** ────────

Find the volumes of these shapes

1)
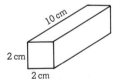

3)
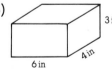

2)
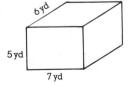

4)
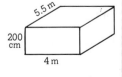

5)
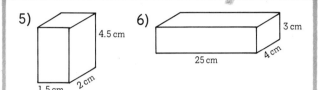

6)

Find the volumes of these cuboids

7) Length 3 m width 2 m height 1.5 m
8) Length 500 cm width 250 cm height 4 m
9) Length 6 mm width 4 mm height 12 mm
10) Length 15 cm width 6 cm height 4 cm
11) Length 35 mm width 1 cm height 5 cm
12) Length 13 cm width 7 cm height 10 cm.
13) Find the volume of a cardboard box measuring 30 cm by 20 cm by 15 cm.
14) The length, width and height of a tin are 15 cm, 10 cm and 8 cm respectively. What is its volume?
15) Calculate the volume of a box with length 35 cm, width 10 cm and height 8 cm.

────────────────────────

EXAMPLE

Find the volume of this podium:

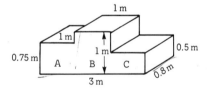

Divide the shape into 3 cuboids (boxes) and label them A, B and C:

A has length 1 m, width 0.8 m and height 0.75 m

B has length 1 m, width 0.8 m and height 1 m

C has length 1 m, width 0.8 m and height 0.5 m

The volume of A $= 1 \times 0.8 \times 0.75 \text{ m}^3$
$$= 0.6 \text{ m}^3$$

The volume of B $= 1 \times 0.8 \times 1 \, m^3$

$\qquad = 0.8 \, m^3$

The volume of C $= 1 \times 0.8 \times 0.5 \, m^3$

$\qquad = 0.4 \, m^3$

The total volume of the podium is $1.8 \, m^3$

—————— **EXERCISE 2** ——————

By dividing these shapes into cuboids (boxes) find their total volumes

1)
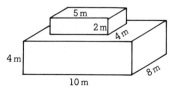

2)
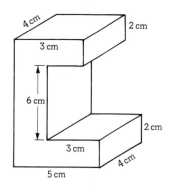

3)
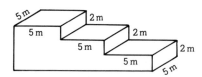

4)
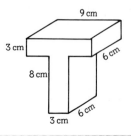

5)
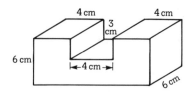

6)
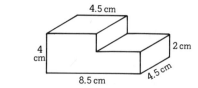

7) A block of wood 10 cm by 10 cm by 9 cm has a square hole drilled into it measuring 8 cm by 8 cm by 9 cm.

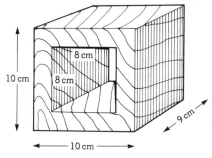

(a) Find the volume of the wood before the hole is drilled.
(b) Find the volume of air in the square hole drilled through the centre of the wood.
(c) By subtracting your answer to (b) from your answer to (a) find the volume of wood remaining after the hole has been drilled.

—————— **VOLUMES OF PRISMS** ——————

A *prism* has the same shape all the way along its length.

It can be shown that:

228

Volume of a prism = Area of the front face
× Length

$$V = A \times l$$

where V is the volume, A is the area of the front face and l is the length.

A new pencil is an example of a *hexagonal prism*. The front face is a hexagon. (It has six sides.)

A *rectangular prism* is another name for a *cuboid* (or box shape)

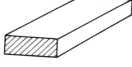

and a *cylinder* is a special name for a *circular prism*. Most tins are made in the shape of a cylinder.

This piece of cheese is in the shape of a *triangular prism*. The front face is a triangle and the cheese has the same shape all the way along its length.

This lean-to greenhouse has the shape of a prism. Its front face is made up from a rectangle plus a triangle.

EXAMPLE

Find the volume of the triangular prism shown below:

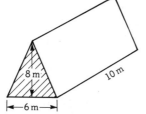

Volume of a prism = Area of front face
× Length

Area of triangle $= \dfrac{1}{2}$ Base × Height

∴ Area of front face $= \dfrac{1}{2} \times 6 \times 8 \, \text{m}^2 = 3 \times 8 \, \text{m}^2$

$$= 24 \, \text{m}^2$$

Volume $= 24 \times 10 \, \text{m}^3 = 240 \, \text{m}^3$

EXAMPLE

Find the volume of this tin:

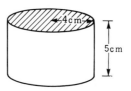

Volume = Area of front face × Length

(If the tin were laid on its side the front face would be a circle, radius 4 cm, and the length of the tin would be 5 cm.)

Area of a circle $= \pi \times$ Radius × Radius

∴ Area of front face $= 3.14 \times 4 \times 4 \, \text{cm}^2$

$$= 12.56 \times 4 \, \text{cm}^2$$

$$= 50.24 \, \text{cm}^2$$

Volume $= 50.24 \times 5 \, \text{cm}^3 = 251.20 \, \text{cm}^3$

—————— **EXERCISE 3** ——————

By finding the area of the front face and multiplying by the length find the volumes of these prisms

1)

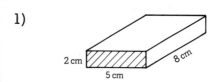

2)

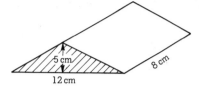

3)

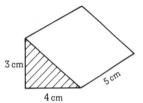

4)

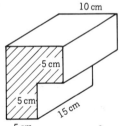

5)

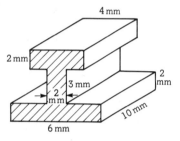

6)

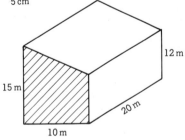

7)

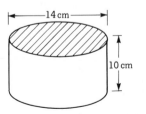

8)

9) Find the volume of a cylinder of length 10 cm and radius 2 cm.

10) Find the volume of a triangular prism with base 10 cm, height 8 cm and length 12 cm.

— VOLUMES OF PYRAMIDS —

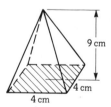

It can be shown that:

$$\text{The volume of a pyramid} = \frac{1}{3} \text{ Area of the base} \times \text{Height}$$

$$V = \frac{1}{3} \times A \times h$$

where V is the volume, A is the area of the base and h is the height.

EXAMPLE

Find the volume of this square based pyramid:

Area of the base $= 4 \times 4 \text{ cm}^2 = 16 \text{ cm}^2$

$$\text{Volume} = \frac{1}{3} \times 16 \times 9 \text{ cm}^3$$

$$= 16 \times 3 \text{ cm}^3$$

$$= 48 \text{ cm}^3$$

A *cone* is a special kind of pyramid with a circular base:

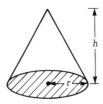

$$\text{The volume of a cone} = \frac{1}{3} \, \text{Area of the base} \times \text{Height}$$

$$= \frac{1}{3} \times \pi \times \text{Radius} \times \text{Radius} \times \text{Height}$$

$$V = \frac{1}{3}\pi r^2 h$$

where V is the volume, r is the radius of the base and h is the height.

EXAMPLE

Find the volume of this cone:

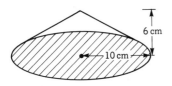

$$\text{Volume} = \frac{1}{3} \times \pi \times \text{Radius} \times \text{Radius} \times \text{Height}$$

$$= \frac{1}{3} \times 3.14 \times 10 \times 10 \times 6 \, \text{cm}^3$$

$$= \frac{1}{3} \times 314 \times 6 \, \text{cm}^3$$

$$= 314 \times 2 \, \text{cm}^3$$

$$= 628 \, \text{cm}^3$$

To find the volume of a pyramid find the area of the base, multiply by the height and divide by 3.

EXERCISE 4

Using the formula *volume of a pyramid* $= \frac{1}{3}$ *area of the base* $\times$ *height* find the volume of these solid figures

1)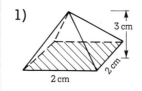
3 cm
2 cm
2 cm

4)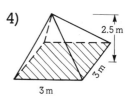
2.5 m
3 m
3 m

2)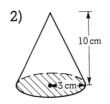
10 cm
3 cm

5)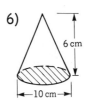
5 cm
6 cm
6 cm

3)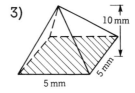
10 mm
5 mm
5 mm

6)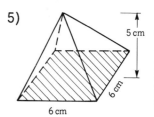
6 cm
10 cm

EXERCISE 5

1) Find the volumes of these boxes

(a)
8 cm
5 cm
2 cm

(b)
1 cm
6 cm
2 cm

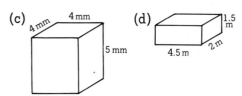

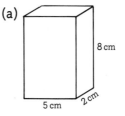

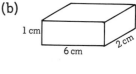

(c)
4 mm
4 mm
5 mm

(d)
1.5 m
4.5 m
2 m

231

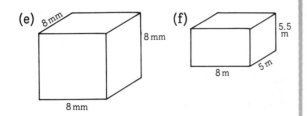

(e) 8 mm, 8 mm, 8 mm

(f) 5.5 m, 8 m, 5 m

2) The measurements of these pieces of wood are as shown. Find their volumes

(a)

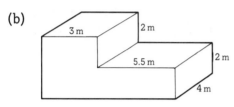

5 m, 3 m, 3 m, 3 m, 1 m, 4 m

(b)

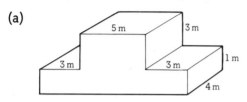

3 m, 2 m, 5.5 m, 2 m, 4 m

3) Find the volumes of these triangular prisms

(a)

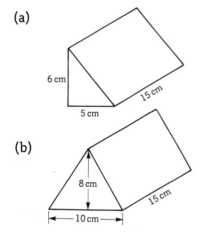

6 cm, 15 cm, 5 cm

(b)

8 cm, 15 cm, 10 cm

4) Using the fact that the volume of a cylinder = area of circle × depth find the volume of this gas holder

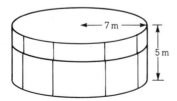

7 m, 5 m

5) The volume of a pyramid is $\frac{1}{3}$ area of the base × height. Use this fact to determine the volume of the pyramid with measurements as shown on this diagram

(The base of this pyramid is a square.)

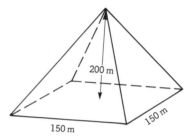

200 m, 150 m, 150 m

Harder questions:

6) The formula for the volume of a cone is $\frac{1}{3}\pi r^2 h$ where r is the radius of the base and h is the height of the cone. Use this formula to find the volume of a cone with base radius 7 cm and height 10 cm.

7) The formula for the volume of a sphere is $\frac{4}{3}\pi r^3$ where r is the radius of the sphere. Use this formula to find the volume of a sphere with radius 2 cm.

8) Another formula for the volume of a cylinder is $\frac{1}{4}\pi d^2 h$ where d is the diameter of the circle and h is the height of the cylinder. Use this formula to find the volume of a cylinder with diameter 10 cm and height 7 cm.

9) The volume of a cuboid is given by $V = l \times w \times h$ where V is the volume, l is the length, w is the width and h is the height. If this formula is rearranged then $h = V \div (l \times w)$. Use this fact to find the height of a box with volume 150 cm³, length 10 cm and width 3 cm.

10) Rearrange the formula $V = l \times w \times h$ to find the width of a box with volume 96 m³, height 8 m and length 4 m.

21

Similarity

Similar figures have the *same* shape.

These two toys are similar in shape:

However, these two trucks are *not* similar in shape:

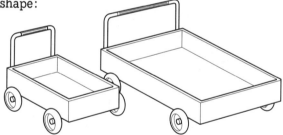

The larger truck is wider and longer but it is not taller. For shapes to be similar they must be wider, longer *and* taller in the *same* proportion.

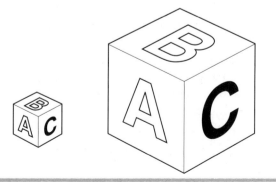

You might say the larger brick is 3 times bigger than the smaller brick and indeed all the *lengths* are 3 times bigger. However, 9 small bricks would sit on top of the larger brick. It would take 27 small bricks to make a shape as big as the larger brick.

If both bricks were made of the same type of wood, the larger brick would be 27 times heavier than the smaller brick.

Notice that the *length* is 3 times bigger.

The *area* is 3×3 times bigger.

The *volume* and the *weight* (if they were made of the same material) are $3 \times 3 \times 3$ times bigger.

The areas vary as the *square* of the lengths.

The volumes (weights) vary as the *cube* of the lengths.

This property applies to all similar figures.

EXAMPLE

The model of a bronze sculpture is 10 times smaller than the actual figure

Model   Sculpture

I need $\frac{1}{2}$ tin of paint for the model. How many tins will I need to paint the actual figure?

If the model weighs 2 kg how much will the actual figure weigh?

> Lengths are 10 times larger.
> Areas are 10×10 times larger.
> Volume and weight are $10 \times 10 \times 10$ times larger.

Painting depends on the area of the figure. The area of the actual figure is 100 times the area of the model.

I will need $100 \times \frac{1}{2}$ tins of paint = 50 tins of paint.

The weight of the sculpture will be 1000 times the weight of the model. The sculpture weighs 1000×2 kg = 2000 kg.

EXERCISE 1

1)

The two tins of beans are similar in shape and content. If the second tin is twice as tall as the first, how much heavier will it be?

2) How many 5 cm tiles will fit on a section of wall measuring 25 cm by 25 cm?

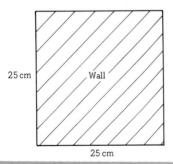

25 cm — Wall — 25 cm

5 cm 5 cm

3)

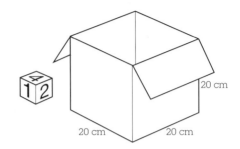

20 cm

20 cm   20 cm

The length of the side of the brick is 5 cm. How many of these bricks will fit inside the box shown?

4) The diagrams show the area of the front of two tunnels. They are similar in shape.

$7\frac{1}{2}$ ft

$2\frac{1}{2}$ ft

If the area of the front of the smaller tunnel is 5 square feet, what is the area of the front of the larger tunnel?

5) How many 1 cm sugar cubes will fit (a) on the first layer, (b) inside a box measuring 6 cm by 6 cm by 6 cm?

6)

The larger bottle of shampoo is similar in shape to the smaller but twice as high. How much more liquid is there in the larger bottle? The price of the smaller is 52 p. The price of the larger is £4.56. Is the larger bottle a good buy?

234

7)

 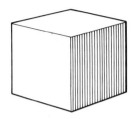

The two cubes are made of the same metal. The larger one is 8 times heavier than the smaller. The length of the side of the smaller cube is 5 cm. What is the length of the side of the larger?

8)

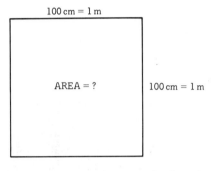

100 cm = 1 m

AREA = ?

100 cm = 1 m

What is the area of a square which is 1 m long and 1 m wide (a) in m^2, (b) in cm^2?

How many square centimetres are there in 1 square metre? (Note carefully that the number of *square centimetres* in one *square metre* is not the same as the number of *centimetres* in one *metre*.)

9)

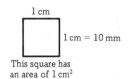

1 cm

1 cm = 10 mm

This square has an area of 1 mm²

This square has an area of 1 cm²

How many mm^2 make $1\ cm^2$?

10)

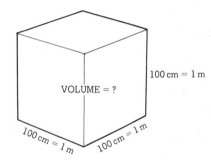

100 cm = 1 m

VOLUME = ?

100 cm = 1 m

100 cm = 1 m

Using the picture above, state how many cm^3 there are in $1\ m^3$.

11)

Costs double!

Why is this diagram misleading?

22

Probability and Flow Charts

In this chapter we look at some of the aspects of mathematics that have been developed relatively recently and are obviously going to continue to be of tremendous importance in the future.

PROBABILITY

This work has great importance in the field of genetics (the study of how human characteristics are passed on from generation to generation), pure science, insurance (the risks have to be carefully calculated before premiums are set), investment in business, and so on.

We measure probability on a scale between 0 and 1.

A probability of 0 means *absolute impossibility* (for example, the probability that we will live to be 200 years old).

A probability of 1 means *absolute certainty* (for example, the probability that all the people alive today were born in either the nineteenth or the twentieth century).

Throwing a 7 on a normal die with six faces is impossible:

probability = 0

We are certain to throw a 1, 2, 3, 4, 5 or 6:

probability = 1

All other probabilities have a value between 0 and 1 which we write as a fraction or a decimal.

We talk of a chance being, say, 1 in 3. As a fraction we would write 1 in 3 as $\frac{1}{3}$.

Two words that often occur with probability are "event" and "random".

An *event* is a particular happening or occurrence that we are interested in. For example, an insurance company may be interested in the number of people unable to work through illness. "Unable to work through illness" would be the event. Another example might be winning the National Lottery. We may say the chance of getting six numbers correct is approximately 1 in 14 million or $\frac{1}{14\,000\,000}$. The event is "winning the National Lottery".

Random means purely by chance in an unbiased way. For example, if we pick a card from a pack at random, then all cards are equally likely to be picked and there is an equal chance of picking any particular card. Winning Premium bonds are selected at random. Every bond has an equal chance of being picked for a prize.

$$\text{Probability of an event} = \frac{\text{Number of ways the event can occur}}{\text{Total possibilities}}$$

Probability is an extremely complicated subject, so it is usually introduced by considering simple (but fairly artificial) events such as "throwing a 5 on a die", or "drawing a queen from a pack of cards" or "picking a green marble at random from a bag".

EXAMPLE

What is the probability of drawing a picture card from a normal pack of cards?

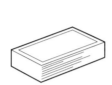

A normal pack of cards has 4 suits of 13 cards each. The suits are hearts, clubs, diamonds and spades with ace, 2 to 10, and the picture cards — Jack, Queen and King — in each suit. Hearts and diamonds are red, spades and clubs are black.

There are 52 cards in a normal pack. 12 of these are picture cards.

The probability of drawing a picture card is $\frac{12}{52} = \frac{3}{13}$ in its lowest terms.

—— TOTAL PROBABILITY ——

The total probability is always 1.

This is useful when we wish to calculate the probability of something *not* happening.

EXAMPLE

What is the probability of drawing a card *other* than a picture card from a normal pack of cards?

The probability of drawing a card other than a picture card is $1 - \frac{3}{13} = \frac{10}{13}$.

EXPERIMENTAL PROBABILITY

When we cannot work out the probability by reason alone, then we have to carry out an experiment on a sample of whatever is involved. We might want to check some light-bulbs made in a factory to see how many of them are faulty. The bigger the sample the more accurate the value for the probability of a light-bulb being faulty. However, if the machine breaks down, it may start to produce all faulty light-bulbs. It is always important to be aware that circumstances may change. Tests should continue to be taken at regular intervals.

EXAMPLE

Tests showed that out of 5000 light-bulbs made in a factory, 25 were found to be faulty. If I pick a light-bulb at random, what is the chance that it is faulty?

Since 25 in 5000 were found to be faulty the chance of picking a faulty light-bulb at random is 25 in 5000 or $\frac{25}{5000}$ which equals $\frac{1}{200}$ in its lowest terms.

—— EXERCISE 1 ——

1) In a family with two children it is possible to have two girls, two boys, a boy then a

girl, or a girl then a boy. This is shown as

GG, BB, BG, GB

with B standing for boy and G for girl. If the chance of having a boy or a girl is the same, write down the probability of having (a) two boys, (b) two girls, (c) a boy *and* a girl in a family of two children.

2) It is known that there are 3 winning raffle tickets and that 150 tickets have been sold. What is the chance of winning with 1 raffle ticket (a) first prize, (b) any prize?

3) It is my turn to throw the die and I need a 6 or a 1 to be able to move. What is the chance that I can move on my next turn?

4) Three boys bet on the colour of marbles which are drawn at random from a bag without looking. There are 6 green marbles, 4 yellow marbles and 2 red marbles. Calculate the probability that the first one drawn will be (a) green, (b) yellow, (c) not red.

5) In a game a coin is placed at random on one of the squares on the board shown below. What is the probability that it is placed on (a) a black square, (b) a shaded square, (c) a white square?

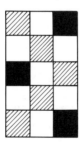

6) A card is drawn at random from a pack of 52 cards. Find the probability that it will be (a) red, (b) a spade, (c) a jack.

7) The probability of a particular couple having a blue-eyed child is $\frac{1}{4}$. What is the probability that the child's eyes will not be blue?

8) 100 ball-bearings are checked in a factory and 15 are found to be defective. If one is picked at random, what is the chance that it will be a good one?

9) A letter is chosen at random from the word POSSESSES. What is the probability that it will be (a) P, (b) S, (c) not E?

10) I have signed 6 letters on my desk but I still have 8 more to sign. If I pick up a letter at random from my desk, what is the chance that it is unsigned?

11) Two dice are thrown together. What is the probability that their total will be (a) 6, (b) less than 6, (c) more than 6?

12) In a survey it is found that 5 out of 150 light-bulbs are faulty. I buy a pack of 10 light-bulbs. What is the chance that the first light-bulb I try is faulty?

13) At a village fête on the tin and bottle stall a prize is won if a raffle ticket ending in 0 or 5 is picked out of a box. If raffle tickets numbered 1 to 200 are placed in the box, what is the initial chance of winning a prize? Describe in words why the chance of winning a prize may vary as the draw proceeds.

14) A hand of cards consists of the king of hearts, the queen of clubs, the ace of spades, the king of diamonds and the 10 of hearts. What is the probability, if I draw a card at random from the hand, of choosing (a) a heart, (b) a spade, (c) a king, (d) a card which is not an ace?

15) The leaves on a certain type of plant may be either striped, variegated or plain. If the chance of their being striped is $\frac{1}{8}$ and the chance of their being variegated is $\frac{1}{4}$, what is the chance of their being plain?

16) I take the four aces, the four kings and the four queens from a pack of cards, shuffle them, and deal three cards to four players. The first player turns up one of

his cards. It is a queen. What is the probability that if he turns up another of his cards it will also be a queen?

———— FLOW CHARTS ————

Flow charts are used in business and computing. Sometimes certain actions have to be carried out in a particular order. Flow charts can show this particularly clearly. A simple example is shown below.

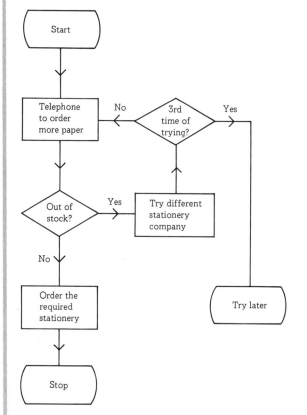

Start and stop boxes are usually shaped like this ⌷. Instruction boxes may be shaped like this ▭ . A diamond-shaped decision box ◇ has two exits, one for "yes" and

one for "no". They allow different paths to be taken through the flow chart. There are usually *loops* which permit certain instructions to be repeated. In the flow chart above, the caller would go round the loop 3 times if three different stationery companies were out of stock.

Quite complicated flow charts might be seen on the wall of an office or on a manager's desk.

A computer programmer would draw a flow chart before writing a program for a computer.

EXAMPLE

The maximum and the minimum measurements of the diameter of a steel rod are entered into a computer. We want the computer to print the larger number but we do not know which is which. If the two measurements are the same we want a message to be printed. Draw a suitable flow chart.

Flow chart to print the larger of two numbers:

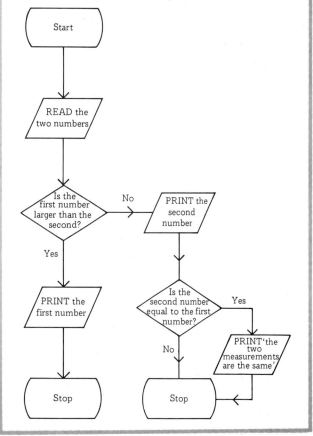

READ and PRINT are called *input* and *output* *instructions* and are in a box shaped like this

This flow chart could be drawn in a shorter way by letting A and B represent the two numbers and by using symbols such as < (less than), > (greater than), or = (equals):

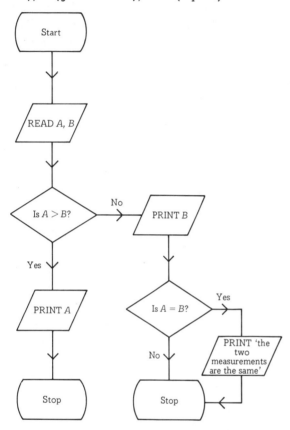

──────── **EXERCISE 2** ────────

Draw a flow chart showing

1) How to make toast and a boiled egg.

2) What to check before going on a holiday which entails a long journey in the car (petrol, water, luggage, map, coffee and sandwiches, etc.).

3) The actions needed to bath a baby (test temperature of water, too hot? too cold? fetch soap, clothes, etc.).

4) Two numbers representing the maximum and minimum age for a particular job are fed into a computer but we do not know which is which. Draw a flow chart to read the two numbers and print the smaller number first and then the larger. Use the signs < or > and letters to represent the numbers.

5) The wages of three people are entered into a computer. Draw a flow chart to print them in ascending order.

────────────────────

USING A FLOW CHART METHOD FOR CALCULATORS WITH ─── A SIMPLE MEMORY ───

In Chapter 2 we showed that multiplication and division must be done before addition and subtraction so that

$$3 + 4 \times 5 = 23 \ (not \ 35)$$
$$\text{and} \quad 15 + 48 \div 12 = 19 \ (not \ 5.25)$$

Some calculators and all computers will do this for you automatically but sometimes, if your calculator is a very simple one, you have to take steps yourself to make sure you get the right answer. (This is very important as it easily leads to errors in calculator work.)

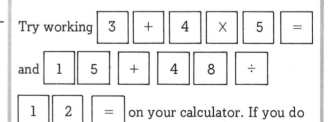

Try working [3] [+] [4] [×] [5] [=] and [1] [5] [+] [4] [8] [÷] [1] [2] [=] on your calculator. If you do

not get the correct answers 23 and 19, then your calculator is one which does not perform arithmetic operations automatically in the correct order. For sums of this type you have to work out the answers by *doing the multi-*

plication and division first yourself. A flow chart can help you to decide the correct order.

If you work out [4] [×] [5] [=] and

then [+] [3] [=] you should get the

right answer 23. The second example is done by

pressing the buttons [4] [8] [÷]

[1] [2] [=] and then [+] [1] [5]

[=] to give the answer 19.

These could be shown on flow charts like this

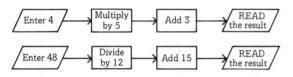

/ Enter 4 / → | Multiply by 5 | → | Add 3 | → / READ the result /

/ Enter 48 / → | Divide by 12 | → | Add 15 | → / READ the result /

(We have written the flow charts across the page to save space.)

──────── **EXERCISE 3** ────────

Suppose you are borrowing a simple calculator. Draw flow charts to show how you would calculate

1) $16 + 84 \times 12$

2) $35 + 112 \div 7$

3) $7 + \dfrac{5}{8}$ as a decimal

4) $5\dfrac{3}{16}''$ in decimal form

5) the price of an electricity bill with fixed charge £10.13 and 415 units at 7.22 p per unit

6) the length of a piece of cloth 1.5 m plus 17 times 0.32 m long

7) the change received from £10 for a tie at £1.25 and 3 pairs of socks at 72 p per pair (you will have to write down an inter-

mediate result or use the [+/−] button to

change the sign in the display)

8) the cost of 4 items at £12.50 each, 17 items at £15.95 each and 9 items at £3.50 each (you will have to write down intermediate results and re-enter them into the calculator for the final total).

9) Use any calculator and the methods shown on your flow charts to work out the correct answers to questions 1) to 8).

────────────────────

FLOW CHART METHOD FOR
── ALGEBRAIC EQUATIONS ──

Flow charts can be used to show algebraic equations. Suppose we wish to flow chart $y = 3x + 5$. We start with x, the unknown letter on the right-hand side of the equation and look to see what is happening to it. Since multiplication is done before addition the first thing is "multiply by 3". We write this in an instruction box and show $3x$ on the exit line. The next thing we do is "add 5" so we write "add 5" in the next instruction box and show $3x + 5$ on the next exit line.

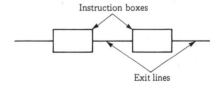

Instruction boxes

Exit lines

If we want to rearrange the equation so that x is on the left-hand side, then we work back again from right to left replacing "add" by "subtract" and "multiply" by "divide".

The flow chart overleaf shows the algebraic equation $y = 3x + 5$. To save space we draw the flow chart across the page and leave out the start and stop boxes.

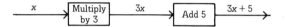

To find what x is in terms of y, we merely reverse the process (working from right to left and starting with y).

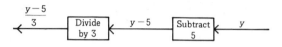

So $x = \dfrac{y-5}{3}$

This method can be useful for rearranging simple formulae.

EXAMPLE

$A = \pi r^2$

Show this on a flow chart and rearrange to find r in terms of A and π.

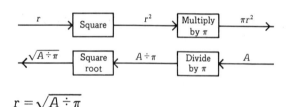

$r = \sqrt{A \div \pi}$

──────── EXERCISE 4 ────────

1) Draw a flow chart to show $y = 7x + 4$ and then draw a second flow chart working from right to left to find x in terms of y.

2) $m = 8n - 6$. Use the flow chart method to find n in terms of m.

3) $C = 2\pi r$. First find r in terms of C and then, using a calculator, find the value of r when $C = 100$.

4) The surface area of a cube is given by $A = 6x^2$ where x is the length of one side. Use the flow chart method to find x in terms of A. What is x when $A = 714$? (Use a calculator.)

5)

```
              b
      ┌──────────────┐
      │              │
    b │              │ b/2
      │        ┌─────┘
      │        │
      └────────┘  b/2
```

The area of this shape is given by $A = \dfrac{5b^2}{4}$.

Find b in terms of A and work out the value of b when $A = 18.75$. (Use a calculator.)

6) Draw a flow chart of $v = u + at$ with t the unknown letter and then reverse it to find t in terms of v, u and a.

7) Show the flow chart of $A = \frac{1}{2}b \times h$ with b the unknown letter. Draw a second flow chart from right to left to find b in terms of A and h. Find b if $h = 10.4$ and $A = 200$. (Use a calculator.) (*Reminder:* multiplying by $\frac{1}{2}$ is the same as dividing by 2.)

242

23

Patterns and Symmetry

NUMBER PATTERNS

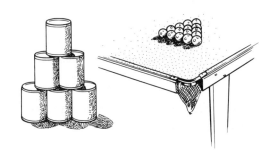

The rows of tins and the snooker balls each contain one more than the row above.

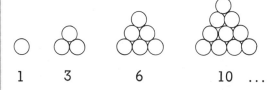

1   3   6   10   ...

is the *sequence* of *triangular* numbers. The dots mean that the sequence goes on for ever. The next triangular number is 15.

A sequence is a series of numbers connected by some definite pattern.

EXAMPLE

The sequence of *multiples of 7* is

7, 14, 21, 28, 35, 42, ...

EXAMPLE

The sequence of *square numbers* is

1   4   9   16   ...

This sequence is formed by multiplying 1×1, 2×2, 3×3, ..., etc.

PRIME NUMBERS

Only 1 and the number itself will go exactly into a prime number. (1 is not a prime number.) The sequence of primes is

2, 3, 5, 7, 11, 13, 17, ...

Numbers which are not prime can be arranged in either a square or a rectangle and are called *rectangular* or *composite* numbers.

1   4   6   8   9   10   ...

1) Give the following 2 numbers in these sequences

(a) 1, 3, 5, 7, 9, 11, ...
(b) 2, 4, 6, 8, 10, 12, ...
(c) 0. 5, 10, 15, 20, 25, ...
(d) 1, 4, 9, 16, 25, ...
(e) 2, 3, 5, 7, 11, ...

2)

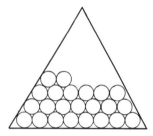

How many table tennis balls would fit into this triangle if there are 8 on the bottom row, 7 on the next and so on?

3) Draw as many different rectangular patterns for the number 24 as possible.

4) Copy and complete the table.

5) Write the next *three* numbers in each sequence

(a) 1, 4, 7, 10, 13, ...
(b) 9, 18, 27, ...
(c) 2, 3, 5, 8, 12, 17, 23, ...
(d) 1, 1, 2, 3, 5, 8, 13, ...
(e) 48, 24, 12, 6, ...

6) Copy the table and complete the last column.

$$1 = 1$$
$$1 + 2 =$$
$$1 + 2 + 3 =$$
$$1 + 2 + 3 + 4 =$$
$$1 + 2 + 3 + 4 + 5 =$$
$$1 + 2 + 3 + 4 + 5 + 6 =$$
$$1 + 2 + 3 + 4 + 5 + 6 + 7 =$$
$$1 + 2 + 3 + 4 + 5 + 6 + 7 + 8 =$$
$$1 + 2 + 3 + 4 + 5 + 6 + 7 + 8 + 9 =$$

What is the name of the sequence of numbers in the last column?

7) Is it possible to fit 29 sugar cubes into a rectangular box if there is more than 1 sugar cube in each row and leaving no gaps?

8) Say whether these statements are true or false

(a) 9 is a square number
(b) 12 is a triangular number
(c) 13 is a prime number
(d) 21 is a multiple of 3
(e) 23 is a rectangular number.

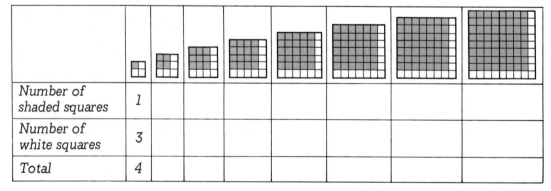

Number of shaded squares	1							
Number of white squares	3							
Total	4							

What is the sequence in (a) the first row,
(b) the middle row, (c) the bottom row of the table?

9) This pattern is called Pascal's Triangle.

```
        1
      1   1
    1   2   1
  1   3   3   1
1   4   6   4   1
1   .   .   .   .   .
```

The numbers in the bottom completed row are found by adding the two numbers above, e.g. $1 + 3$ and $3 + 3$ etc.
$$= 4 \qquad = 6,$$

Find the next two rows of Pascal's Triangle. Write down the sum of the numbers in (a) the third row, (b) the fourth row, (c) the fifth row, (d) the sixth row. What do you think the sum of the numbers in the seventh row will be?

10)

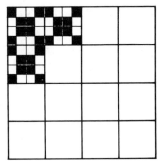

How many black tiles will be needed to *complete* this pattern? How many white tiles are there in the whole pattern?

REFLECTIONS AND LINE SYMMETRY

In nature we see many examples of the property called *symmetry.*

Oak leaves are roughly symmetrical.

So is the human face.

Although the leaf "looks the same" on both sides, in nature the symmetry is not usually quite precise.

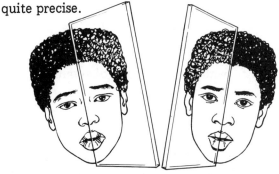

If you use a mirror to reflect one half of a photograph of a person then the slight difference between each side of the face immediately becomes obvious.

Children make symmetrical patterns with ink blots by folding blotted paper in half.

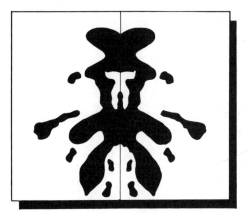

The left-hand pattern is a reflection of the right-hand pattern, and similarly, the right-hand pattern is a reflection of the left-hand pattern.

The line about which the pattern is symmetrical is known as a *line of symmetry,* or an *axis of symmetry,* or a *mirror line.*

After reflection the shape appears "turned over" but otherwise is the same shape and size.

EXAMPLE

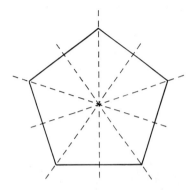

A regular pentagon has 5 lines of symmetry.

--- **EXERCISE 2** ---

1) Copy the shapes below and complete them so they have the dotted line as a line of symmetry.

(a) 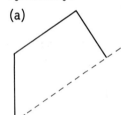 (b)

2) How many axes of symmetry does this shape have?

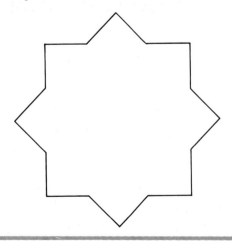

3)

Which of the above letters has (a) no line of symmetry, (b) one line of symmetry, (c) two lines of symmetry?

4) Copy the drawings below and draw on each all the lines of symmetry.

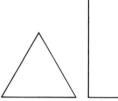

5) Plot the shapes drawn below on graph paper and reflect them in the mirror line $y = x$.

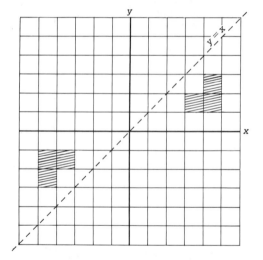

6) Reflect the word shown below in the x-axis. (This means that the x-axis is to be the mirror line.)

SUSAN

246

7) How many lines of symmetry do these shapes have?

(a)
A square

(b)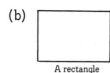
A rectangle

(c)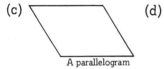
A parallelogram

(d)
A kite

(e)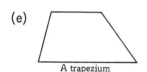
A trapezium

(f)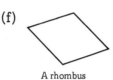
A rhombus

8) Give the coordinates of the point, P(3, 2), after reflecting it in (a) the x-axis, (b) the y-axis, (c) the line $x = 1$, (d) the line $y = 1$.

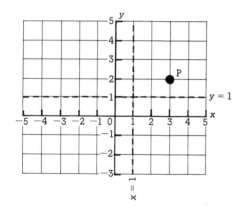

9) Plot the points given below and join them up in alphabetical order. Reflect this shape in (a) the x-axis, (b) the y-axis and then complete your drawing so it has four lines of symmetry.

A(0, 0), B(0, 2), C(2, 4), D(2, 6),
E(4, 6), F(4, 4), G(6, 4), H(6, 2),
I(4, 2), J(2, 0), K(0, 0)

10) Copy and complete this figure so that it has two axes of symmetry as shown.

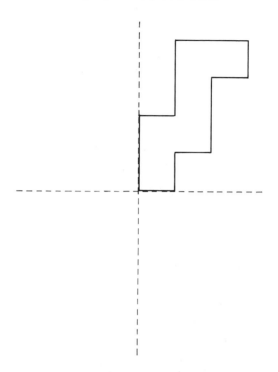

Repeat using three *different* pairs of lines of symmetry which either touch or pass through the shape.

ROTATIONS AND THE ORDER
− OF ROTATIONAL SYMMETRY −

A flywheel and a humming top have one property in common. They both spin or *rotate* about an axis passing through their centre.

If the above shape was cut out of cardboard and a pin was put through the centre it could also be rotated. It would appear to be in the same position 3 times in one complete turn. It has *order of rotational symmetry* 3. Its centre is called the *point of symmetry.*

The shape shown below has order of rotational symmetry 5.

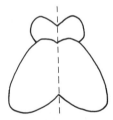

Note carefully that the butterfly shape below has line symmetry but does not have rotational symmetry. Its order of rotational symmetry is 1 because it is only in the same position once during one complete turn.

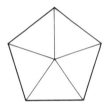

A rotation is given by an *angle of rotation* and the *centre of rotation.* (Rotations are measured anticlockwise.)

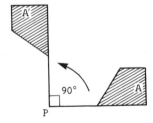

The diagram shows a rotation of 90° anticlockwise about the point P. After a rotation the shape remains exactly the same size.

1) Give the order of rotational symmetry of these figures.

(a)

(b)

(c)

(d)

(e)

2) Copy and complete this figure so it has order of rotational symmetry 3.

3) For each of the letters drawn below decide whether they have line symmetry and/or rotational symmetry. Draw a sketch of each letter showing any axes of symmetry or points of symmetry.

4)

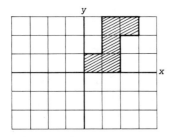

Copy the shape drawn above and rotate it through 90° anticlockwise. Repeat twice more until you have drawn a figure with order of rotational symmetry 4.

5) What is the order of rotational symmetry of

(a)

A square

(b)

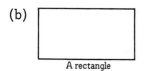

A rectangle

(c)

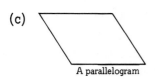

A parallelogram

(d)

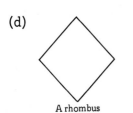

A rhombus

(e)

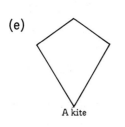

A kite

(f)

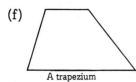

A trapezium

6)

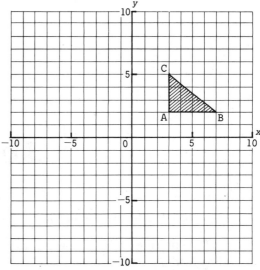

Rotate triangle ABC 90° anticlockwise about (0, 0) and label the new triangle A'B'C'. Rotate triangle A'B'C' 90° anticlockwise about (0, 0) and label the new triangle A"B"C". Rotate A"B"C" 90° anticlockwise about (0, 0) and label the new triangle A'''B'''C'''.

7)

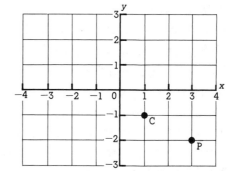

Write down the coordinates of the point P after an anticlockwise rotation of 90° about the point C.

8) Copy the shape below and rotate it about C, 60° in an anticlockwise direction, six times to give a figure with order of rotational symmetry 6.

249

Aural Test

The questions should be read *twice* slowly.

Allow 30 seconds for each of these questions including reading time.

1) Add together sixty-three and forty-seven.
2) Take seventeen away from thirty.
3) Multiply seven by nine.
4) What is the cost of five articles at forty-five pence each?
5) Write three point one four correct to one decimal place.
6) From six times eight take away nine.
7) Multiply six point two by ten.
8) Write the fraction four-fifths as a decimal.
9) Multiply seven point six by one hundred.
10) Write the number eight thousand, five hundred and forty to the nearest hundred.
11) How many centimetres are there in two and a half metres?
12) How many quarters are there in two and three-quarters?
13) Write in figures the number seventy thousand.
14) Express forty marks out of fifty as a percentage.
15) A rectangle has an area of fifteen square centimetres. How wide is it if the length is five centimetres?
16) How many minutes are there in one and a half hours?
17) Twenty-five per cent of a class are boys. What percentage are girls?
18) What is the time eighteen minutes after eight fifty-five?
19) Divide sixteen point eight by four.
20) What is one-fifth of two pounds fifty?
21) Multiply nought point three by five.
22) Using the twenty-four hour clock write down the time seven minutes after fourteen fifty-eight.
23) What is the area of a triangle of base three metres and height six metres?
24) Write down the average of nine, eleven and sixteen.
25) How much change is left from a pound note after buying three articles at twelve pence each?
26) What is the perimeter of a square whose sides are each two and a half feet long?
27) A motorist travels fifty miles in two hours. What is his average speed?
28) Multiply nought point four by nought point three.
29) How many feet are there in three yards?
30) Give the cost of ten twenty-five pence stamps.

Miscellaneous Questions

1) (a) 3020×407
   (b) 47×38
   (c) $250 \div 8$ (to the nearest whole number)
   (d) Give the remainder when 250 is divided by 8.

2) (a) 250 g of pepper are shared equally into 8 boxes. What is the weight in each box? (Give the answer first as a whole number and a fraction and then as a whole number and a decimal.)
   (b) The surface temperature of a pond changes from $-3°$ to $5°$ from one day to the next. By how many degrees has the temperature risen?
   (c) A man has £35 in his bank account. He writes a cheque for £50. By how much will he be overdrawn?

3) (a) $145 - (24 + 92)$
   (b) $15 + 4 \times 3$
   (c) $2 \times 9 + 3 \times 5$
   (d) $10 - 36 \div 6$
   (e) $84 \div 12 - 49 \div 7$

4) (a) A woman saves £70 in order to redecorate a room. If the wallpaper costs £41.94 and the paint £24.27, how much has she left over for other items?
   (b) A carton contains 840 g of soap powder when full. If 2460 g of powder are shared equally between 3 cartons how much more must be added to each carton to fill it?

5) Work out
   (a) 9^2

   (b) the square root of 64
   (c) 4^3
   (d) $3^2 + 4^2$
   (e) $10^2 - 5^2$

6) What is the value of 3 in
   (a) 22 103   (b) 30.6   (c) 1.3
   (d) 2.035?

7) Write in standard form
   (a) 150 000   (b) 0.000 15.

8) Express as single numbers
   (a) 2.4×10^5 (b) 2.4×10^{-2}.

9) Which is larger 1.5×10^4 or 16×10^3?

10) (a) 2.8×0.3 (b) 0.5×0.1
    (c) $307.2 \div 4$ (d) $6.4 \div 0.8$
    (e) $2.1 \times 3.6 + 0.83 \times 1.2$

11) (a) How many pieces of wool for rug-making can be cut from a piece of wool 1.56 m long if each piece of wool must be 7.8 cm long?
    (b) Find the total cost of 1.2 kg bacon at £2.50 per kg and 0.5 kg sausages at £1.28 per kg.

12) (a) $\dfrac{3}{16} + \dfrac{1}{4}$ (b) $2\dfrac{5}{6} + 3\dfrac{7}{12}$

    (c) $3\dfrac{5}{16} - 2\dfrac{9}{16}$ (d) $3\dfrac{1}{2} \times 1\dfrac{1}{7}$

    (e) $5\dfrac{1}{3} \times 3$ (f) $\dfrac{7}{16} \div 2\dfrac{5}{8}$

    (g) $3\dfrac{3}{4} \div \dfrac{1}{2}$

13) (a) Find the cost of $3\frac{1}{2}$ dozen eggs at 42 p per $\frac{1}{2}$ dozen.

(b) If a piece of wood $4\frac{1}{4}$ m long is divided into 17 equal lengths, how long is each piece?

(c) Three people share a flat. One pays $\frac{1}{3}$ of the rent, the second $\frac{5}{12}$ and the third the rest. If the rent is £60 a week how much does each pay?

(d) Basic wages are £160 a week. How much is this per hour if a basic week is 40 hours?

(e) Overtime is paid at time and a quarter. How much is paid for 3 hours overtime if the basic wage is £8 per hour?

14) VAT is charged at $17\frac{1}{2}$% on a £50 coat. What is the price of the coat plus VAT?

15) A salesman earns 2% commission on a sale worth £2500. How much commission does he get?

16) A girl earns 14 marks out of 25. What percentage is this?

17) Using a calculator find (a) $\sqrt{0.52}$ (b) $\sqrt{520}$.

18)

Temperature		Temperature	
°C	°F	°C	°F
0°	32°	50°	122°
10°	50°	60°	140°
20°	68°	70°	158°
30°	86°	80°	176°
40°	104°	90°	194°
		100°	212°

Use the above table to convert
(a) 15°C to °F
(b) 140°F to °C.

19) Use the 24 hour clock to convert
(a) 9.00 a.m.
(b) 4.15 p.m.
to 24 hour clock time.

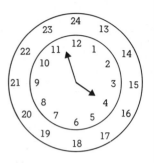

20) How many minutes before midnight is 2333 hours?

21) Give the fares from
(a) Linden Avenue to the Bus Station
(b) Princess Avenue to High Street
(c) Queen Street to The Bullring.

Linden Avenue

20	The Bullring						
30	20	High Street					
30	30	20	Weston Road				
34	30	30	20	Knight Lane			
40	34	34	30	20	Princess Av.		
50	40	34	34	30	20	Queen Street	
60	50	40	34	30	30	20	Bus Station

All fares in this chart are in pence.

22) Plot these points on a graph, choosing a suitable scale.

x	0	1	2	3	4	5
y	2	5	8	11	14	17

Draw a line through the points.
(a) Give the value of y when x = 2.5.
(b) Give the value of x when y = 15.5.
(c) Write down the equation of the straight line.

23) Change to fractions
    (a) 0.75   (b) 0.35   (c) 1.75
    (d) 60%   (e) 36%   (f) 125%.

24) Change to decimals
    (a) $\dfrac{1}{10}$ (b) $\dfrac{5}{8}$ (c) $2\dfrac{4}{5}$

    (d) 80%   (e) $12\dfrac{1}{2}\%$ (f) 175%.

25) Change to percentages
    (a) $\dfrac{1}{5}$ (b) $\dfrac{1}{3}$ (c) $2\dfrac{5}{8}$
    (d) 0.5   (e) 0.6̇6̇   (f) 1.625.

26) (a) What fraction of 400 is 25?
    (b) What percentage of 400 is 25?

27) Which is greater, 0.3 or $\dfrac{1}{3}$?

28) Write down in the correct order the buttons that you would press on a calculator to add

    £1.25 + 12 p + £12.50

    What answer should you expect to see in the display and what would this mean in pounds and pence?

29) Approximately how many centimetres are there in 1 foot?

30) Two swimming pools have shallow ends of depth 3 ft 6 in and 1 metre respectively. Which is deeper?

31) A boy in Britain buys 4 oz sweets while on the same day his French penfriend buys 100 g sweets. Who has more?

32) The marks obtained in 10 different tests are
    2, 4, 6, 2, 3, 5, 5, 2, 3 and 4
    (a) find the mean, mode and median
    (b) give the range.

33) The cricket scores for 5 different innings are 30, 110, 0, 50 and 70
    (a) What is the average score?
    (b) What must the batsman score in the next innings to improve his average by 3 runs?

34) A motorist travels 60 miles at an average speed of 40 mph and 30 miles at an average speed of 60 mph. Give the total time, the total distance and the overall average speed.

35) The curved surface area of a cylinder is $2\pi r l$ where r is the radius and l is the length. Find the area when $r = 5$ cm and $l = 21$ cm. Take π as $\frac{22}{7}$.

36) The volume of a sphere is given by
    $V = \dfrac{4}{3}\pi r^3$ where r is the radius. Find the volume, V, when $r = 10$ cm. Take π as 3.14. Give your answer correct to 3 significant figures.

37) The radius of a cylindrical piece of wood is given by $r = \sqrt{V \div (\pi \times h)}$ where V is the volume and h is the height. Find r when $h = 7$ cm and $V = 88$ cm³. Take π as $\frac{22}{7}$.

38) How many right angles are there in 270°?

39) Find (a) x (b) y (c) z.

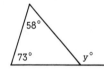

40) There are 60 minutes in 1 degree, so 6 minutes = 0.1°. (60 minutes may be written as 60′.)
    (a) Express 23.5° in degrees and minutes.
    (b) Write 54°12′ in decimal form.

41) Following is a sketch of a small picture in a frame.
    (a) What is the total area of the picture and frame?
    (b) What is the area of the picture only?
    (c) What is the area of the frame alone?
    (d) If the frame cost 5p per cm² what is the cost of the frame?

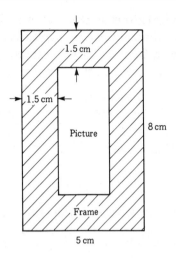

1.5 cm
1.5 cm
8 cm
Picture
Frame
5 cm

42) A ladder 13 ft long leans against a vertical wall so that it reaches to a height of 12 ft. How far from the wall is the base of the ladder?

43) The cardboard pieces in a game are made in the shape of a triangle and a semi-circle as shown. Find the area of cardboard in each shape and the total area of six such pieces.

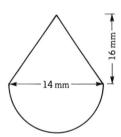

16 mm

14 mm

44) What is the reading on (a) the scales and (b) the gas-meter dials?

(a)

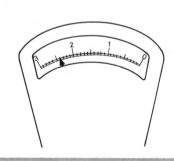

(b)

45) Plot these points, choosing a suitable scale and draw a smooth curve through them.

Temperature in °C	Time in minutes
86°	0
68°	1
54°	$2\frac{1}{2}$
44°	4
38°	$5\frac{1}{2}$
32°	8
30°	10

(a) From your graph estimate the time at which the temperature is 50°.
(b) What is the probable temperature after 5 minutes?

46)

Inches	Millimetres
1	25.4
2	50.8
3	76.2
4	101.6
5	127.0
10	254.0
50	1270.0
100	2540.0

Using this conversion table change
(a) 3   (b) 8
(c) 14   (d) 56
(e) 98
inches to millimetres.

47) The chart below shows how index numbers have changed. -2 means down 2, $+7$ means up 7. One of the numbers in the right-hand column is wrong. Which one is it and what should the correct number be?

Index	1st week	2nd week	Overall change	Final value
500	-2	-4	-6	494
512	$+4$	$+3$	$+7$	519
511	-6	$+5$	-1	510
514	$+7$	-4	$+3$	517
517	-7	$+6$	-1	516
520	-2	-5	-7	513
519	$+3$	-2	-1	518
516	$+5$	-1	$+4$	520

48) (a) Profit (P) is worked out by taking the cost price (C) from the selling price (S). Write an equation for calculating profit.

(b) A motorist travels s miles at an average speed of u miles/hour. He takes t hours. Write an equation for s using u and t.

(c) An electricity bill is worked out by multiplying the number of units used (N) by the cost per unit (Q) and adding the fixed charge (F). Write an equation for the cost of electricity (C) in terms of N, Q and F.

Multiple Choice Revision Exercise

Write down the letter corresponding to the correct answer for each question.

1) The fraction of the circle which has been shaded in the diagram is

A $\frac{5}{18}$ B $\frac{1}{2}$ C $\frac{5}{8}$

D $\frac{13}{18}$

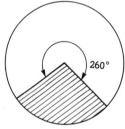

260°

2) What is the median of the numbers: 8, 9, 8, 12, 10, 7, 6, 9, 10 and 8?
A 9   B 8.5   C 8   D 7.5

3) $78.75 \div 0.35$ is
A 2.25   B 22.5   C 225   D 2250

4) $\frac{7}{40}$ as a percentage is

A 7%   B $17\frac{1}{2}$%   C 40%   D 70%

5) During a sale a shop reduced the price of everything by 10%. What was the sale price of an article originally priced at £12.90?
A £1.29   B £11.61   C £11.91   D £14.19

6) $2\frac{1}{4} \times 3\frac{1}{2}$ is

A $6\frac{1}{8}$ B $6\frac{7}{8}$ C $7\frac{1}{8}$ D $7\frac{7}{8}$

7) 350×60 is
A 2.1×10^5 B 2.1×10^4 C 2.1×10^3
D 2.1×10^2

8) A boy scored 60% in a test. If the maximum mark was 50, then the boy's score was
A 5   B 15   C 30   D 35

9) A girl obtains 60 marks out of 75 in a test. This is equivalent to a percentage of
A 30%   B 60%   C 80%   D 90%

10) The perimeter of the triangle ABC shown in the diagram is
A 70 cm
B 80 cm
C 82 cm
D 90 cm

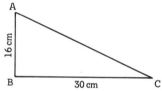

11) If the exchange rate for Swedish kronor is 10 kronor for £1, then 2876 kronor changed into pounds would be
A £28.76   B £287.60   C £2876
D £28 760

12) A transistor radio can be bought for £37.90 or on hire purchase for 9 monthly payments of £4.74 each. How much more does the hire purchase method cost?
A £4.76   B £5.90   C £6.62   D £7.42

13) $10 - 0.005$ equals
A 0.005   B 9.005   C 9.995   D 10.005

14) What is $\frac{1}{2} + \left(\frac{1}{2} \times \frac{1}{2}\right)$?

A $\frac{1}{8}$ B $\frac{1}{2}$ C $\frac{3}{4}$ D 1

15) The number 0.075 538 correct to two decimal places is
A 0.07   B 0.075   C 0.076   D 0.08

16) Which of the following does *not* equal $\frac{1}{2}xy$?

A $\frac{xy}{2}$ B $x \times \frac{y}{2}$ C $\frac{1}{2}xy$ D $\frac{1}{2x} \times y$

17)
Hockey	⊥⊦⊤
Baseball	⊥⊦⊤ ‖
Rugby	⊥⊦⊤ ‖‖ (⊥⊦⊤ equals 5)
Cricket	⊥⊦⊤ ‖‖
Football	⊥⊦⊤ ⊥⊦⊤ ‖

The tally marks above show the distribution in the sports preferred by a sample of schoolboys. What fraction of the sample preferred rugby?

A $\frac{1}{8}$ B $\frac{7}{40}$ C $\frac{1}{5}$ D $\frac{1}{4}$

18) What is 0.0089 correct to two decimal places?
A 0.008   B 0.009   C 0.01   D 0.09

19) £189 is divided into two parts so that the first part is $\frac{3}{4}$ of the second. What is the value of the larger part?
£108   B £84   C £81   D £72

20) In an examination, 240 candidates out of 720 results were awarded Grade C. On a pie chart showing all the grades, the angle at the centre for Grade C would be
A 135°   B 120°   C 105°   D 90°

21) The value of 0.7×0.05 is
A 0.035   B 0.35   C 0.75   D 3.5

22) In the number 350.47, the actual value represented by the digit 7 is
A 70   B 7   C $\frac{7}{10}$ D $\frac{7}{100}$

23) If £240 is divided in the ratio 2 to 3 then the smaller share is
A £80   B £96   C £120   D £144

24) If 20% of a sum of money is £200 then the whole sum of money is
A £20   B £40   C £400   D £1000

25) If $4M = 20$, then the value of M is
A 80   B 24   C 16   D 5

26) The total surface area of a cube of edge 2 cm is
A 8 cm²   B 8 cm³   C 24 cm   D 24 cm²

27) Calculate the mean of 1, 2, 5, 7 and 15.
A 4   B 6   C 7   D 30

28) Find the value of $13^2 - 12^2$.
A 1^2 B 5   C 25   D 125

29) The pie chart in the diagram illustrates the subjects taken by a group of pupils in the sixth form of a school. What percentage take French?
A 20%   B 25%   C 30%   D 50%

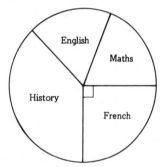

30) Find the value of x in the equation $5x + 1 = 61$.
A 10   B 12   C 50   D 62

31) Find the size of the angle marked Q in the diagram.
A 63°
B 107°
C 117°
D 127°

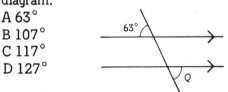

32) Find the area of a triangle whose sides are 3 cm, 4 cm and 5 cm.
A 6 cm²   B 10 cm²   C 12 cm²   D 15 cm²

33) The sum of eleven thousand and eleven hundred is
A 11 100   B 11 110   C 11 111   D 12 100

34) $8 + 5 \times (4 - 2)$ is equal to
A 2   B 18   C 22   D 50

35) Which is the largest of the following fractions?
A $\frac{1}{2}$ B $\frac{2}{3}$ C $\frac{5}{6}$ D $\frac{7}{10}$

36) $0.1 \times 0.2 \times 0.3$ equals
A 0.006   B 0.06   C 0.6   D 6

37) Which of the following is the largest?
A 0.22   B 2.20   C 2.02   D 0.022

38) The cost of 2 metres of material at £2.40 per metre and 6 metres at £1.50 per metre is
A £5.40   B £13.20   C £13.80   D £14.20

39) 0.9 as a percentage is
A 0.9%   B 9%   C 90%   D 99%

40) For her holidays a woman saved 10% of her weekly wage of £150 for 40 weeks in the year. How much did she save for her holiday?
A £150   B £300   C £450   D £600

41) The perimeter of a square is 36 cm. Its area, in square centimetres, is
A 6   B 9   C 36   D 81

42) 42 600 m expressed in kilometres is
A 4.26   B 42.6   C 426   D 4260

43) Four packets have weights of 2 kg, 250 g, $3\frac{1}{2}$ kg and 500 g. Their total weight in kilograms is
A 5.25   B 6   C 6.25   D 6.5

44) The number 63 700 written in standard form is
A 637×10^2 B 63.7×10^3 C 6.37×10^4 D 0.637×10^5

45) If $2q = 8$ then the value of q is
A $\frac{1}{4}$ B 4   C 6   D 16

46) The square of 0.12 is
A 0.0144   B 0.14   C 0.144   D 0.24

47) $2\frac{1}{5}$ when written in decimal form is
A 2.01   B 2.02   C 2.1   D 2.2

48) $6\frac{1}{5} - 2\frac{3}{4} + 1\frac{1}{6}$ equals
A $2\frac{17}{60}$ B $4\frac{17}{60}$ C $4\frac{37}{60}$ D $5\frac{22}{60}$

49) Correct to 2 significant figures, 3.0394 is equal to
A 3.0   B 3.03   C 3.039   D 3.4

50) A man's taxable income is £1800. If tax is levied at 20% how much does he pay in income tax?
A £36 000   B £1620   C £360   D £1440

Answers

Answers

CHAPTER 1

Exercise 1

1) (a) (i) 35 (ii) XXXV
   (b) (i) 305 (ii) CCCV
   (c) (i) 350 (ii) CCCL
2) (a) (i) 79
   (ii) LXXVIIII or LXXIX
   (b) (i) 709
   (ii) DCCVIIII or DCCIX
   (c) (i) 790
   (ii) DCCLXXXX (or DCCXC)
3) (a) (i) 65 (ii) LXV
   (b) (i) 605 (ii) DCV
   (c) (i) 650 (ii) DCL
4) (a) nine hundred and eighty
   (b) ninety-eight
   (c) nine hundred and eight
5) (a) eight hundred (b) eighty
   (c) eight
6) (a) eighty-nine
   (b) eight hundred and ninety
   (c) eight hundred and nine
7) (a) one hundred and ten
   (b) one hundred and one
   (c) eleven
8) (a) seventy-eight
   (b) seven hundred and eight
   (c) seven hundred and eighty
9) (a) one hundred and thirty
   (b) one hundred and three
   (c) thirteen
10) (a) fourteen
    (b) one hundred and forty
    (c) one hundred and four
11) (a) 26   (b) 206   (c) 260
12) (a) 55   (b) 550   (c) 505
13) (a) 12   (b) 120   (c) 102
14) (a) 21   (b) 210   (c) 201

15) (a) 700   (b) 70   (c) 7
16) Number in words | figures

Number in words	H	T	U
twenty-four		2	4
nineteen		1	9
seventy-six		7	6
eight			8
one hundred and sixty-four	1	6	4
seven hundred and eight	7	0	8
three			3
sixty-one		6	1
two hundred and fifty-four	2	5	4
fifty-eight		5	8

17) 5 hundreds 4 tens 2 units
18) 9 hundreds 0 tens 2 units
19) (a) 5 hundreds (b) 5 tens
    (c) 5 units   (d) 5 tens
20) (a) 3 units   (b) 3 hundreds
    (c) 3 tens   (d) 3 units
21) (a) 9 tens   (b) 9 tens
    (c) 9 units   (d) 9 hundreds
22) (c) and (d)
23) (b) and (d)
24) (b) and (c)
25) 123, 132, 321
26) 35, 53, 305, 350, 503, 530
27) 13, 31, 103, 130, 301, 310
28) 510, 501, 150, 105, 51, 15
29) 321, 312, 231, 213, 132, 123

Exercise 2

1) 2451
2) 5311
3) 6112
4) 1014

5) 6009
6) 5300
7) 14 000
8) 24 000
9) 70 000
10) 16 208
11) 75 142
12) 40 064
13) 300 000
14) 195 580
15) 114 613
16) 351 645
17) 712 201
18) five thousand, seven hundred and sixty-three
19) six thousand, seven hundred and twenty-one
20) nine thousand, eight hundred and eighty-four
21) sixteen thousand
22) sixty-five thousand
23) eight hundred and nine thousand
24) seventy-two thousand and fifty
25) seven thousand and nine
26) eight thousand and two
27) seven thousand and eighty
28) eight thousand and ninety
29) six thousand and fifty
30) eight thousand nine hundred
31) seven thousand eight hundred
32) sixty-five thousand, three hundred and thirty-two
33) fourteen thousand five hundred
34) one hundred and sixty-three thousand, four hundred and four
35) nine hundred thousand and nine
36) four hundred and fifty-two thousand, nine hundred and eighty-six

37) three hundred and four thousand and twenty-one

38) . . . .

```
      5  0  0  0
   6  2  0  0  0
3  0  0  0  0  0
   7  4  0  0  9
6  0  0  2  0  0
   7  0  0  5  0
9  9  0  0  0
   6  0  0  0
```

Exercise 3

1) 8 000 000
2) 19 000 000
3) 600 000 000
4) 324 000 567
5) 3 700 030
6) 4 000 040
7) 64 300 666
8) 999 000 000
9) 464 330 206
10) seven million
11) fifty-seven million
12) one hundred and eighty-nine million
13) seven million, twelve thousand, five hundred and six
14) five million, eight hundred and sixty-nine thousand and fourteen
15) one hundred and sixty-seven million, one hundred and eighty-nine thousand, one hundred and twelve
16) nine hundred million, three thousand and forty
17) . . . .

```
1 610 000 000
    8 030 050
  706 000 000
```

Exercise 4

1) 60
2) 10
3) 80
4) 80
5) 70
6) 800
7) 800
8) 100
9) 500
10) 900
11) 4000
12) 1000
13) 7000
14) 8000
15) 7000

Exercise 5

1) (a) yes (b) yes (c) no
   (d) no   (e) yes (f) yes
   (g) yes (h) no
2) (a) no   (b) yes (c) yes
   (d) no   (e) yes (f) no
   (g) no   (h) yes
3) (a) yes (b) no   (c) no
   (d) no   (e) yes (f) yes
   (g) yes (h) no
4) (a) yes (b) no   (c) yes
   (d) yes (e) no   (f) yes
   (g) yes (h) no
5) (a) no   (b) yes (c) no
   (d) yes (e) no   (f) yes
   (g) no   (h) yes
6) (b) and (c)
7) (a), (b), (e) and (h)
8) (a), (c), (d) and (e)
9) (a) and (b)

Exercise 6

1) (a) 1, 2, 5, 10
   (b) 1, 2, 7, 14
   (c) 1, 3, 5, 15
   (d) 1, 2, 4, 8, 16
   (e) 1, 2, 4, 5, 10, 20
   (f) 1, 5, 25
   (g) 1, 2, 3, 4, 6, 9, 12, 18, 36
   (h) 1, 2, 3, 6, 7, 14, 21, 42
2) (a) 1 not prime
   (b) 1, 3 prime
   (c) 1, 5 prime
   (d) 1, 2, 3, 6 not prime
   (e) 1, 11 prime
   (f) 1, 17 prime
   (g) 1, 19 prime
   (h) 1, 11, 22 not prime
   (i) 1, 2, 3, 4, 6, 8, 12, 24 not prime
   (j) 1, 2, 13, 26 not prime
   (k) 1, 31 prime
   (l) 1, 2, 4, 8, 16, 32 not prime

Exercise 7

1) (a), (b), (e), (h), (i), (k), (l), (n) and (o)

Exercise 8

1) (a) 8 (b) 7 (c) 7 (d) 6
2) (a) 2 (b) 5 (c) 4 (d) 2
3) (a) 1 (b) 3 (c) 9 (d) 1
4) (a) 6 (b) 9 (c) 6 (d) 3

Exercise 9

1) (b) and (d)
2) (b) and (d)
3) (b) and (c)
4) (a), (b) and (d)
5) (c)
6) (a), (b) and (c)
7) (a) and (c)

Exercise 10

1) (b)
2) (a), (b) and (c)
3) (c)
4) (a) and (c)

Exercise 11

1) (a), (b) and (c)
2) (a), (b) and (c)
3) (a) and (c)
4) (a) and (b)
5) (a), (b) and (c)

Exercise 12

1) (a), (b), (d), (f), (g), (i), (k), (m) and (n)
2) (i) (b), (f), (h), (j), (k), (l) and (o)
   (ii) (b), (h), (j) and (o)
3)

(i)	(ii)	(iii)	(iv)
4	9	25	7
8	33	35	19
16	321	215	
34			
104			
206			

Exercise 13

1) (c), (d), (f) and (h)
2) (a), (c) and (e)

Exercise 14

1) (b), (c), (d) and (e)
2) (b), (c), (d) and (e)
3) (b)

Exercise 15

1) negative whole number
2) positive fraction
3) positive decimal
4) positive whole number
5) negative fraction
6) negative decimal

7) positive whole number
8) negative decimal
9) positive fraction

Exercise 16

1) $6\frac{1}{2}$ 2) $13\frac{1}{2}$ 3) $509\frac{1}{2}$
4) $\underline{4}.1$ 5) $\underline{45}.2$ 6) $\underline{390}.12$

Exercise 17

1) $1°C$ 6) $1°C$
2) $-2°C$ 7) $8°C$
3) $0°C$ 8) $-2°C$
4) $-5°C$ 9) $-5°C$
5) $-2°C$ 10) $8°C$

Exercise 18

1) 240
2) six hundred and five
3) six thousands
4) no
5) 40 018
6) fifty thousand and twenty
7) yes
8) no
9) yes
10) yes
11) yes
12) yes
13) yes
14) (a) 1, 3, 9 not prime
    (b) 1, 3, 7, 21 not prime
    (c) 1, 23 prime
    (d) 1, 29 prime
15) no
16) no
17) yes
18) yes
19) yes
20) yes
21) 12, 14, 16, 18, 20
22) 6, 9, 12, 15, 18, 21, 24
23) yes
24) yes
25) yes
26) yes
27) $2\frac{1}{2}$, $\underline{12}.5$
28) yes
29) yes
30) sixty thousand and fifty
31) 3 hundreds 1 ten 2 units
32) 3 thousands 2 units
33) (a) 8   (b) 1
34) (a) and (b)

35) $-4°C$
36) $-2°C$
37) $11°C$
38) fallen by $4°C$
39) 7
40) 4

CHAPTER 2

Exercise 1

1) 136   31) 133 036
2) 209   32) 1 362 951
3) 363   33) 523 745
4) 160   34) 58 330
5) 3977   35) 197 135
6) 8701   36) 915 997
7) 2012   37) 2180
8) 9956   38) 4498
9) 9113   39) 298
10) 25 976   40) 397
11) 221   41) 1834
12) 7798   42) 10 713
13) 54 772   43) 5071
14) 89 043   44) 630
15) 93 972   45) 5458
16) 31 368   46) 152
17) 138 103   47) 959
18) 902 028   48) 148
19) 18 538   49) 85
20) 10 972   50) 169
21) 1 109 676   51) 5656
22) 1 900 716   52) 17 210
23) 10 777   53) 1 015 021
24) 469   54) 104 813
25) 666 679   55) 27 751
26) 9099   56) 4 110 286
27) 1 683 830   57) 279 639 414
28) 98 199   58) 14 008
29) 581 587   59) 31 987
30) 2 124 689   60) 224 851

Exercise 3

1) 32   12) 151
2) 13   13) 233
3) 33   14) 171
4) 33   15) 68
5) 33   16) 471
6) 43   17) 749
7) 18   18) 6918
8) 66   19) 4943
9) 26   20) 2939
10) 46   21) 7911
11) 114   22) 4937

23) 1465   36) 1598
24) 4020   37) 2483
25) 1451   38) 2889
26) 977   39) 1070
27) 5957   40) 53 625
28) 9646   41) 4094
29) 2609   42) 9421
30) 5278   43) 75 096
31) 4337   44) 1921
32) 6551   45) 41 931
33) 493   46) 2
34) 8202   47) 24 992
35) 9862   48) 50 902

Exercise 4

1) 235   7) 473
2) 183   8) 175
3) 178   9) 2023
4) 379   10) 4780
5) 4558   11) 2919
6) 501   12) 3708

Exercise 6

1) 8   9) 16
2) 7   10) 6
3) 6   11) 5
4) 23   12) 13
5) 4   13) 8
6) 12   14) 4
7) 9   15) 11
8) 13

Exercise 8

1) 56   9) 24
2) 30   10) 36
3) 21   11) 40
4) 63   12) 30
5) 32   13) 16
6) 48   14) 81
7) 42   15) 49
8) 45

16) (a) (i) 4   (ii) 6   (iii) 8   (iv) 12
    (b) (i) 6   (ii) 9   (iii) 12 (iv) 18
    (c) (i) 8   (ii) 12 (iii) 16 (iv) 24
    (d) (i) 10 (ii) 15 (iii) 20 (iv) 30
17) (a) (i) 30 (ii) 36 (iii) 54
    (b) (i) 35 (ii) 42 (iii) 63
    (c) (i) 40 (ii) 48 (iii) 72
    (d) (i) 45 (ii) 54 (iii) 81

	6) 729
69	7) 296
144	8) 415
4) 260	9) 324
5) 448	10) 392

Exercise 10

1) 4536	11) 675
2) 1215	12) 5306
3) 2233	13) 5376
4) 3372	14) 5418
5) 3400	15) 7016
6) 1356	16) 2169
7) 792	17) 4208
8) 8334	18) 3012
9) 2149	19) 4182
10) 2736	20) 4672

Exercise 11

1) (i) 80	(ii) 800
2) (i) 650	(ii) 6500
3) (i) 170	(ii) 1700
4) (i) 3310	(ii) 33 100
5) (i) 1510	(ii) 15 100
6) (i) 40 000	(ii) 400 000
7) (i) 49 900	(ii) 499 000
8) (i) 511 990	(ii) 5 119 900

Exercise 12

1) (i) 680	(ii) 1020
(iii) 1700	(iv) 2720
2) (i) 1440	(ii) 2160
(iii) 3600	(iv) 5760
3) (i) 3020	(ii) 4530
(iii) 7550	(iv) 12 080
4) (i) 15 780	(ii) 23 670
(iii) 39 450	(iv) 63 120
5) (i) 900	(ii) 1350
(iii) 2250	(iv) 3600
6) (i) 1200	(ii) 1800
(iii) 3000	(iv) 4800
7) (i) 6020	(ii) 9030
(iii) 15 050	(iv) 24 080
8) (i) 131 060	(ii) 196 590
(iii) 327 650	(iv) 524 240
9) (i) 1120	(ii) 1680
(iii) 2800	(iv) 4480
10) (i) 620	(ii) 930
(iii) 1550	(iv) 2480

11) (i) 13 620	(ii) 20 430
(iii) 34 050	(iv) 54 480
12) (i) 40 160	(ii) 60 240
(iii) 100 400	(iv) 160 640

Exercise 13

1) 11 352	11) 2730
2) 17 608	12) 7222
3) 13 468	13) 22 824
4) 8192	14) 17 950
5) 56 238	15) 15 570
6) 11 046	16) 25 912
7) 8730	17) 35 168
8) 9975	18) 28 350
9) 19 965	19) 10 710
10) 21 975	20) 28 728

Exercise 14

1) 46 008	6) 158 080
2) 127 484	7) 226 171
3) 370 230	8) 722 160
4) 316 050	9) 404 670
5) 128 512	10) 27 468

Exercise 15

1) 4953
2) 3 846 828
3) 107 898
4) 227 169
5) 13 198 185
6) 6 769 815
7) 4 628 547
8) 15 140 432

Exercise 16

1) 3564	17) 427 r 2
2) 158	18) 1365
3) 128	19) 697 r 5
4) 61 r 2	20) 715 r 3
5) 130 r 3	21) 1135 r 7
6) 49	22) 2050 r 1
7) 43	23) 565 r 6
8) 101 r 4	24) 610
9) 125	25) 101 r 1
10) 35	26) 1129 r 1
11) 91	27) 320
12) 61	28) 1049
13) 29	29) 469
14) 25	30) 432
15) 93 r 2	31) 257 r 6
16) 112 r 3	32) 241 r 5

Exercise 17

1) 27	8) 19
2) 16	9) 1500
3) 23	10) 2600
4) 13 r 24	11) 750
5) 11	12) 130 r 3
6) 48 r 2	13) 2101
7) 33	14) 371

Exercise 19

1) 32	35) 61
2) 40	36) 72
3) 36	37) 10
4) 49	38) 6
5) 42	39) 11
6) 63	40) 15
7) 72	41) 13
8) 45	42) 17
9) 27	43) 11
10) 32	44) 14
11) 88	45) 17
12) 96	46) 10
13) 13	47) 12
14) 16	48) 13
15) 20	49) 81
16) 31	50) 9
17) 10	51) 8
18) 31	52) 32
19) 40	53) 7
20) 17	54) 7
21) 11	55) 35
22) 10	56) 2
23) 10	57) 14
24) 12	58) 52
25) 17	59) 5
26) 19	60) 22
27) 15	61) 13
28) 36	62) 45
29) 27	63) 18
30) 77	64) 32
31) 53	65) 14
32) 45	66) 12
33) 23	67) 5
34) 43	68) 24

Exercise 20

1) should be 10 099
2) correct
3) correct
4) should be 27 292
5) should be 1171

6) should be 70
7) should be 9004
8) should be 9002

Exercise 21

1) 11 067	7) 6732
2) 3666	8) 20 532
3) 4680	9) 17 884
4) 11 937	10) 6562
5) 3468	11) 263 504
6) 11 152	12) 397 350

Exercise 22

1) 19	11) 9
2) 26	12) 540
3) 82	13) 31
4) 102	14) 400
5) 33	15) 31
6) 28	16) 42
7) 9	17) yes
8) 60	18) 32
9) 432	19) 4
10) 5	20) 1420

Exercise 23

1) 9340, 15 458, 1990
2) (a) 344   (b) 277
    (c) 5778   (d) 4348
    (e) 492   (f) 1725
3) (a) 4578   (b) 1152
    (c) 4128   (d) 2409
    (e) 5334
4) (a) 49   (b) 91
    (c) 36   (d) 170
5) (a) 2560   (b) 17 700
    (c) 30 458   (d) 7665
    (e) 376 012
6) (a) 22   (b) 11
    (c) 18   (d) 96
    (e) 23
7) 22
8) 7
9) 6
10) 10
11) 9
12) 40
13) 26
14) 10

Exercise 24

1) 382 mm
2) 3104 ohms

3) 154 mm
4) 759 litres
5) 4950 grams
6) 129 kilograms
7) 30 mm
8) 15 000 grams
9) 109 metres
10) 612 millimetres
11) 18 millimetres
12) £13
13) 355 grams
14) 83 kilograms
15) £1339
16) 29 remainder 8 millimetres
17) 78 patterns
18) 203
19) 422
20) 582 millimetres
21) (a) 267   (b) 148
    (c) 435
22) 10 hours
23) 100 pages
24) 36 litres
25) 7866

Exercise 26

1) 43 min
2) (a) £760   (b) £893
3) 3 m
4) 31 min
5) 30 min
6) (a) 400   (b) 1600
7) (a) 3 mm   (b) 87 mm
8) 30

CHAPTER 3

Exercise 1

1) $\frac{3}{5}$
2) (a)

     (b)

     (c)

     (d)
  $\frac{4}{5}$

     (e)
  $\frac{5}{12}$

     (f)
  $\frac{5}{8}$

3) (a)   (b)
 $\frac{1}{3}$ $\frac{2}{3}$

     (c)
 $\frac{5}{6}$

Exercise 2

1)
  $\frac{4}{5}$ $\frac{8}{10}$

2)
  $\frac{1}{3}$ $\frac{3}{9}$

3)
  $\frac{2}{3}$ $\frac{4}{6}$

4)
  $\frac{1}{5}$ $\frac{2}{10}$

5)
  $\frac{3}{4}$ $\frac{9}{12}$

6)
  $\frac{1}{3}$ $\frac{4}{12}$

7)
  $\frac{3}{5}$ $\frac{6}{10}$

8)
  $\frac{1}{3}$ $\frac{2}{6}$

9)
  $\frac{2}{3}$ $\frac{8}{12}$

10) (a) $\frac{1}{8}$ (b) $\frac{5}{6}$ (c) $\frac{6}{12}$ or $\frac{1}{2}$
     (d) $\frac{2}{3}$

Exercise 3

1) (a) $\frac{1}{4} = \frac{5}{20}$ (b) $\frac{2}{3} = \frac{4}{6}$ (c) $\frac{5}{6} = \frac{10}{12}$ (d) $\frac{4}{5} = \frac{16}{20}$
2) (a) $\frac{3}{4} = \frac{12}{16}$ (b) $\frac{1}{2} = \frac{5}{10}$ (c) $\frac{3}{8} = \frac{9}{24}$ (d) $\frac{3}{5} = \frac{6}{10}$
3) (a) $\frac{5}{6} = \frac{15}{18}$ (b) $\frac{1}{4} = \frac{6}{24}$ (c) $\frac{6}{7} = \frac{36}{42}$ (d) $\frac{9}{11} = \frac{72}{88}$
4) (a) $\frac{4}{9} = \frac{12}{27}$ (b) $\frac{6}{11} = \frac{12}{22}$ (c) $\frac{5}{8} = \frac{35}{56}$ (d) $\frac{3}{4} = \frac{15}{20}$
5) (a) $\frac{5}{6} = \frac{30}{36}$ (b) $\frac{6}{7} = \frac{18}{21}$ (c) $\frac{3}{5} = \frac{15}{25}$ (d) $\frac{7}{8} = \frac{28}{32}$
6) (a) $\frac{3}{8} = \frac{15}{40}$ (b) $\frac{2}{3} = \frac{12}{18}$ (c) $\frac{5}{9} = \frac{25}{45}$ (d) $\frac{3}{5} = \frac{18}{30}$
7) (a) $\frac{5}{11} = \frac{15}{33}$ (b) $\frac{3}{8} = \frac{24}{64}$ (c) $\frac{1}{2} = \frac{20}{40}$ (d) $\frac{2}{3} = \frac{18}{27}$
8) (a) $\frac{4}{5} = \frac{20}{25}$ (b) $\frac{7}{8} = \frac{49}{56}$ (c) $\frac{3}{7} = \frac{21}{49}$ (d) $\frac{5}{12} = \frac{25}{60}$

Exercise 4

1) (a) $\frac{6}{9}$, $\frac{2}{3}$ (b) $\frac{6}{8}$, $\frac{3}{4}$ (c) $\frac{4}{12}$, $\frac{1}{3}$
2) $\frac{1}{3}$ 15) $\frac{1}{2}$
3) $\frac{1}{4}$ 16) $\frac{3}{4}$
4) $\frac{3}{5}$ 17) $\frac{2}{3}$
5) $\frac{1}{3}$ 18) $\frac{3}{4}$
6) $\frac{1}{2}$ 19) $\frac{3}{5}$
7) $\frac{1}{3}$ 20) $\frac{3}{5}$
8) $\frac{6}{7}$ 21) $\frac{1}{6}$
9) $\frac{6}{7}$ 22) $\frac{1}{7}$
10) $\frac{3}{4}$ 23) $\frac{1}{3}$
11) $\frac{7}{8}$ 24) $\frac{1}{2}$
12) $\frac{3}{5}$ 25) $\frac{2}{3}$
13) $\frac{1}{4}$ 26) $\frac{1}{2}$
14) $\frac{4}{5}$

Exercise 5

1) 4   6) 4
2) 5   7) 3
3) 3   8) 5
4) 2   9) 12
5) 7   10) 5

Exercise 6

1) $2\frac{1}{4}$ 19) $3\frac{1}{4}$
2) $3\frac{2}{3}$ 20) $3\frac{3}{5}$
3) $2\frac{1}{7}$ 21) $9\frac{1}{9}$
4) $2\frac{1}{8}$ 22) $11\frac{2}{5}$
5) $2\frac{6}{7}$ 23) $9\frac{1}{4}$
6) $6\frac{3}{8}$ 24) $4\frac{5}{12}$
7) $2\frac{1}{2}$ 25) $1\frac{7}{8}$
8) $2\frac{1}{3}$ 26) $2\frac{1}{5}$
9) $2\frac{3}{4}$ 27) $8\frac{3}{5}$
10) $2\frac{5}{8}$ 28) $3\frac{2}{7}$
11) $6\frac{2}{3}$ 29) $5\frac{1}{4}$
12) $6\frac{1}{4}$ 30) $4\frac{1}{4}$
13) $2\frac{1}{9}$ 31) $5\frac{1}{5}$
14) $5\frac{5}{6}$ 32) $8\frac{3}{8}$
15) $6\frac{3}{4}$ 33) $6\frac{1}{2}$
16) $5\frac{3}{8}$ 34) $4\frac{4}{5}$
17) $2\frac{7}{11}$ 35) $1\frac{1}{2}$
18) $7\frac{1}{8}$ 36) $4\frac{1}{8}$

Exercise 7

1) $\frac{7}{2}$ 19) $\frac{29}{9}$
2) $\frac{23}{4}$ 20) $\frac{58}{5}$
3) $\frac{19}{3}$ 21) $\frac{19}{7}$
4) $\frac{14}{3}$ 22) $\frac{31}{5}$
5) $\frac{15}{2}$ 23) $\frac{28}{5}$
6) $\frac{5}{2}$ 24) $\frac{89}{10}$
7) $\frac{11}{3}$ 25) $\frac{31}{7}$
8) $\frac{22}{5}$ 26) $\frac{68}{7}$
9) $\frac{16}{3}$ 27) $\frac{10}{3}$
10) $\frac{53}{6}$ 28) $\frac{62}{7}$
11) $\frac{48}{5}$ 29) $\frac{75}{8}$
12) $\frac{55}{8}$ 30) $\frac{31}{2}$
13) $\frac{44}{7}$ 31) $\frac{13}{5}$
14) $\frac{16}{5}$ 32) $\frac{37}{4}$
15) $\frac{59}{6}$ 33) $\frac{39}{5}$
16) $\frac{60}{7}$ 34) $\frac{29}{5}$
17) $\frac{15}{7}$ 35) $\frac{20}{3}$
18) $\frac{32}{9}$

Exercise 8

1) 6   13) 16
2) 15   14) 30
3) 8   15) 72
4) 6   16) 42
5) 10   17) 20
6) 20   18) 18
7) 18   19) 12
8) 10   20) 24
9) 12   21) 30
10) 14   22) 8
11) 30   23) 12
12) 30   24) 90

Exercise 9

1) (a) (b) (d) (c)
2) (b) (a) (c) (d)
3) (c) (d) (a) (b)
4) (b) (c) (a) (d) (e)

Exercise 10

1) $1\frac{1}{4}$ 25) $8\frac{1}{4}$
2) $\frac{5}{6}$ 26) $7\frac{7}{8}$
3) $\frac{7}{8}$ 27) $3\frac{9}{20}$
4) $\frac{7}{10}$ 28) $9\frac{5}{8}$
5) $1\frac{1}{12}$ 29) $8\frac{7}{10}$
6) $1\frac{1}{2}$ 30) $6\frac{14}{15}$
7) $\frac{7}{8}$ 31) $5\frac{5}{8}$
8) $1\frac{7}{15}$ 32) $5\frac{17}{20}$
9) $\frac{17}{21}$ 33) $5\frac{1}{2}$
10) $\frac{7}{12}$ 34) 5
11) $\frac{11}{12}$ 35) 5
12) $1\frac{1}{12}$ 36) $13\frac{3}{4}$
13) $1\frac{3}{8}$ 37) $6\frac{1}{6}$
14) $1\frac{1}{16}$ 38) $7\frac{5}{12}$
15) $\frac{11}{15}$ 39) $6\frac{1}{9}$
16) $1\frac{1}{8}$ 40) $\frac{15}{56}$
17) $\frac{8}{15}$ 41) $1\frac{8}{63}$
18) $\frac{16}{21}$ 42) $5\frac{1}{2}$
19) $1\frac{1}{6}$ 43) $5\frac{6}{7}$
20) $1\frac{3}{20}$ 44) $7\frac{7}{8}$
21) $\frac{5}{8}$ 45) $4\frac{1}{2}$
22) $1\frac{11}{20}$ 46) $4\frac{7}{15}$
23) $1\frac{1}{30}$ 47) $\frac{1}{2}$
24) $1\frac{1}{9}$ 48) $4\frac{1}{10}$

Exercise 11

1) $\frac{1}{10}$ 6) $\frac{5}{22}$
2) $\frac{3}{14}$ 7) $\frac{1}{2}$
3) $\frac{1}{8}$ 8) $\frac{1}{12}$
4) $\frac{1}{2}$ 9) $\frac{3}{10}$
5) $\frac{1}{8}$ 10) $\frac{1}{12}$

Exercise 12

1) $1\frac{1}{2}$ 7) $\frac{5}{6}$
2) $2\frac{1}{2}$ 8) $\frac{7}{12}$
3) $\frac{5}{6}$ 9) $\frac{3}{8}$
4) $3\frac{1}{4}$ 10) $\frac{5}{12}$
5) $1\frac{11}{12}$ 11) $\frac{1}{2}$
6) $\frac{7}{8}$ 12) $2\frac{1}{2}$

Exercise 13

1) $2\frac{5}{8}$
2) 1
3) $\frac{3}{20}$
4) $\frac{11}{12}$
5) $\frac{7}{8}$
6) $1\frac{1}{12}$
7) $\frac{7}{12}$
8) $\frac{5}{12}$
9) $1\frac{5}{8}$

Exercise 14

1) $\frac{1}{2}$
2) $\frac{12}{35}$
3) $\frac{3}{20}$
4) $\frac{1}{3}$
5) 2
6) $\frac{4}{5}$
7) $14\frac{1}{2}$
8) 1
9) $\frac{1}{2}$
10) 3
11) $\frac{1}{6}$
12) $\frac{1}{6}$
13) $\frac{4}{5}$
14) $\frac{1}{4}$
15) 2
16) $\frac{5}{8}$
17) $\frac{2}{15}$
18) $\frac{3}{4}$
19) $\frac{3}{4}$
20) $\frac{1}{32}$
21) $1\frac{1}{3}$
22) $\frac{4}{9}$
23) $\frac{1}{2}$
24) $\frac{3}{10}$
25) 1
26) 6
27) 2
28) $\frac{2}{3}$

Exercise 15

1) $1\frac{1}{2}$
2) $1\frac{1}{4}$
3) $2\frac{4}{5}$
4) $\frac{2}{3}$
5) 2
6) 12
7) 12
8) 2
9) $\frac{5}{6}$
10) 4
11) $\frac{1}{6}$
12) $\frac{2}{3}$
13) $1\frac{1}{2}$
14) $\frac{3}{4}$
15) $2\frac{2}{3}$
16) $3\frac{1}{2}$
17) 5
18) 14
19) $\frac{1}{2}$
20) $1\frac{1}{3}$
21) $\frac{5}{6}$
22) $\frac{1}{4}$
23) 6
24) $1\frac{1}{2}$

Exercise 16

1) $5\frac{1}{3}$
2) 2
3) 9
4) 4
5) $\frac{1}{4}$
6) $\frac{3}{4}$
7) $\frac{2}{9}$
8) $\frac{2}{5}$

Exercise 17

1) 12
2) 12
3) 15
4) 21
5) 60
6) 180
7) 90
8) 120

Exercise 18

1) (a) $\frac{1}{6}$ (b) $\frac{2}{7}$ (c) $\frac{2}{5}$ (d) $\frac{1}{3}$
2) (a) $\frac{1}{4}$ (b) $\frac{7}{12}$
   (c) $\frac{3}{8}$
3) $\frac{5}{6}$ $\frac{10}{12}$
4) $\frac{5}{7}$
5) (a) $\frac{6}{8}$ (b) $\frac{10}{20}$ (c) $\frac{2}{12}$ (d) $\frac{9}{15}$
6) (a) $\frac{1}{2}$ (b) $\frac{1}{5}$ (c) $\frac{3}{4}$ (d) $\frac{1}{8}$
7) (a) $5\frac{2}{5}$ (b) $3\frac{5}{5}$ (c) $5\frac{1}{6}$ (d) $7\frac{1}{6}$
8) (a) $\frac{5}{3}$ (b) $\frac{20}{3}$ (c) $\frac{59}{8}$ (d) $\frac{23}{2}$
9) 40
10) $\frac{1}{2}, \frac{3}{5}, \frac{5}{8}, \frac{3}{4}$
11) (a) $5\frac{3}{4}$ (b) $1\frac{1}{10}$ (c) $6\frac{11}{12}$
12) (a) $\frac{9}{10}$ (b) $1\frac{11}{12}$ (c) $\frac{7}{16}$ (d) $\frac{3}{20}$
13) $1\frac{1}{8}$
14) $2\frac{3}{4}$
15) (a) $\frac{1}{2}$ (b) $\frac{1}{2}$ (c) 2 (d) $1\frac{1}{2}$
16) (a) $\frac{24}{25}$ (b) $\frac{8}{15}$
17) (a) $\frac{3}{4}$ (b) $1\frac{1}{3}$ (c) $1\frac{1}{8}$ (d) $\frac{1}{6}$
18) (a) 420 (b) 280

Exercise 19

1) $\frac{1}{2}$ of £32
2) $15\,\text{m}\ell$
3) £12 000
4) 10
5) 18
6) $\frac{1}{4}$
7) $\frac{3}{8}$
8) $\frac{1}{10}$, £24

Exercise 20

1) £1.80
2) £1400
3) 14 yards
4) 412
5) 160
6) $5\frac{1}{4}$ metres
7) $43\frac{7}{8}$ minutes
8) 95
9) $146\frac{1}{2}$ litres
10) 340
11) 630 grams copper, 162 grams tin, 108 grams zinc
12) $727\frac{1}{2}$ tonnes
13) 38 metres
14) 935
15) £8700

CHAPTER 4

Exercise 1

1) 0 units 2 tenths
2) 3 units 4 tenths
3) 1 ten 0 units 0 tenths 4 hundredths
4) 1 ten 6 units 0 tenths 2 hundredths
5) 8 units 2 tenths 4 hundredths
6) 1 ten 5 units 1 tenth 6 hundredths
7) 3 units 8 tenths 1 hundredth 7 thousandths
8) 0 units 0 tenths 8 hundredths 2 thousandths
9) 5 tens 0 units 2 tenths 0 hundredths 7 thousandths
10) 0 units 3 tenths 0 hundredths 4 thousandths
11) 6 tenths
12) 2 hundredths
13) 7 tens
14) 5 thousandths

Exercise 2

13) 0.7 is larger
14) 5.8 is larger
15) 7.0 is larger
16) 7.0 is larger
17) 5.2 is larger
18) 2.5 is larger
19) 6.8 is larger
20) 4.4 is larger
21) 2.1 is larger
22) 5.3 is larger

Exercise 3

1) (a) (i) 5.0 (ii) 5.00 (iii) 5.000
   (b) (i) 8.0 (ii) 8.00 (iii) 8.000
   (c) (i) 9.0 (ii) 9.00 (iii) 9.000
2) (a) yes (b) yes (c) yes (d) no
3) (a) yes (b) yes (c) yes (d) no
4) (a) 2 (b) 56 (c) 342 (d) 15

Exercise 4

1) 48.8
2) 4
3) 6
4) 8
5) 0.8
6) 0.7

7) 0.045
8) 1.103
9) 5070
10) 300
11) 0.272
12) 11
13) 2.02
14) 0.055
15) 6
16) no
17) yes
18) no
19) 6.5

Exercise 5

1) 0.6
2) 0.03
3) 0.004
4) 0.25
5) 0.127
6) 0.3
7) 0.08
8) 0.009
9) 0.14
10) 0.628
11) 0.5
12) 0.07
13) 0.006
14) 0.75
15) 0.954
16) 0.2
17) 0.06
18) 0.003
19) 0.15
20) 0.205
21) 0.011
22) 0.015
23) 0.062
24) 0.058

Exercise 6

1) 103.45
2) 8.6
3) 49.93
4) 236.81
5) 1549.8
6) 13.98
7) 13.55
8) 57.4
9) 34.48
10) 178.1
11) 9.851
12) 28.759
13) 29.508
14) 36.47
15) 45.25
16) 7.99
17) 26.3
18) 20.67
19) 19.706
20) 62.9
21) 92.448
22) 19.054
23) 413.043
24) 48.90
25) 842.23
26) 740.092
27) 189.642
28) 61.985
29) 184.58
30) 386.12
31) 78.013
32) 98.20
33) 29.09
34) 22.26

Exercise 7

1) 4.32
2) 1.39
3) 5.1
4) 56.49
5) 314.82
6) 1.4
7) 0.04
8) 3.8
9) 113.1
10) 5.5
11) 12.88
12) 799.84
13) 35.7
14) 6947.5
15) 5273.6
16) 14.5
17) 194.03
18) 467.492
19) 12.46
20) 153.302
21) 6.34
22) 40.61

23) 132.71
24) 76.06
25) 10.12
26) 298.5
27) 388.88
28) 12.361
29) 0.0054
30) 0.098
31) 12.34
32) 192.4
33) 0.15
34) 8.055
35) 93.1
36) 308.9

Exercise 8

1) 37.17
2) 9.28
3) 12.14
4) 256.017
5) 77.547
6) 137.4
7) 47.51
8) 9.68
9) 30.04
10) 94.11

Exercise 9

1) (i) 57 (ii) 570
   (iii) 5700
2) (i) 62.1 (ii) 621
   (iii) 6210
3) (i) 0.23 (ii) 2.3
   (iii) 23
4) (i) 4.6 (ii) 46
   (iii) 460
5) (i) 2.178 (ii) 21.78
   (iii) 217.8
6) (i) 0.028 54 (ii) 0.2854
   (iii) 2.854
7) (i) 5.823 (ii) 58.23
   (iii) 582.3
8) (i) 0.07 (ii) 0.7
   (iii) 7
9) (i) 1600.7 (ii) 16 007
   (iii) 160 070
10) (i) 1.4729 (ii) 14.729
    (iii) 147.29

Exercise 10

1) 26.8
2) 10.5
3) 33.6
4) 73.2
5) 124.6
6) 318.6
7) 7.5
8) 23.87
9) 46.64
10) 457.5
11) 614.4
12) 51.85
13) 0.0572
14) 0.008
15) 0.3213
16) 2.014
17) 17.4
18) 41.72
19) 0.576
20) 0.813 72
21) 23.3
22) 40.28
23) 16.371
24) 10.881
25) 401.28
26) 15.225
27) 0.072
28) 82.34
29) 11.475
30) 3.914

31) 20.259
32) 0.2416
33) 0.0234
34) 239.4
35) 0.0261
36) 0.732
37) 8.64
38) 79.8
39) 6.351
40) 0.1512

Exercise 11

1) (i) 0.61 (ii) 0.061
   (iii) 0.0061
2) (i) 7.25 (ii) 0.725
   (iii) 0.0725
3) (i) 0.005 (ii) 0.0005
   (iii) 0.000 05
4) (i) 26.005 (ii) 2.6005
   (iii) 0.260 05
5) (i) 0.0469 (ii) 0.004 69
   (iii) 0.000 469
6) (i) 0.128 (ii) 0.0128
   (iii) 0.001 28
7) (i) 0.000 26 (ii) 0.000 026
   (iii) 0.000 002 6
8) (i) 32.55 (ii) 3.255
   (iii) 0.3255
9) (i) 462.89 (ii) 46.289
   (iii) 4.6289
10) (i) 71.3598 (ii) 7.135 98
    (iii) 0.713 598

Exercise 12

1) 17.5
2) 9.7
3) 2.03
4) 0.184
5) 0.51
6) 11.33
7) 8.16
8) 0.223
9) 0.36
10) 0.903
11) 72.1
12) 201.1
13) 178.9
14) 0.478
15) 0.091
16) 0.2
17) 1.544
18) 2.166
19) 1.037
20) 0.357
21) 1.713
22) 0.176
23) 0.21
24) 6.49
25) 10.35
26) 1.844
27) 8.04
28) 2.2
29) 1.744
30) 1.206

Exercise 14

1) 10.6
2) 10.5
3) 69
4) 6900
5) 80
6) 8100
7) 0.21
8) 0.6
9) 13
10) 68 500
11) 70
12) 0.4
13) 8.5
14) 10.3

15) 314 100
16) 172.1
17) 1970
18) 10.24
19) 80
20) 501
21) 1031
22) 112.4
23) 0.223
24) 72
25) 106.8

26) 20.2
27) 40.6
28) 0.9
29) 12.07
30) 14.7
31) 4.08
32) 0.009
33) 0.14
34) 0.17
35) 8.64

Exercise 15

1) 2
2) 4
3) 3
4) 3
5) 8

6) 1
7) 5
8) 3
9) 1
10) 4

Exercise 16

1) 2.2
2) 5.8
3) 9.4
4) 8.1
5) 1.2

6) 2.3
7) 5.7
8) 9.3
9) 8.2
10) 1.3

Exercise 17

1) 0.25
2) 0.36
3) 0.68
4) 0.79
5) 0.11
6) 0.12

7) 0.80
8) 0.79
9) 5.23
10) 5.24
11) 7.22
12) 7.22

Exercise 18

1) 0.86
2) 0.76
3) 0.56
4) 5.66
5) 0.25
6) 0.19
7) 0.39
8) 0.49

9) 3.70
10) 4.60
11) 8.30
12) 11.20
13) 102.18
14) 84.19
15) 15.78
16) 14.29

Exercise 19

1) 0.653
2) 0.652
3) 0.653
4) 0.653
5) 0.652
6) 5.512

7) 5.513
8) 5.512
9) 5.513
10) 5.513
11) 0.780
12) 0.779

13) 0.780
14) 3.000
15) 3.000
16) 2.999

17) 2.999
18) 3.000
19) 3.000
20) 2.999

Exercise 20

1) 1.733
2) 68.889
3) 3.971
4) 7.889
5) 0.729
6) 117.333

7) 13.667
8) 0.093
9) 70.167
10) 77.143
11) 75.167
12) 5.722

Exercise 21

1) 0.25
2) 0.5
3) 0.125
4) 0.875
5) 0.2

6) 0.4
7) 0.6
8) 0.05
9) 0.04
10) 0.15

Exercise 22

1) 3.75
2) 2.5
3) 1.5
4) 1.25
5) 2.75
6) 4.75
7) 5.5
8) 1.375
9) 2.625
10) 3.5

11) 15.5
12) 3.8
13) 2.4
14) 2.125
15) 5.6
16) 2.2
17) 6.25
18) 10.5
19) 11.75
20) 4.875

Exercise 23

1) (a) 1/2   (b) 0.5
2) (a) 1/4   (b) 0.25
3) (a) 1/10   (b) 0.1
4) (a) 1/50   (b) 0.02
5) (a) 1/100   (b) 0.01
6) (a) 1/200   (b) 0.005
7) (a) 1/1000   (b) 0.001
8) (a) 1/1 000 000
    (b) 0.000 001
9) 0.7407
10) 0.042
11) 0.002 01
12) 14.286
13) 111.11
14) 13.021
15) 3.1646
16) 0.8
17) 0.4
18) 4

19) 2
20) 10
21) 100
22) 1000
23) 10 000
24) 2
25) 5
26) 8
27) 10
28) 40
29) 75
30) (a) 2/5   (b) 0.4
31) (a) 3/2   (b) 1.5
32) (a) 7/10   (b) 0.7
33) (a) 11/100   (b) 0.11
34) (a) 7/8   (b) 0.875
35) (a) 3/8   (b) 0.375

Exercise 24

1) $\frac{3}{5}$
2) $\frac{7}{10}$
3) $\frac{9}{100}$
4) $\frac{1}{50}$
5) $\frac{1}{25}$
6) $\frac{1}{20}$
7) $\frac{2}{25}$
8) $\frac{17}{100}$

9) $\frac{31}{50}$
10) $\frac{11}{20}$
11) $\frac{3}{20}$
12) $\frac{23}{100}$
13) $\frac{29}{40}$
14) $\frac{123}{1000}$
15) $\frac{513}{1000}$

Exercise 25

1) $2\frac{1}{2}$
2) $3\frac{3}{4}$
3) $5\frac{1}{4}$
4) $6\frac{1}{5}$
5) $18\frac{1}{10}$

6) $27\frac{1}{2}$
7) $3\frac{3}{10}$
8) $5\frac{1}{4}$
9) $14\frac{1}{2}$
10) $22\frac{1}{5}$

Exercise 26

1) $0.\dot{4}$
2) $0.\dot{3}\dot{4}$
3) $0.4\dot{6}75\dot{3}$
4) $7.\dot{8}$
5) $30.7\dot{8}$

6) $12.50\dot{3}0\dot{9}$
7) $0.\dot{2}$
8) $0.1\dot{9}$
9) $2.314\dot{5}$

Exercise 27

1) 31.7
2) 0.26
3) 21
4) 6
5) 4.98
6) 0.06

7) 20
8) 0.8
9) 9.8
10) 21.885
11) 0.03
12) 2.3

Exercise 28

1) (a) 4 tens   (b) 4 tenths
   (c) 4 hundredths (d) 4 units
2) (a) 17.0   (b) 17.000
3) (a) no   (b) yes
4) (a) 72.5   (b) 63.1
   (c) 402.05
5) (a) 0.3   (b) 0.06
   (c) 0.007   (d) 0.35
6) (a) 9.3   (b) 1.12
7) (a) 16.338   (b) 98.04
8) 70.85
9) (a) 7.2   (b) 72
10) (a) 2.6   (b) 10.4
11) (a) 14.85   (b) 0.2412
   (c) 1.548
12) (a) 0.1756   (b) 0.017 56
13) 72.6
14) (a) 5.12   (b) 0.122
15) 51.2
16) 1038
17) 500
18) (a) 4.3   (b) 4.2
19) (a) 14.630   (b) 14.629
20) (a) 59   (b) 59.3
   (c) 59.34   (d) 59.342
   (e) 59.3416
21) 1.667
22) (a) 0.8   (b) 0.7
   (c) $0.\dot{3}$
23) (a) 2.25   (b) 5.1
   (c) 4.2
24) (a) $\frac{1}{5}$ (b) $\frac{2}{25}$
   (c) $\frac{9}{20}$
25) (a) $6\frac{1}{2}$ (b) $4\frac{1}{4}$
   (c) $9\frac{1}{10}$ (d) $1\frac{3}{4}$
26) (a) $0.\dot{1}$ (b) $0.\dot{2}\dot{9}$
   (c) $0.\dot{5}01\dot{2}$
27) (a) 2.27   (b) 2.79
   (c) 1.78

Exercise 29

1) 0.8 kg
2) 0.22 m
3) £2
4) £5.64
5) 10 litres
6) £11.25
7) 24 dollars
8) 15 litres
9) 2.2 tonnes
10) 7
11) 0.33 kg
12) 2.1
13) 92 p
14) £4360
15) 80 p
16) £333.34
17) 60, £2348.50
18) (a) 0.184 p
   (b) 0.26 p
19) 1.6 km
20) 11.81 in

Exercise 30

1) 7.1 tonnes
2) 765 tonnes
3) 554.4 litres
4) 147.84 kilograms
5) 85.97 mm
6) 950.4 millimetres
7) 280.46 minutes (4 h 40.46 min)
8) 54.9 ohms per metre
9) 2916
10) 68.15 metres
11) 856.8p (£8.57 to nearest penny)
12) 0.3 millimetres
13) 0.882, by 0.007 or $\frac{7}{1000}$
14) 700

CHAPTER 5

Exercise 1

1) 3 kg
2) 5 kg
3) 9 kg
4) 1.5 kg
5) 6.5 kg
6) 7.5 kg
7) 3.85 kg
8) 5.67 kg
9) 4.5 kg
10) 1.654 kg
11) 3.045 kg
12) 6.723 kg
13) 0.423 kg
14) 0.56 kg
15) 0.305 kg
16) 25 kg
17) 40 kg
18) 50 kg

Exercise 2

1) 5000 g
2) 7000 g
3) 9000 g
4) 2340 g
5) 9160 g
6) 2080 g
7) 500 g
8) 900 g
9) 400 g
10) 5100 g
11) 4150 g
12) 3200 g
13) 25 000 g
14) 50 000 g
15) 60 000 g
16) 150 g
17) 350 g
18) 950 g

Exercise 3

1) 2000
2) 4000
3) 10 000
4) 500
5) 3500
6) 3000
7) 5000
8) 500
9) 8500
10) 250

Exercise 4

1) (a) 4   (b) 40   (c) 80
2) (a) 10   (b) 60   (c) 110
3) (b)

4) more
5) (b)

Exercise 5

1) 20
2) 40
3) 60
4) 90
5) 15
6) 5
7) 55
8) 95
9) 2400
10) 3500
11) 5000

Exercise 6

1) 200
2) 400
3) 700
4) 900
5) 40
6) 60
7) 80
8) 90
9) 4000
10) 5000
11) 8500
12) 9500

Exercise 7

1) 4 cm
2) 2 cm
3) 2.5 cm
4) 3.7 cm
5) 6.5 cm
6) 4.5 cm
7) 2.4 cm
8) 0.9 cm
9) 5.6 cm
10) 5 cm

Exercise 8

1) 2 km
2) 4 km
3) 7 km
4) 9 km
5) 2.5 km
6) 3.55 km
7) 4.5 km
8) 7.5 km
9) 0.5 km
10) 0.65 km
11) 0.7 km
12) 0.9 km

Exercise 9

1) 9
2) 880
3) 42
4) 14 ft
5) less
6) 16
7) longer
8) (a) 4   (b) 2   (c) 8
9) (a) 90   (b) 45   (c) 15
10) (a) 16   (b) 24
11) (b)
12) less

Exercise 10

1) 50
2) 2

3) more
4) (d)
5) less
6) (a) 20   (b) 6
7) 75 cℓ
8) yes
9) 2.5 litres
10) 23
11) 40
12) 5
13) 500
14) 2500
15) (b)

Exercise 11

1) the 1530 train
2) the 1700 train
3) 1 h 55 min
4) 30 min
5) no

Exercise 12

1) 135
2) 210
3) 2 h 30 min
4) 1 h 40 min
5) (a) 31   (b) 30   (c) 29
6) 26
7) £500

Exercise 13

1) £5.74   6) £1.12
2) £3.97   7) £30.08
3) £57.40   8) £49.47
4) £4.83   9) £30.53
5) £2.05   10) £342

Exercise 14

1) £72.31   6) £541.65
2) £334.18   7) £29.95
3) £201.90   8) £0.75
4) £59.50   9) £20.48
5) £20.11   10) £33.50

Exercise 15

1) £9   9) £1084
2) £9.90   10) £4.05
3) £22.88   11) £280
4) £5.25   12) £169.05
5) £83   13) £125
6) £68.20   14) £248.50
7) £30.15   15) £4.00
8) £105   16) £13.50

Exercise 16

1) £1.12   11) £0.55
2) £1.20   12) £2.04
3) £2.03   13) £1.67
4) £3.12   14) £1.13
5) £1.20   15) £3.50
6) £3.35   16) £6.75
7) £3.55   17) £15.50
8) £0.27   18) £5.10
9) £1.04   19) £5.11
10) £1.50   20) £1.07

Exercise 17

1) 2.2 kg   6) 1.25 kg
2) 1.05 km   7) 500 m
3) 1.7 litres   8) 7.5 litres
4) 12.7 kg   9) 1250 g
5) 2.1 m   10) 5 cm

Exercise 18

1) 2250 g   6) 600 m
2) 670 cm   7) 3450 g
3) 3550 g   8) 100 cm
4) 38 mm   9) 750 g
5) 6.7 km   10) 30 mm

Exercise 19

1) 800 g   11) 750 g
2) 1 km   12) 50 g
3) 150 mm   13) 3.5 litres
4) 17.5 kg   14) 500 m
5) 0.5 litres 15) 6 km
6) 150 km   16) 7 mm
7) 1500 g   17) 10.5 kg
8) 15 cm   18) 80 cm
9) 4 kg   19) 28 m
10) 50 cm   20) 30 cℓ

Exercise 20

1) 56 p
2) (a) 14   (b) 35   (c) 49
   (d) 63
3) £40
4) £17.50
5) £1.95
6) £1.26
7) £2.48
8) (a) 8   (b) 20   (c) 28
9) £8
10) £4
11) £30, £20 and £10
12) £3.07

13) £1.92
14) £16.50
15) £12.25
16) £3.84
17) 2200 hours, 10 p.m.
18) £2.80
19) £6.74
20) 110 litres
21) $13\frac{1}{2}$ min

CHAPTER 6

Exercise 1

1) £3   5) 55 p
2) 4 cm   6) 7.5 km
3) 2 p   7) 9 m
4) 6 min   8) 32 p

Exercise 2

1) $\frac{1}{5}$ 12) $\frac{3}{4}$
2) $\frac{3}{10}$ 13) $\frac{17}{20}$
3) $\frac{2}{5}$ 14) $\frac{1}{25}$
4) $\frac{1}{2}$ 15) $\frac{1}{50}$
5) $\frac{3}{5}$ 16) $\frac{1}{100}$
6) $\frac{7}{10}$ 17) $\frac{1}{40}$
7) $\frac{4}{5}$ 18) $\frac{1}{200}$
8) $\frac{9}{10}$ 19) $\frac{3}{8}$
9) $\frac{11}{20}$ 20) $\frac{5}{8}$
10) $\frac{7}{20}$ 21) $1\frac{1}{2}$
11) $\frac{3}{20}$

Exercise 3

1) £9   9) £9
2) 9 cm   10) 9 kg
3) 2 km   11) £18.28
4) 1 p   12) £60
5) 6 mm   13) £60
6) 15 m   14) 38 p
7) £20   15) £1
8) £2

Exercise 4

1) £18   9) £28
2) £20   10) £4.57
3) £40   11) £52.50
4) £110   12) £450
5) £35   13) £30
6) £58.50   14) £12
7) £1.60   15) £30
8) £7.50   16) £5

Exercise 5

1) £63
2) £132.40
3) £165.50
4) £126
5) £61.50
6) £258

Exercise 6

1) 20%
2) 80%
3) 25%
4) 50%
5) 75%
6) $33\frac{1}{3}$%
7) $66\frac{2}{3}$%
8) $12\frac{1}{2}$%
9) $37\frac{1}{2}$%
10) $62\frac{1}{2}$%
11) 40%
12) $87\frac{1}{2}$%
13) 5%
14) $2\frac{1}{2}$%
15) 70%
16) 90%
17) 10%
18) 1%
19) 3%
20) 2%

Exercise 7

Percentage	Fraction	Decimal
50%	$\frac{1}{2}$	—
75%	—	0.75
—	$\frac{1}{3}$	$0.3\dot{3}$
$66\frac{2}{3}$%	—	$0.6\dot{6}$
$12\frac{1}{2}$%	$\frac{1}{8}$	—
—	$\frac{1}{4}$	0.25
10%	$\frac{1}{10}$	—
—	$\frac{1}{5}$	0.2
30%	$\frac{3}{10}$	—
—	$\frac{5}{8}$	0.625

Exercise 8

1) 20%
2) 60%
3) 10%
4) 16%
5) 30%
6) 50%
7) 80%
8) $16\frac{2}{3}$%
9) 20%
10) 40%
11) 15%
12) $12\frac{1}{2}$%, $52\frac{1}{2}$%, $47\frac{1}{2}$%
13) 75%, $62\frac{1}{2}$%
14) 4%
15) 65%

Exercise 9

1) 0.4, $\frac{1}{2}$, 60%
2) $\frac{1}{5}$, 0.22, 25%
3) 45%, 0.54, $\frac{4}{5}$
4) 66%, $\frac{2}{3}$, 0.67
5) all the same
6) $12\frac{1}{2}$%, $\frac{3}{20}$, 0.2
7) 9%, $\frac{1}{9}$, 0.9
8) 0.33, 35%, $\frac{3}{8}$
9) $\frac{5}{6}$, 85%, 0.875
10) $1\frac{1}{4}$, 150%, 1.75

Exercise 10

1) £324
2) £520
3) £4.86
4) £18, £102
5) £10
6) £21
7) 13%
8) $33\frac{1}{3}$%
9) $12\frac{1}{2}$%
10) 528
11) 10 500
12) 46%
13) least 9.8 mm, most 10.2 mm
14) 60%

Exercise 11

1) (a) £4   (b) £20
   (c) £3.65   (d) £1.82
   (e) 40 g
   (f) 0.25 kg (or 250 g)
   (g) 34 m   (h) 2 s
2) (a) $\frac{1}{8}$ (b) 0.125
3) (a) £2   (b) £4
4) (a) $\frac{1}{3}$ (b) 0.33
5) (a) £2   (b) £5
6) (a) $\frac{19}{20}$ (b) 0.95
7) (a) £6   (b) £66.50
8) £181.50
9) (a) £82.75   (b) £332.75
10) (a) 17%   (b) 95%
11) (a) 0.3   (b) 30%
12) £52.87
13) £1500
14) £84
15) £518.75
16) £24
17) 30
18) 59 400
19) 80%, 88% and 96%
20) maximum 2.04 cm,
   minimum 1.96 cm

21) 80%
22) 10%
23) (a)

3 5 0 × 1 7 . 5 % or

1 7 . 5 × 3 5 0 ÷ 1 0 0 or

0 . 1 7 5 × 3 5 0 or others

   (b) £61.25   (c) 14 p
24) (a) £87.50   (b) £587.50
25) £7
26) (a) £17.50   (b) £35
   (c) £8.75   (d) £70
27) £450, £78.75, £528.75

CHAPTER 7

Exercise 1

1) 4
2) 25
3) 36
4) 49
5) 64
6) 81

Exercise 2

1) 169
2) 676
3) 1225
4) 2304
5) 3249
6) 4761
7) 5184
8) 5776
9) 5929
10) 6561
11) 7744
12) 9025

Exercise 3

1) $\frac{1}{16}$
2) $\frac{25}{49}$
3) $\frac{16}{81}$
4) $\frac{4}{25}$
5) $\frac{25}{64}$
6) $\frac{9}{49}$
7) $\frac{16}{121}$
8) $\frac{49}{144}$
9) $2\frac{7}{9}$
10) $1\frac{7}{9}$
11) $6\frac{1}{4}$
12) $5\frac{1}{16}$
13) $20\frac{1}{4}$
14) $30\frac{1}{4}$
15) $11\frac{1}{9}$

Exercise 4

1) 2.25
2) 12.25
3) 17.64
4) 24.01
5) 26.01
6) 30.25
7) 33.64
8) 39.69
9) 0.25
10) 0.49

Exercise 5

1) 8
2) 7
3) 9
4) 3
5) 4
6) 5
7) 10
8) 6

Exercise 6

1) $\frac{3}{5}$
2) $\frac{4}{7}$
3) $\frac{7}{8}$
4) $\frac{6}{11}$
5) $\frac{7}{10}$
6) $\frac{5}{6}$
7) $\frac{5}{9}$
8) $\frac{3}{10}$
9) $\frac{4}{5}$

Exercise 7

1) 3.2
2) 5.5
3) 6.7
4) 7.1
5) 7.9
6) 8.7
7) 9.2
8) 9.3
9) 9.5
10) 5.9
11) 5.4
12) 7.7

Exercise 8

1) 289
2) 65 025
3) 422 500
4) 0.0625
5) 6.5536
6) 475.24
7) 1.18
8) 3.79
9) 3.74
10) 29.92
11) 9.46
12) 0.95
13) 1.44
14) 576
15) 1764
16) 0.8
17) 0.1
18) 0.4

Exercise 9

1) (a) 9   (b) 100
2) 49, 64, 81, 100, 121, 144
3) (a) 196   (b) 400
    (c) 22 500
4) (a) $\frac{1}{4}$ (b) $\frac{9}{25}$
    (c) $2\frac{1}{4}$
5) (a) 0.04   (b) 1.21
    (c) 0.0016
6) (a) 25   (b) 40
7) (a) 11   (b) 12
8) (a) 3.5   (b) 5.3
9) (a) $\frac{2}{3}$ (b) $\frac{5}{8}$
10) (a) 5   (b) 4
11) (a) 2916   (b) 0.73
12) (a) 0.323   (e) 3.464
    (b) 2.890   (f) 4.899
    (c) 12.674   (g) 2.254
    (d) 0.608   (h) 3
13) (a) $\boxed{4. \quad 10}$ (b) $\boxed{4. \quad -10}$

CHAPTER 8

Exercise 1

1) £600
2) £600
3) £700
4) £900
5) £1500
6) £23 900
7) £50
8) £150
9) £200
10) £250
11) £800
12) £15 600
13) £80
14) £180
15) £180
16) £290
17) £480
18) £46
19) £73
20) £31
21) £174
22) £274

Exercise 2

1) 20 m, 17.1 m
2) £2, £1.89
3) 4 kg, 3.978 kg
4) 80 cm, 89 cm
5) £9, £9.01
6) £25, £25.02
7) 40, each child has 44 sweets and 3 are left over
8) 800 kg, 842.1 kg
9) 3 m, 2.871 m
10) 30 g, 29 g

Exercise 3

1) 30, 33
2) 1500, 1341
3) 100, 90
4) 7, 7.2
5) 10, 9.45
6) 80, 81.7
7) 20, 23.2
8) 110, 108
9) 2, 2.31
10) 21, 21.1702
11) 100, $104\frac{1}{2}$
12) 4, 4.455

Exercise 4

1) 3^2
2) 4^3
3) 6^4
4) 5^6
5) 10^2
6) 9^{10}
7) 11^3
8) $3 \times 3 \times 3 \times 3$
9) 7×7
10) $8 \times 8 \times 8$
11) $10 \times 10 \times 10 \times 10 \times 10 \times 10$
12) $12 \times 12 \times 12$
13) 8
14) 625
15) 216

16) 81
17) 512
18) 24
19) 54
20) 11

Exercise 5

1) 100
2) 1
3) 0.1
4) 1000
5) 1 000 000
6) 10
7) 0.01
8) 0.001
9) 0.0001
10) 10 000

Exercise 6

1) 2.48×10^2
2) 6.8×10^3
3) 7.81×10^2
4) 5.63×10^2
5) 7.8×10^3
6) 4.1×10^3
7) 5.2×10^3
8) 3.43×10^2
9) 2.3×10^2
10) 4.56×10^3

Exercise 7

1) 6×10^{-1}
2) 2×10^{-1}
3) 9×10^{-2}
4) 5×10^{-2}
5) 4.5×10^{-1}
6) 5.6×10^{-1}
7) 5.4×10^{-2}
8) 6.1×10^{-2}
9) 1×10^{-1}
10) 8.7×10^{-2}
11) 8×10^{-2}
12) 3×10^{-1}

Exercise 8

1) 5.8×10^4
2) 3.4×10^3
3) 1.56×10^5
4) 6.3×10^1
5) 5.97×10^2
6) 7.8×10^{-3}
7) 5.43×10^{-2}
8) 4.52×10^{-4}
9) 5.67×10^8
10) 5.9×10^{-7}
11) 4.97×10^5
12) 3.5×10^{-4}
13) 2.6×10^{-5}
14) $6.351\,24 \times 10^8$

Exercise 9

1) 5300
2) 340
3) 690 000
4) 0.92
5) 0.943
6) 0.0311
7) 4.7
8) 56.6
9) 0.4505
10) 3 200 000
11) 0.014
12) 0.008

Exercise 10

1) 3.26	9) 62.8
2) 3.27	10) 62.9
3) 3.26	11) 62.9
4) 3.26	12) 62.9
5) 3.27	13) 62.8
6) 3.26	14) 62.8
7) 3.27	15) 62.9
8) 62.9	16) 62.8

Exercise 11

1) 1350	11) 5560
2) 6730	12) 4330
3) 5670	13) 6070
4) 6790	14) 4520
5) 7130	15) 3400
6) 24 600	16) 1200
7) 146 000	17) 4600
8) 3010	18) 23 400
9) 4030	19) 568 000
10) 5420	20) 4560

Exercise 12

1) 0.412	9) 0.0891
2) 0.413	10) 0.0892
3) 0.413	11) 0.0809
4) 0.412	12) 0.0810
5) 0.006 11	13) 0.005 20
6) 0.006 12	14) 0.005 19
7) 0.006 12	15) 0.006 10
8) 0.006 11	16) 0.006 09

Exercise 13

1) 5.6	7) 280
2) 3.9	8) 0.37
3) 6.2	9) 1500
4) 7.3	10) 0.51
5) 2.3	11) 64
6) 5.4	12) 5.8

Exercise 14

1) £44
2) £800
3) £6.70
4) 2 m
5) 500 g
6) 4 h
7) 25 cm
8) 8 litres
9) 6 kg
10) 4 km

11) 2000 g, 1710 g
12) 4 m, 3.9 m
13) £13, £13.02
14) 50 cm, 70 cm
15) 60, 63
16) 1200, 1071
17) 100, 101.25
18) (a) 100 (b) 0.1
    (c) 1 (d) 10
19) (a) 3.72×10^2 (b) 4.5×10^1
    (c) 6.5×10^1 (d) 9×10^{-2}
20) (a) 5.7×10^4 (b) 5×10^{-5}
21) (a) 2.56 (b) 0.659
    (c) 8.08 (d) 5720
22) (a) 0.13 (b) 7300
    (c) 0.037 (d) 4.5

Exercise 15

1) £2.17, give 2p to charity
2) 3×5 needs to be worked out
   first, | 4 | 5 | ÷ | 3 |
   | ÷ | 5 | = | or others
3) 54.0 cm^2
4) (a) £9 (b) £10
5) (a) 3 (b) 3.14
   (c) 3.142
6) (a) £4.40 (b) 2.53
   (c) 8 p
7) 73.5
8) 2.8 metres, 2 whole metres
9) (a) 1.6×10^{-9} (b) 1.6×10^{11}
   (c) | 1.6 | −09 |
       | 1.6 | 11 |
   (d) no ×10 in display
10) 1.98
12) 20

CHAPTER 9

Exercise 1

1) 1:2	9) 6:7
2) 1:4	10) 5:6
3) 1:2	11) 5:6
4) 1:3	12) 2:3
5) 2:5	13) 5:4
6) 1:3	14) 3:2
7) 1:3	15) 4:5
8) 1:5	16) 5:6

17) 7:8	27) 2:3
18) 7:8	28) 1:2
19) 5:9	29) 7:8
20) 3:4	30) 2:3
21) 4:5	31) 9:10
22) 6:7	32) 7:8
23) 5:6	33) 5:6:7
24) 9:10	34) 5:6:7
25) 2:3	35) 4:5:6:7
26) 2:3	

Exercise 2

1) 3:1	16) 4:5
2) 1:2	17) 11:4
3) 1:25	18) 3:5
4) 3:1	19) 5:1
5) 100:1	20) 6:5
6) 4:1	21) 4:1
7) 1:5	22) 10:9
8) 1:3	23) 3:2
9) 3:1	24) 6:1
10) 3:4	25) 5:1
11) 1:10	26) 2:1
12) 5:2	27) 1:10
13) 1:2	28) 8:3
14) 1:10	29) 10:1
15) 2:1	30) 10:1

Exercise 3

1) 9:4	9) 7:3
2) 2:1	10) 1:2
3) 4:3	11) 4:1
4) 2:1	12) 2:1
5) 9:1	13) 11:6
6) 6:1	14) 13:9
7) 9:2	15) 1:10
8) 3:1	16) 16:15

17) (a) 6 oz butter
       5 oz caster sugar
       2 eggs separated
       12 oz SR flour
       2 pinches salt
       3 oz currants
       1 oz mixed peel
    (b) 18 oz butter
       15 oz caster sugar
       6 eggs separated
       2 lb 4 oz SR flour
       6 pinches salt
       9 oz currants
       3 oz mixed peel

(c) $1\frac{1}{2}$ oz butter

$1\frac{1}{4}$ oz caster sugar

$\frac{1}{2}$ egg separated

3 oz SR flour

small pinch salt

$\frac{3}{4}$ oz currants

$\frac{1}{4}$ oz mixed peel

18) (a) 4 onions

8 carrots

8 sticks celery

1 kg cooked beef

four 340 g cans oxtail soup

salt and pepper

16 tsp curry powder

200 g raisins

(b) 10 onions

20 carrots

20 sticks celery

2.5 kg cooked beef

ten 340 g cans oxtail soup

salt and pepper

40 tsp curry powder

500 g raisins

(c) $\frac{1}{2}$ onion

1 carrot

1 stick celery

125 g cooked beef

half a 340 g can oxtail soup

salt and pepper

2 tsp curry powder

25 g raisins

Exercise 4

1) (a) 1.5,   (b) 3,
   ratio 1:2

2) (a) 3,   (b) 2,
   ratio 3:2

3) (a) 1.5,   (b) 3.5,
   ratio 3:7

4) (a) 2,   (b) 2.5,
   ratio 4:5

5) (a) 1,   (b) 2,
   ratio 1:2

6) (a) 1.75,   (b) 2,
   ratio 7:8

7) (a) 2.5,   (b) 1.75,
   ratio 10:7

8) (a) 3,   (b) 1.75,
   ratio 12:7

9) (a) 2.75,   (b) 1.5,
   ratio 11:6

10) (a) 1.25   (b) 2.75,
   ratio 5:11

Exercise 5

1) (a) $\frac{1}{5}$ (b) 20%
2) (a) $\frac{1}{2}$ (b) 50%
3) (a) $\frac{2}{3}$ (b) $66\frac{2}{3}$%
4) (a) $\frac{7}{10}$ (b) 70%
5) (a) $\frac{4}{5}$ (b) 80%
6) (a) $\frac{5}{8}$ (b) $62\frac{1}{2}$%
7) (a) $\frac{3}{20}$ (b) 15%
8) (a) $\frac{1}{3}$ (b) $33\frac{1}{3}$%
9) (a) $\frac{1}{8}$ (b) $12\frac{1}{2}$%
10) (a) $\frac{2}{3}$ (b) $66\frac{2}{3}$%
11) (a) $\frac{11}{20}$ (b) 55%
12) (a) $\frac{1}{20}$ (b) 5%
13) $3:5;\frac{3}{5};60\%$
14) $16:40;2:5;2:7;28\frac{4}{7}\%$
15) 40%; £1.20; £1.80

Exercise 6

1) £10:£5
2) £12:£4
3) £25:£100
4) £15:£5
5) 180 g:120 g
6) 1000 g:2000 g
7) 250 g:150 g
8) £20:£15
9) 45 min:15 min
10) 20 min:25 min
11) 5 km:1 km
12) 28 p:14 p
13) 10 p:40 p
14) 25 g:15 g
15) £200 and £300
16) 200 g copper, 300 g tin and 200 g nickel
17) 60 cm and 15 cm

Exercise 7

1) 18 p
2) 70 p
3) 24 p
4) £6.24
5) 48 p
6) 40 p
7) £1.08
8) £750
9) £3.75
10) 99 p
11) £4.20
12) 36 p
13) £1.05
14) £3.30

15) (a) 120 miles   (b) 30 miles
16) 20 kg
17) (a) 24   (b) 48
   (c) 144
18) (a) £1.50   (b) £15
19) 4 kg
20) £1.12
21)

50 kg £1.00	8 kg £9.60
100 kg £2.00	4 kg £4.80
10 kg 20 p	24 kg 28.80
20 kg 40 p	1 kg £1.20

7 kg £8.40	15% £3.00
1 kg £1.20	5% £1.00
$3\frac{1}{2}$ kg £4.20	1% 20 p
14 kg £16.80	30% £6.00

5 kg £5.15	20% £1500
1 kg £1.03	10% £750
20 kg £20.60	1% £75
40 kg £41.20	$\frac{1}{2}$% £37.50

15 kg £3.00	50% 250 g
5 kg £1.00	100% 500 g
1 kg 20 p	10% 50 g
30 kg £6.00	1% 5 g

$3\frac{1}{2}$ kg £7.00	100% 80 g
7 kg £14.00	10% 8 g
1 kg £2.00	5% 4 g
21 kg £42.00	$2\frac{1}{2}$% 2 g

Exercise 8

1) 36 days   6) 15 min
2) 3 days   7) 3 h 36 min
3) 5 sweets   8) 4 sweets
4) 20 min   9) 4 h
5) 10 min

Exercise 9

1) 50 g @ £1.00
2) 150 mℓ @ 42 p
3) 300 g @ £3.18
4) 850 g @ £1.02
5) yes
6) 20 p
7) 62 p; I save 10 p

Exercise 10

1) (a) 2 cm   (b) 150 cm
2) 2 km, 5 km, 2.5 km

3) (a) 216 cm (or 2.16 m)
   (b) 36 cm
   (c) 1440 cm (or 14.4 m)
   (d) 1800 cm (or 18 m)
4) (a) 10 km   (b) 2.5 km
   (c) 25 km   (d) 55 km
   (e) 2 cm   (f) 6 cm
   (g) 0.5 cm   (h) 1.5 cm

Exercise 11

1) 16 French francs
2) 6 German marks
3) 18 000 lire
4) 1600 pesetas
5) 30 Swiss francs
6) £10
7) 50 p
8) £4
9) £2.50
10) £5
11) £20
12) £25
13) £12.50

Exercise 12

1) (a) 1:5   (b) 5:4
   (c) 3:4   (d) 3:1
2) (a) 3:5   (b) 5:1
   (c) 3:10   (d) 2:1
3) (a) 10:1   (b) 1:5
4) (a) 9:1   (b) 2:3
5) (a) 8 egg whites
       400 g caster sugar
   (b) 12 egg whites
       600 g caster sugar
   (c) 2 egg whites
       100 g caster sugar
6) $2\frac{1}{4}$ cm : $1\frac{1}{2}$ cm, 3:2
7) £30:£20
8) 10:6
9) 120 g:30 g
10) $\frac{1}{2}$ h : $1\frac{1}{2}$ h
11) £10:£20:£10
12) 24 p:28 p
13) 42 p
14) 35 p
15) 24 p
16) 1 h 36 min
17) 35 g for 70 p
18) 10 g for 20 p
19) 40 marks
20) £24
21) 28 km
22) 4 m

CHAPTER 10

Exercise 1

1) 4   18) 2.80
2) 6   19) 7.3
3) 2   20) 5.8
4) 5   21) $2\frac{1}{2}$
5) 5   22) $3\frac{1}{3}$
6) 3   23) $2\frac{5}{6}$
7) 3   24) $1\frac{3}{5}$
8) 12   25) $\frac{1}{4}$
9) 14   26) $3\frac{3}{4}$
10) 3   27) $\frac{5}{12}$
11) 23   28) $3\frac{1}{7}$
12) 67   29) 3.14
13) 0.5   30) 1.78
14) 1.6   31) 4.333
15) 0.6   32) 0.106
16) 1.2   33) 1.4
17) 86   34) 6.0

Exercise 2

1) 2
2) 58 runs
3) 93 g
4) 52
5) £3.58
6) 11 years 2 months
7) 12 cm
8) (a) 1200 g   (b) 240 g
9) 12.6
10) 10

Exercise 3

1) 200 runs   5) 32
2) 34 years   6) 1012 g
3) £21.70   7) 800 g
4) 78   8) £2.50

Exercise 4

1) 34, 39
2) 55 years 10 months,
   11 years 8 months
3) £98, £104
4) 17
5) 45 g
6) 130 m
7) 13, 78, 20
8) 4
9) 90 g

Exercise 5

1) 18 p   5) 29 p
2) 29 p   6) £1.70
3) 28 p   7) 77 p
4) 96 p   8) 72 p

Exercise 6

1) 3   4) 10.4
2) 23   5) 36.1
3) 4   6) 62

Exercise 7

1) (a) £700   (b) £70
2) (a) 73 g   (b) 7.3 g
3) 290 cm
4) (a) 19 goals   (b) 1.9 goals
5) 64 g
6) 24 p
7) 25.1 cm
8) 545 g
9) 6.66
10) £17 181, 200 people, £86

Exercise 8

1) 4 h   19) 150 km
2) 3 h   20) 130 km
3) 5 h   21) 2 mph
4) 3 h   22) 3 mph
5) 3 h   23) 5 mph
6) 3 h   24) 2 mph
7) 4 h   25) 2 mph
8) 4 h   26) 4 mph
9) 3 h   27) 7 mph
10) 2 h 30 min 28) 30 mph
11) 2 h 24 min 29) 40 mph
12) 2 h 30 min 30) 50 mph
13) 2 h 15 min 31) 3 km/h
14) 8 miles   32) 9 km/h
15) 10 miles   33) 50 km/h
16) 24 miles   34) 2.5 km/h
17) 60 miles   35) 12.5 km/h
18) 75 miles

Exercise 9

1) 80 miles   6) 40 mph
2) 4 h   7) 32 mph
3) 30 mph   8) $1\frac{1}{2}$ h
4) 105 miles   9) $4\frac{1}{2}$ h
5) 100 miles   10) 60 mph

Exercise 10

1) 48 mph
2) $33\frac{1}{3}$ mph
3) (a) 40 mph   (b) 60 mph
    (c) 48 mph
4) 400 km/h   (a) 5400 km
    (b) 8 h   (c) 675 km/h
5) (a) 5 h   (b) 140 miles
    (c) 28 mph
6) 2800 km, 560 km/h
7) 36 mph

Exercise 11

1) 8 km/h
2) 16 km/h
3) 32 km/h
4) 40 km/h
5) 64 km/h
6) 80 km/h
7) 88 km/h
8) 104 km/h
9) 112 km/h
10) 128 km/h
11) 19.2 km/h
12) 41.6 km/h
13) 51.2 km/h
14) 76.8 km/h
15) 86.4 km/h
16) 37.6 km/h
17) 65.92 km/h
18) 87.2 km/h
19) 192 km/h
20) 1600 km/h
21)

mph	15	30	45	60	75	90
km/h	24	48	72	96	120	144

Exercise 12

1) mean 3.5, median 3, mode 3
2) mean 13, median 13, mode 13
3) mean 5, median 3, mode 1
4) mean 16.8, median 17, mode 18
5) mean 36.3, median 34.5, mode 34
6) median 5, mode 5
7) mean 4.6̇, median 5, mode 5
8) mean 9.5, median 11
9) mean 2.4, median 2, mode 2
10) mean 91.6̇ g, median 92 g,
    mode 92 g

Exercise 13

1) 9   4) 7
2) 9   5) 7
3) 8   6) 6

Exercise 14

1) 3
2) 16.25 cm
3) 28
4) 2048 g
5) 51 runs

6) (a) 140 miles   (b) 4 h
    (c) 35 mph
7) 1 h
8) 35p
9) 95 p
10) (a) 45   (b) 42
    (c) 45   (d) 11

11) 7, 1
12) 175 miles, 325 miles, 6 hours,
    $54\frac{1}{6}$ mph
13) 0
14) 5
15) (a) 4.2   (b) 8

CHAPTER 11

Exercise 1

1) Village Fete

Income

15 Aug	Donation	£75
24 Aug	Ticket sale	124
24 Aug	Raffle	11
24 Aug	Stall sales	50
		£260

Expenses

1 Aug	Lorry hire	£30
3 Aug	Field hire	35
3 Aug	Stall hire	15
	Balance c/d	180
		£260

2) Sports Club Gymnastic Display

Income

11 May	Ticket sale	£40
17 May	Ticket sale	50
20 May	Programme sales	14
20 May	Ticket sale	50
		£154

Expenses

3 May	Hall hire	£50
7 May	Ticket printing	3.75
15 May	Equipment hire	10
	Balance c/d	90.25
		£154

3) Annual Works Outing

Income

5 Jul	Receipts from workers	£140
10 Jul	Subsidy	35
		£175

Expenses

10 Jul	Coach hire	£75
15 Jul	Teas	72
15 Jul	Entertainment	25
	Balance c/d	3
		£175

4) School Sale of Work

Income

17 Jan	Raffle	£15
17 Jan	Ticket sale	12
17 Jan	Good-as-new stall profit	56
17 Jan	Other stalls profit	75
		£158

Expenses

10 Jan	Stall hire	£15
15 Jan	Decorations	7
	Balance c/d	136
		£158

Exercise 2

1)

Receipts	Date	Particulars	Payments	Postage	Stationery	Travel expenses	Office expenses	Cleaning
40.00	5 Jan	Cash in hand						
	6 Jan	Postage	3.10	3.10				
	6 Jan	Window cleaner	3.50					3.50
	6 Jan	String	0.65		0.65			
	7 Jan	Ballpoint pens	0.35		0.35			
	7 Jan	Rail fares	4.50			4.50		
	8 Jan	Cleaners' wage	10.00					10.00
	8 Jan	Taxi fares	4.50			4.50		
	9 Jan	Office coffee	0.95				0.95	
	9 Jan	Telegram	2.10	2.10				
	9 Jan	Envelopes	0.65		0.65			
		Totals	30.30	5.20	1.65	9.00	0.95	13.50
30.30	9 Jan	Reimbursement						
		Balance c/d	40.00					
70.30			70.30					

2)

Receipts	Date	Particulars	Payments	Postage	Stationery	Travel expenses	Office expenses	Cleaning
30.00	4 May	Cash in hand						
	4 May	Envelopes	0.60		0.60			
	4 May	Office tea	0.70				0.70	
	5 May	Pencils	0.35		0.35			
	5 May	Window cleaner	2.50					2.50
	5 May	Taxi fare	1.75			1.75		
	5 May	Postage	1.35	1.35				
	6 May	Typing ribbon	1.10		1.10			
	6 May	Cleaners' wage	15.00					15.00
	6 May	Bus fares	1.45			1.45		
	7 May	Parcel post	1.35	1.35				
	7 May	Office coffee	0.92				0.92	
	8 May	Photocopying	1.04				1.04	
	8 May	Ballpoint pens	0.55		0.55			
		Totals	28.66	2.70	2.60	3.20	2.66	17.50
28.66	8 May	Reimbursement						
		Balance c/d	30.00					
58.66			58.66					

Exercise 3

1)		2)	
	50.00		2.40
	7.50		6.00
	15.00		3.00
	10.00		1.00
	82.50		12.40

Exercise 4

1)			2)		
		8.00			105.00
		40.00			200.00
		15.00			47.50
	Total gross	63.00			360.00
	Less 10%	6.30		Total gross	712.50
		56.70		Less 10%	71.25
					641.25

3)

R. Evans		Tel:	
101 Dark St		Bevington 3055	
BEVINGTON	INVOICE		
Quantity	Description	Unit Cost	Gross
33	Jars coffee	0.50	16.50
20	Tins soup	0.15	3.00
10	Boxes biscuits	2.00	20.00
50	Tins fruit	0.20	10.00
40	Pkts sugar	0.30	12.00
		Total gross	61.50
		Less 10% discount	6.15
VAT zero rated			55.35

4)

J. Saunders		Tel:	
45 James St		Oxminster 1139	
OXMINSTER	INVOICE		
Quantity	Description	Unit cost	Gross
10	Maths books	2.10	21.00
30	French books	3.00	90.00
15	English books	2.00	30.00
10	Geography books	4.35	43.50
25	History books	2.50	62.50
		Total gross	247.00
		Less 10% discount	24.70
VAT zero rated			222.30

CHAPTER 12

Exercise 1

1) £240
2) £5
3) (a) £152   (b) £171
   (c) £233.70   (d) £275.50
4) (a) £8   (b) £4
   (c) £4.50   (d) £5
5) £148
6) (a) £187.50   (b) £206.25
   (c) £150   (d) £168.75

Exercise 2

1) (a) £6   (b) £10.50
   (c) £9.30   (d) £5.40
2) (a) £8   (b) £12
   (c) £4.40   (d) £5.60
3) (a) £8.40   (b) £5.25
4) (a) £5   (b) £10
   (c) £2.50   (d) £3.75
5) (a) £11.34   (b) £8.97
   (c) £5.69   (d) £6.86

Exercise 3

1) £19	4) £5.20
2) £28	5) £18.50
3) £11	6) £19

Exercise 4

1) (a) £2   (b) £4   (c) £5
   (d) £6   (e) £11.50
2) £60
3) £12
4) £20
5) £56
6) £8

Exercise 5

1) £716.67	5) £1000
2) £750	6) £1333.33
3) £791.67	7) £1500
4) £833.33	8) £2000

Exercise 6

1) £4890, £407.50
2) £350
3) £11.76
4) £60.25
5) £598.72
6) £390
7) £1120
8) £71.02
9) £355.86
10) £320, £16

Exercise 7

1) £215.20
2) £4
3) (a) £9.90
   (b) £8.80
4) £280.60
5) £6
6) £1.10
7) £12
8) £7
9) £1600
10) £77.62

CHAPTER 13

Exercise 1

1) (a) A
   (b) C
   (c) H
2) $\frac{8}{9}$
3) $\frac{7}{9}$
4) (a) £540
   (b) £720
   (c) £1170
5) £957
6) £600
7) £600
8) (a) £522
   (b) £391.50
9) £66
10) £1.75

Exercise 2

£3750	£33 750	£2916
£4250	£38 250	£3324
£5300	£47 700	£4128
£5500	£49 500	£4296
£6000	£54 000	£4680
£7550	£67 950	£5856
£10 300	£92 700	£8076
£12 500	£112 500	£9744

Exercise 3

1) £4.50
2) £12
3) £80
4) £96
5) (a) £2   (b) £20   (c) £8
6) (a) £1.50   (b) £10.50

Exercise 4

1) (a) £1256   (b) £1253
   (c) £1251   (d) £1246
2) (a) £11 283   (b) £11 283
   (c) £11 270   (d) £11 251
3) (a) £1255   (b) £2129
4) £1200
5) (a) £6624   (b) £11 241
6) £6000

Exercise 5

1) £400
2) £1800, £360
3) £4360, £3640, £728,
4) £3820, £1780, £356, £4908
5) £604, £6431

Exercise 6

1) (a) £780
   (b) £1200
   (c) £1980
2) (a) £780
   (b) £1440
   (c) £2220
3) (a) £780
   (b) £1680
   (c) £2460
4) (a) £780
   (b) £1920
   (c) £2700
5) (a) £780
   (b) £2088
   (c) £2868
6) (a) £780
   (b) £2328
   (c) £3108
7) (a) £780
   (b) £2808
   (c) £3588
8) (a) £780
   (b) £3840
   (c) £4620

Exercise 7

1) £51
2) £60
3) £42750
4) £3840
5) £11.25
6) £100
7) £1800
8) £630
9) £5580

CHAPTER 14

Exercise 1

1) (a) £56.28   (b) 92
   (c) 2913
2) 2800 p   £28
   7250 p   £72.50
   4431 p   £44.31
   7450 p   £74.50
   5101 p   £51.01
3) (a)

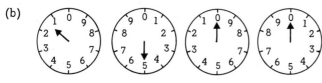

   (b)

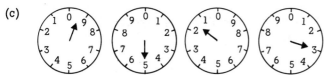

   (c)

4) (a) 2502   (b) 3308
   (c) 2000
5) 1212

Exercise 2

1) £35.66
2) 317
3) 7.22 p
4) £10.13
5) £22.89
6) £10.13 + £22.89 + £2.64 =
   £35.66
7) £72.20, £82.33, £6.59, £88.92
8) 11 March 1994

Exercise 3

1) (a) £40.04, £7.01, £47.05
   (b) £144.03, £25.21, £169.24
2) £30
3) £21
4) £163
5) £39

Exercise 4

1) 41.0 p
2) 141.6 p
3) 167.1 p
4) 56.6 p
5) 116.9 p
6) 127.3 p
7) 12 minutes
8) 10 minutes
9) 424.8 p
10) (a) 29.5 p   (b) 82.1 p
    (c) 198.2 p   (d) 309 p

Exercise 5

1) £2   6) £312
2) £1   7) £7
3) £179.88   8) £364
4) £8.91   9) £869.78
5) £6

Exercise 6

1) £2152
2) (a) £950   (b) £480
   (c) £1960
3) £9

4) 5 p
5) £500
6) 40
7) £16.38
8) £624
9) 200 miles, 12 gallons, £27.60;
   240 miles, 10 gallons, £23;
   £4.60

Exercise 7

1) (a) £4.80   (b) £32
   (c) £48.96   (d) 9.6 p
2) £129.60
3) £1.60
4) £58.75
5) £5.41
6) £4.23
7) £42
8) (a) £5.25   (b) £52.50
   (c) £525
9) £94
10) £258.50

Exercise 8

1) £60   4) £100
2) £40   5) £80
3) £2   6) £200

Exercise 9

1) £144
2) (a) £100   (b) £10
   (c) £11   (d) £130
3) (a) £260   (b) £60
4) £10
5) £60

Exercise 10

1) £45
2) (a) £25.13, £2.01, £27.14
   (b) £57.27, £4.58, £61.85
3) (a) 2010   (b) 9500
4) £64, £75.20
5) £13
6) £83.05, £14.53, £97.58
7) £2120
8) £500
9) £12.10
10) £40, £540
11) £2000
12) (a) £100   (b) £600
    (c) £60

CHAPTER 15

Exercise 1

1) A(1, 2)   B(3, 4)   C(3, 1)
   D(4, 3)   E(5, 4)   F(5, 0)
2) A(10, 5)   B(30, 10)   C(40, 25)
   D(50, 40)   E(70, 45)
   F(80, 45)   G(90, 35)
   H(100, 20)   I(110, 10)
   J(120, 5)

Exercise 2

1) (a) 108 cm   (b) 9 years old
2) (a) 38.5 miles per gallon
   (b) 15 mph, 78 mph
   (c) 50 mph
3) (a) 68°F   (b) 37°C
4) (a) 25.6 inches   (b) 95 cm
5) (a) 10 oz   (b) 350 g
6) (a) 50 mph   (b) $7\frac{1}{2}$ min
   (c) $12\frac{1}{2}$ min   (d) 10 min
7) (a) 20 years   (b) 200 cm
8) (a) 120 miles   (b) 1 hour
   (c) 60 mph   (d) 30 mph

Exercise 3

1) £10 per year
2) 40 metres/second per second
3) $\frac{1}{2}$
4) 20 mph

Exercise 4

1) 60 p   5) 60 p
2) 80 p   6) 25 p
3) 60 p   7) 16 p
4) 64 p

Exercise 5

1) 5 p
2) 39 p
3) 19 p
4) 87 p
5) 20 p
6) evenings and night-time
7) 32 p
8) 65 p
9) 39 p
10) 35 min
11) 10 min
12) daytime
13) 5 p
14) £19.70

Exercise 6

1) (a) women
   (b) (i) £180 (ii) £120 (iii) £60
2) 32p, 32p, 16p, 48p
3)

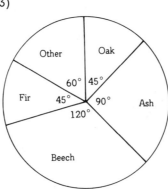

Exercise 7

1)

Scale: 1 cm to 20 people

4)

2)

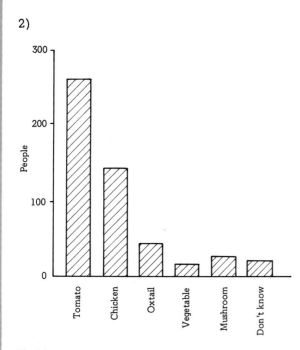

3) (a) type A   (b) type B
    (c) 11 000   (d) 8000
    (e) yes

Exercise 8

1)

Measurement	Tally Marks	Frequency
35.92	|||	3
35.93	ⅢⅠⅠ ||	7
35.94	ⅢⅠⅠ ⅢⅠⅠ		11
35.95	ⅢⅠⅠ |||	8
35.96	||||	4
35.97	||	2
35.98			1

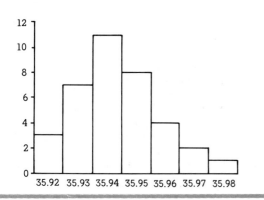

2)

Score	Tally marks	Frequency
1					3
2	ⅢⅠⅠ ||	7
3	ⅢⅠⅠ ⅢⅠⅠ		11
4	ⅢⅠⅠ ⅢⅠⅠ |||	13
5	ⅢⅠⅠ ⅢⅠⅠ ||	12
6	ⅢⅠⅠ |||	8
7	ⅢⅠⅠ ||	7
8	ⅢⅠⅠ	5
9						4
10				2

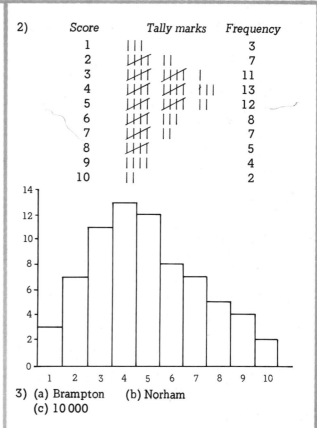

3) (a) Brampton   (b) Norham
    (c) 10 000

Exercise 9

1) 1500, 2000, 2250
2) (a) 135   (b) 120
    (c) 90
3) (a) 1988   (b) 35 000
    (c) 15 000   (d) 10 000, 25%

Exercise 10

1) (a) 5 kg   (b) 16 months
2) (a) 19 km
    (b) 19 miles; 1.6 km/mile
3) 6 km per hour
4) (a) 51 p   (b) 30 p
    (c) 51 p
5)

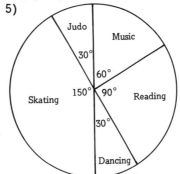

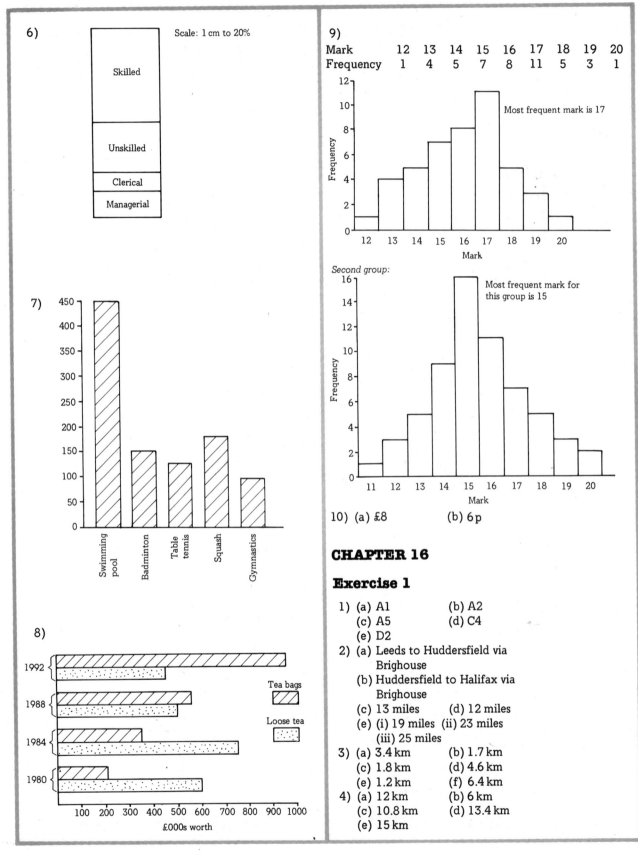

6) Scale: 1 cm to 20%

Skilled

Unskilled

Clerical

Managerial

7)

8)

1992
1988
1984
1980

Tea bags

Loose tea

£000s worth

9)

Mark	12	13	14	15	16	17	18	19	20
Frequency	1	4	5	7	8	11	5	3	1

Most frequent mark is 17

Mark

Second group:

Most frequent mark for this group is 15

Mark

10) (a) £8   (b) 6p

CHAPTER 16

Exercise 1

1) (a) A1   (b) A2
   (c) A5   (d) C4
   (e) D2
2) (a) Leeds to Huddersfield via Brighouse
   (b) Huddersfield to Halifax via Brighouse
   (c) 13 miles   (d) 12 miles
   (e) (i) 19 miles (ii) 23 miles
       (iii) 25 miles
3) (a) 3.4 km   (b) 1.7 km
   (c) 1.8 km   (d) 4.6 km
   (e) 1.2 km   (f) 6.4 km
4) (a) 12 km   (b) 6 km
   (c) 10.8 km   (d) 13.4 km
   (e) 15 km

Exercise 2

1) 49 minutes
2) 1010
3) 10 minutes
4) 0912
5) 421
6) There is a railway connection at Stroud
7) 438, 426
8) yes
9) Sapperton, Daglingworth
10) 1217

Exercise 3

1) 9.15 a.m.
2) $1\frac{1}{2}$ h
3) 7.58 a.m., 2 min
4) 42 min
5) 1107
6) 1431 from Kemble (1317 from Worcester)
7) 0840, 4 h 2 min
8) (a) 2022   (b) 1757
    (c) 1818; 2 h 52 min, 2 h 42 min, 2 h 50 min

Exercise 4

1) 91
2) Wednesday
3) 4
4) 20 February
5) 4
6) 2 September
7) 35
8) 25
9) 6 October
10) Sunday

Exercise 5

1) 1400
2) 1900
3) 1200
4) 0630
5) 0000 or 2400
6) 1625
7) 1100
8) 1415
9) 1805
10) 0745
11) 2325
12) 0200
13) 25
14) 90
15) 40
16) 111
17) 7.15 a.m.
18) 7.35 p.m.
19) 11.14 p.m.
20) 6.48 a.m.

Exercise 6

1) 0500
2) 0846
3) 0420
4) 1805
5) 2100
6) 0812
7) 1756
8) 1640
9) 0555
10) 2205

CHAPTER 17

Exercise 1

1) $x = 2$
2) $x = 3$
3) $x = 7$
4) $x = 7$
5) $x = 8$
6) $x = 7$
7) $x = 7$
8) $x = 3$
9) $x = 7$
10) $x = 10$
11) $x = 4$
12) $x = 3$
13) $x = 1$
14) $x = 12$
15) $x = 9$
16) $x = 14$
17) $x = 8$
18) $x = 2$
19) $x = 16$
20) $x = 13$

Exercise 2

1) $3z$
2) $2y$
3) $2t$
4) $6p$
5) $4q$
6) $3r$
7) $4n$
8) $5t$
9) $2m$
10) $3s$
11) $3p$
12) $3w$

Exercise 3

1) $x = 4$
2) $p = 3$
3) $w = 4$
4) $h = 5$
5) $y = 6$
6) $m = 4$
7) $z = 7$
8) $t = 8$
9) $z = 3$
10) $t = 12$
11) $y = 12$
12) $s = 7$

Exercise 4

1) $9x$
2) $8d$
3) $4c$
4) $16y$
5) $12h$
6) $7c$
7) $6z$
8) $6s$
9) $7t$
10) $8j$
11) will not simplify further
12) $3n$
13) $6p$
14) 0
15) $8e$
16) $5c + 6d$
17) $9m + 5n$
18) $15w + 5v$
19) $4z + 7q$
20) $5a + 3b$
21) $5a + 2b$
22) $7x + 9$
23) $5y + 9$
24) $11p + 6q$
25) $3a + 4b + 4c$
26) $8a + 3b + 1$
27) will not simplify further
28) $2c$
29) 6

Exercise 5

1) 14
2) 15
3) 14
4) 27
5) 6
6) 5
7) 29
8) 13
9) 74
10) 37
11) 5
12) 12
13) 14
14) 6
15) 13
16) 16
17) 9
18) 15
19) 136
20) 25

Exercise 6

1) b^3
2) m^4
3) n^2
4) p^3
5) s^2
6) q^6
7) r^3
8) w^5
9) z^3
10) 4
11) 16
12) 81
13) 121
14) 1000
15) 64
16) 20
17) 21
18) 202
19) 77

Exercise 7

1) pqr
2) $5st$
3) mn
4) abc
5) $8xyz$
6) $8pqr$
7) $14ab$
8) $12pqr$
9) $3pqr$
10) $11ab$
11) $3abc$
12) $2pq$

13) 15
14) 40
15) 24
16) 45
17) 200
18) 168
19) 120
20) 120
21) 120
22) 480
23) 480
24) 480
25) 26
26) 184
27) 124
28) 46
29) 288
30) 500
31) 264
32) 50
33) 49
34) 14
35) 20
36) 22
37) 7
38) 10
39) 13

Exercise 8

1) $x = 1$
2) $x = 3$
3) $x = 3$
4) $x = 3$
5) $x = 6$
6) $x = 2$
7) $x = 4$
8) $x = 8$
9) $x = 12$
10) $x = 8$
11) $x = 3$
12) $x = 7$

Exercise 9

1) $x = 2$
2) $x = 2$
3) $x = 8$
4) $x = 9$
5) $x = 5$
6) $x = 3$
7) $x = 4$
8) $x = 2$
9) $x = 3$
10) $x = 5$
11) $x = 7$
12) $x = 3$

Exercise 10

1) $x = 3$
2) $x = 6$
3) $x = 6$
4) $x = 14$
5) $x = 9$
6) $x = 4$
7) $x = 5$
8) $x = 7$
9) $x = 3$
10) $x = 25$
11) $x = 121$
12) $x = \frac{1}{4}$

Exercise 11

1) $x = 1$
2) $x = 2$
3) $x = 3$
4) $x = 1$
5) $x = 5$
6) $x = 1$
7) $x = 5$
8) $x = 4$
9) $x = 4$
10) $x = 12$
11) $x = 5$
12) $x = 5$
13) $x = 4$
14) $x = 6$
15) $x = 2$
16) $x = 3$
17) $x = 2$
18) $x = 3$
19) $x = 10$
20) $x = 1$

Exercise 12

1) $A = 63$
2) $C = 62.8$
3) $A = 314$
4) $v = 20$
5) $f = 4$
6) $s = 88$
7) $s = 3$
8) $V = 628$
9) $I = 100$
10) 242

Exercise 13

1) (a) gradient 2
   (b) y-intercept is 7
2) (a) gradient 3
   (b) $y = 5$ when $x = 0$
3) (a) gradient 2
   (b) y-intercept is 3
   (c) $y = 2x + 3$
4) (a) gradient 3
   (b) $y = -7$ when $x = 0$
5) (a) gradient 1
   (b) line cuts y-axis at $y = 6$
   (c) equation of line $y = x + 6$

Exercise 14

1) 20, 30
2) 4, 12
3) 4, 8, 2
4) 4, 20, 96, 1.5
5) 1, 2, 3, 4, 5
6) 3, 6, 9, 12, 15, 18, 21, 24, 27, 30
7) 1, 8, 27, 64, 125, 216, 343, 512
8) HELLO, HELLO, HELLO, HELLO, HELLO, HELLO
9) 99, 98, 97, 96, 95
10) 3, 5, 7, 9, 11, 13, 15, 17, 19, 21
11) 2, 5, 8, 11, 14, 17, 20, 23
12) 0.5, 1, 1.5, 2, 2.5, 3, 3.5, 4, 4.5, 5

CHAPTER 18

Exercise 1

1) (a) 3 (b) 6
   (c) 4 (d) 5
   (e) 4 (f) 8
   (g) 4
2) (a) 12 cm (b) 15.8 cm
   (c) 22 mm (d) $7\frac{1}{2}$ cm
3) (a) 6 (b) 5
   (c) 5 (d) 8
   (e) 4

4) (a) 8 (b) 12
   (c) 9 (d) 6
5) (a) 8 (b) 5
   (c) 6 (d) 16
6) Examples:
   (a)
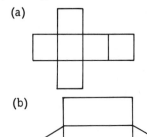
   (b)
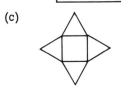
   (c)

7) (b), (f) and (g) would, the rest would not

Exercise 2

Acute	Obtuse	Reflex
17°	173°	190°
65°	135°	240°
54°	155°	230°
72°	169°	185°
85°	120°	270°
32°	110°	315°
88°	95°	198°
12°	103°	350°
51°	114°	210°
63°		

Exercise 3

1) 60° 4) 48°
2) 130° 5) 143°
3) 82° 6) 90°

Exercise 4

a: BÂC or CÂB
b: AB̂C or CB̂A
c: AĈB or BĈA
d: GD̂E or ED̂G
e: DÊF or FÊD
f: EF̂G or GF̂E
g: DĜF or FĜD
p: QP̂R or RP̂Q
q: PQ̂R or RQ̂P
r: PR̂Q or QR̂P

Exercise 5

1)

CÂB	60°
AB̂C	60°
BĈA	60°
Sum	180°

2)

CÂB	130°
AB̂C	20°
BĈA	30°
Sum	180°

3)

CÂB	45°
AB̂C	90°
BĈA	45°
Sum	180°

4)

CÂB	70°
AB̂C	50°
BĈA	60°
Sum	180°

5)

CÂB	50°
AB̂C	50°
BĈA	80°
Sum	180°

6)

CÂB	60°
AB̂C	80°
BĈA	40°
Sum	180°

7)

CÂB	125°
AB̂C	30°
BĈA	25°
Sum	180°

Exercise 6

1) $x = 75$ (alternate angles)
2) $a = 68$ (alternate angles)
   $b = 112$ (angles on a straight line)
3) $x = 110$ (corresponding angles)
   $y = 70$ (angles on a straight line)
   $z = 110$ (vertically opposite angles or angles on a straight line)
4) $x = 50$ (angle sum of a triangle)
5) $x = 48$ (angle sum of a triangle)
   $y = 132$ (angles on a straight line)
6) $a = 53$ (angles on a straight line)
   $b = 58$ (alternate angles)
   $c = 69$ (alternate angles or angle sum of a triangle)
7) $x = 122$ (angles on a straight line)
8) $x = 92$ (angles on a straight line)
   $y = 40$ (angles on a straight line)
9) $x = 28$ (angles on a straight line)
   $y = 152$ (vertically opposite angles)
10) $x = 57$ (angles on a straight line)
    $y = 57$ (alternate angles)
    $z = 57$ (vertically opposite angles)
    $w = 62$ (angles on a straight line)
    $t = 61$ (alternate angles)
11) $x = 90$ (angles on a straight line)
    $w = 36$ (angle sum of a triangle)
    $t = 126$) (angles on a straight line)
12) $y = 115$ (angles at a point add up to 360°)
13) $w = 62$ (alternate angles)
    $x = 46$ (angle sum of a triangle)
    $y = 72$ (angles on a straight line or alternate angles)
    $z = 108$ (angles on a straight line)

Exercise 7

1) 5 cm
2) 13 mm
3) (a) 8.6 cm   (b) 8.94 cm
   (c) 9 mm   (d) 8 cm
4) 26 cm
5) 7.07 cm
6) 2.24 cm

Exercise 8

1) 3 cm   5) 5.74 mm
2) 1.73 cm   6) 7.48 cm
3) 8 cm   7) 9.22 cm
4) 5.20 m   8) 6.93 cm

Exercise 9

1) 6.28 cm   6) 44 m
2) 15.7 cm   7) 62.8 mm
3) 22 mm   8) 22 cm
4) 66 cm   9) 220 cm
5) 18.8 cm   10) 12.6 m

Exercise 10

2) 5.2 cm, 20.8 cm
3) Perimeter = 40.5 cm
4) (a) and (d) will fold to make a cuboid

Exercise 11

1) (a) 225°   (b) 315°
2) (a)

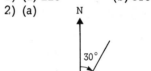

(b)

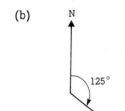

(c)

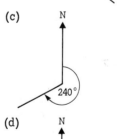

(d)

287

3) (a) 53°   (b) 114°
   (c) 208°
4) Distance AQ = 32 km, bearing
                    061° from A
5) (a) 10.5 km   (b) 054°
   (c) 19 km   (d) 117°

Exercise 12

1) square
2) equilateral triangle
3) two straight lines
4) regular pentagon
5) letter T
6) (a) diamond or rhombus
   (b) FD 100   LT 330
        FD 100   LT 60
        FD 100   LT 120
        FD 100   LT 60
        FD 100
   (c) as above with LT 330
       replaced by RT 30

CHAPTER 19

Exercise 1

1) 10 cm^2
2) 18 cm^2
3) 56 cm^2
4) 60 mm^2
5) 270 ft^2
6) 165 cm^2
7) 28 cm^2
8) 17 cm^2
9) 8 m^2
10) 7.5 ft^2
11) 27.5 cm^2
12) 9 cm^2
13) 22.5 cm^2
14) 59.5 yd^2
15) 33.25 in^2
16) 80 mm^2
17) 1.5 m^2
18) 14 cm^2
19) 10.35 cm^2
20) 116.1 cm^2
21) 7 m^2
22) 225 cm^2
23) 28 sq. ft
24) 6 m^2, £13.50
25) £9
26) 1.5 m^2
27) 150, £25

Exercise 2

1) 25 cm^2 7) 93 cm^2
2) 103 cm^2 8) 52 cm^2
3) 72 cm^2 9) 36 cm^2
4) 36 cm^2 10) 39 cm^2
5) 42 cm^2 11) 49 cm^2
6) 64 cm^2 12) 52 cm^2

Exercise 3

1) 14 cm^2
2) 12 cm^2
3) 18 cm^2
4) 20 cm^2
5) 18 cm^2
6) 24 cm^2
7) 66 cm^2
8) (a) 108 cm^2 (b) 70 m^2
   (c) 38 m^2 (d) 76 p
9) (a) 3 m^2 (b) 1.5 m^2
   (c) 1.5 m^2 (d) £3

Exercise 4

1) 28.26 cm^2
2) 78.5 cm^2
3) 314 cm^2
4) 615.4 cm^2 (or 616 cm^2 if $\pi = \frac{22}{7}$
   is used)
5) 38.5 cm^2 (or $38\frac{1}{2} \text{ cm}^2$)
6) 9.62 cm^2 (or $9\frac{5}{8} \text{ cm}^2$)

Exercise 5

1) 113 cm^2
2) 201 mm^2
3) 1260 m^2
4) 1.54 m^2
5) 11.6 cm^2
6) 11.1 cm^2
7) 164 or 165 cm^2
8) 26.3 cm^2
9) 98.5 cm^2
10) 12.3 cm^2
11) (a) 100 cm^2 (b) 78.5 cm^2
   (c) 21.5 cm^2
12) (a) 0.25 cm^2 (b) 0.196 cm^2
   (c) 0.054 cm^2
13) (a) 16 cm^2 (b) 12.6 cm^2
   (c) 3.4 cm^2
14) (a) 196 cm^2 (b) 154 cm^2
   (c) 42 cm^2
15) (a) 64 cm^2 (b) 12.6 cm^2
   (c) 51.4 cm^2

16) (a) 49 cm^2 (b) 19.2 cm^2
   (c) 29.8 cm^2

Exercise 6

1) 30 cm^2 11) 14 cm^2
2) 15 cm^2 12) 52.8 cm^2
3) $17\frac{1}{2} \text{ cm}^2$ 13) 56 mm^2
4) 21 cm^2 14) 6.4 m^2
5) 36 cm^2 15) 104 mm^2
6) 30 m^2 16) 37.5 cm^2
7) 21.5 cm^2 17) 60 m^2
8) 28 cm^2 18) 3 cm^2
9) $4\frac{1}{2} \text{ cm}^2$ 19) 225 mm^2
10) 66 cm^2 20) 184 mm^2

Exercise 7

1) 20 cm^2 6) 62.5 mm^2
2) 39 cm^2 7) 51.9 cm^2
3) 59.2 cm^2 8) 48 cm^2
4) 28.75 cm^2 9) 85 cm^2
5) 22 m^2 10) 48 m^2

Exercise 8

1) 8 cm
2) 6 m
3) (a) 6 m   (b) 8 mm
4) 4 mm
5) 2 cm
6) 14 cm
7) 5 mm
8) 6 cm
9) 24 mm
10) 14 cm

Exercise 9

1) (a) 15 cm^2 (b) 16 cm^2
   (c) 20 cm^2 (d) 133 cm^2
   (e) 9 cm^2 (f) 42.5 cm^2
2) 130 cm^2
3) (a) 24 cm^2 (b) 14 mm^2
   (c) 44 cm^2
4) (a) 28   (b) 60
   (c) 88
5) 7.5 m^2, 14 m^2, 6.5 m^2, 26 p
6) (a) 12.56 cm^2 (b) 3850 cm^2
7) (a) 72 cm^2 (b) 9 m^2
8) (a) 32.13 cm^2 (b) 48 cm^2
9) (a) 12.86 cm^2 (b) 67.5 cm^2
10) (a) 10 cm   (b) 4 mm
11) (a) 17 cm^2 (b) 22 cm^2

CHAPTER 20

Exercise 1

1) $40\,\text{cm}^3$
2) $210\,\text{yd}^3$
3) $72\,\text{in}^3$
4) $44\,\text{m}^3$
5) $13.5\,\text{cm}^3$
6) $300\,\text{cm}^3$
7) $9\,\text{m}^3$
8) $50\,\text{m}^3$
9) $288\,\text{mm}^3$
10) $360\,\text{cm}^3$
11) $1.75\,\text{cm}^3$ (or $1750\,\text{mm}^3$)
12) $910\,\text{cm}^3$
13) $9000\,\text{cm}^3$
14) $1200\,\text{cm}^3$
15) $2800\,\text{cm}^3$

Exercise 2

1) $360\,\text{m}^3$
2) $112\,\text{cm}^3$
3) $300\,\text{m}^3$
4) $306\,\text{cm}^3$
5) $360\,\text{cm}^3$
6) $117\,\text{cm}^3$
7) (a) $900\,\text{cm}^3$ (b) $576\,\text{cm}^3$
    (c) $324\,\text{cm}^3$

Exercise 3

1) $80\,\text{cm}^3$ 4) $30\,\text{cm}^3$
2) $240\,\text{cm}^3$ 5) $260\,\text{mm}^3$
3) $9.42\,\text{cm}^3$ 6) $1125\,\text{cm}^3$

---

7) $2700\,\text{m}^3$ 9) $125.6\,\text{cm}^3$
8) $1540\,\text{cm}^3$ 10) $480\,\text{cm}^3$

Exercise 4

1) $4\,\text{cm}^3$ 4) $7.5\,\text{m}^3$
2) $94.2\,\text{cm}^3$ 5) $60\,\text{cm}^3$
3) $83\frac{1}{3}\,\text{mm}^3$ 6) $157\,\text{cm}^3$

Exercise 5

1) (a) $80\,\text{cm}^3$ (b) $12\,\text{cm}^3$
    (c) $80\,\text{mm}^3$ (d) $13.5\,\text{m}^3$
    (e) $512\,\text{mm}^3$ (f) $220\,\text{m}^3$
2) (a) $104\,\text{m}^3$ (b) $92\,\text{m}^3$
3) (a) $225\,\text{cm}^3$ (b) $600\,\text{cm}^3$
4) $770\,\text{m}^3$
5) $1\,500\,000\,\text{m}^3$
6) $513\frac{1}{3}\,\text{cm}^3$
7) $33.5\,\text{cm}^3$
8) $550\,\text{cm}^3$
9) $5\,\text{cm}$
10) $3\,\text{m}$

CHAPTER 21

Exercise 1

1) 8 times heavier
2) 25
3) 64
4) 45 square feet
5) (a) 36   (b) 216
6) 8 times; no, since $8 \times 52\,\text{p}$
    $= £4.16$.

---

7) $10\,\text{cm}$
8) $10\,000\,\text{cm}^2 = 1\,\text{m}^2$
9) $100\,\text{mm}^2 = 1\,\text{cm}^2$
10) $1\,000\,000\,\text{cm}^3 = 1\,\text{m}^3$
11) The second diagram with £10 notes is 4 times larger in area, and the wad is 8 times larger in volume.

CHAPTER 22

Exercise 1

1) $\frac{1}{4}, \frac{1}{4}, \frac{1}{2}$
2) (a) $\frac{1}{150}$ (b) $\frac{1}{50}$
3) $\frac{1}{3}$
4) $\frac{1}{2}$ (b) $\frac{1}{3}$ (c) $\frac{5}{6}$
5) (a) $\frac{1}{5}$ (b) $\frac{1}{3}$ (c) $\frac{7}{15}$
6) (a) $\frac{1}{2}$ (b) $\frac{1}{4}$ (c) $\frac{1}{13}$
7) $\frac{3}{4}$
8) $\frac{17}{20}$
9) (a) $\frac{1}{9}$ (b) $\frac{5}{9}$ (c) $\frac{7}{9}$
10) $\frac{4}{7}$
11) (a) $\frac{5}{36}$ (b) $\frac{5}{18}$ (c) $\frac{7}{12}$
12) $\frac{1}{30}$
13) $\frac{1}{5}$; as tickets are taken out of the box, the proportion of winning tickets might not remain the same, especially near the end of the draw
14) (a) $\frac{2}{5}$ (b) $\frac{1}{5}$ (c) $\frac{2}{5}$
    (d) $\frac{4}{5}$
15) $\frac{5}{8}$
16) $\frac{3}{11}$

Exercise 2

1)

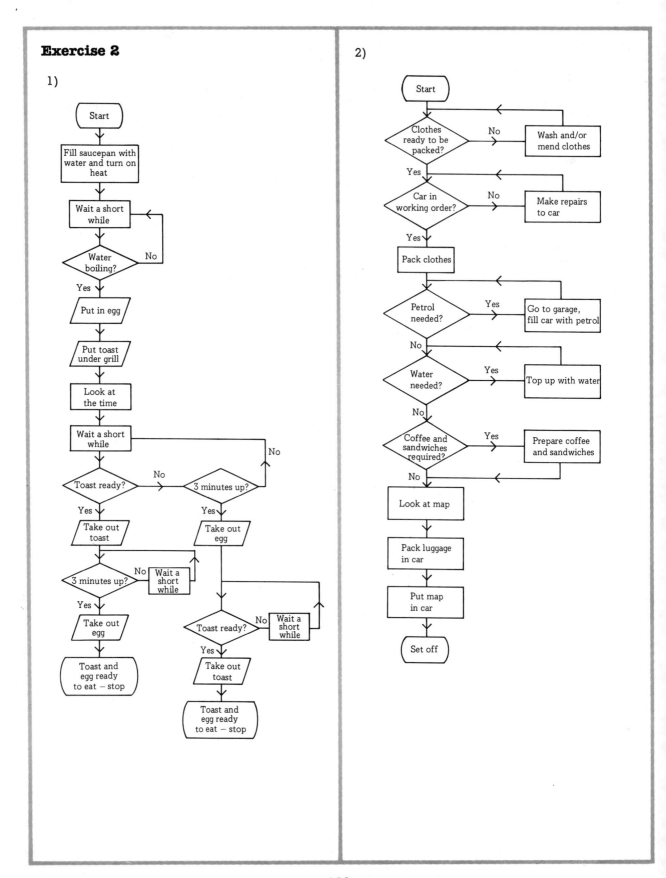

2)

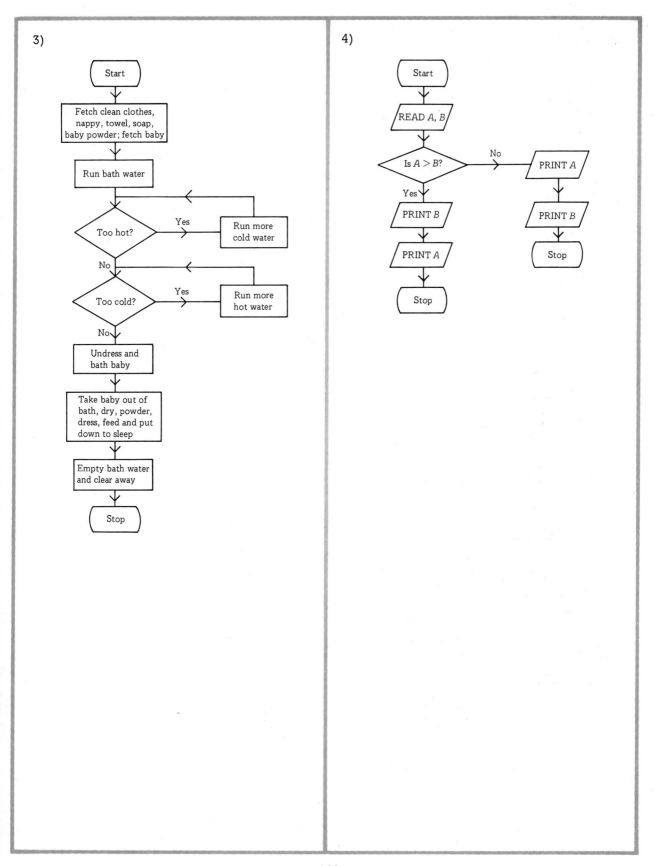

3)

Start

Fetch clean clothes, nappy, towel, soap, baby powder; fetch baby

Run bath water

Too hot? — Yes → Run more cold water

No

Too cold? — Yes → Run more hot water

No

Undress and bath baby

Take baby out of bath, dry, powder, dress, feed and put down to sleep

Empty bath water and clear away

Stop

4)

Start

READ A, B

Is A > B? — No → PRINT A

Yes

PRINT B

PRINT A

Stop

PRINT B

Stop

5)

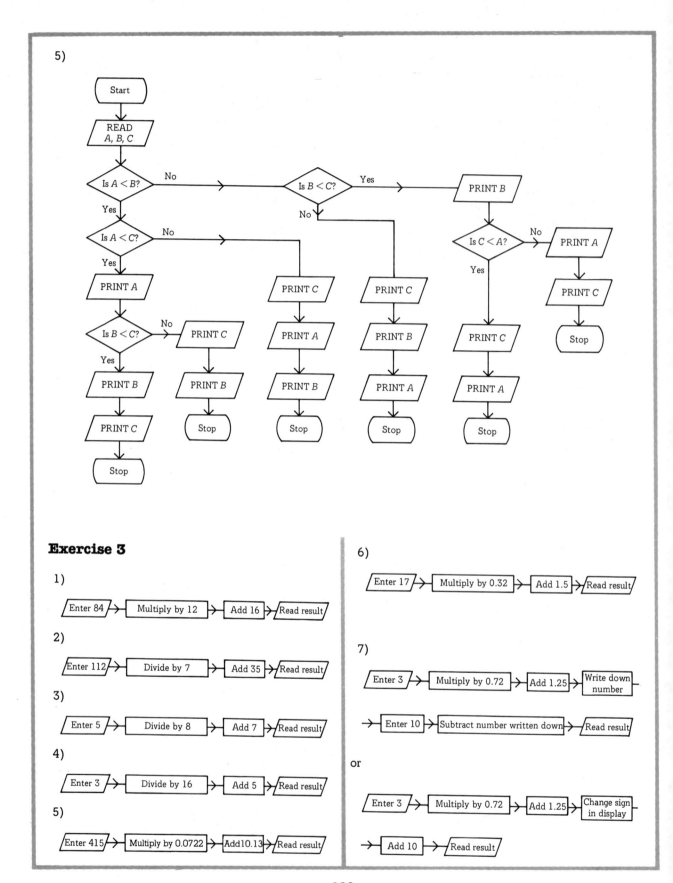

Exercise 3

1)

Enter 84 → Multiply by 12 → Add 16 → Read result

2)

Enter 112 → Divide by 7 → Add 35 → Read result

3)

Enter 5 → Divide by 8 → Add 7 → Read result

4)

Enter 3 → Divide by 16 → Add 5 → Read result

5)

Enter 415 → Multiply by 0.0722 → Add10.13 → Read result

6)

Enter 17 → Multiply by 0.32 → Add 1.5 → Read result

7)

Enter 3 → Multiply by 0.72 → Add 1.25 → Write down number

→ Enter 10 → Subtract number written down → Read result

or

Enter 3 → Multiply by 0.72 → Add 1.25 → Change sign in display

→ Add 10 → Read result

8)

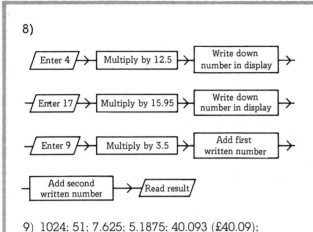

9) 1024; 51; 7.625; 5.1875; 40.093 (£40.09);
6.94 (6.94 m); 6.59 (£6.59); 352.65 (£352.65)

Exercise 4

1)

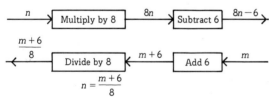

$$x = \frac{y-4}{7}$$

2)

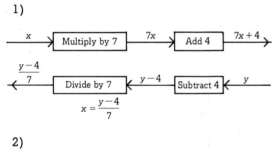

$$n = \frac{m+6}{8}$$

3)

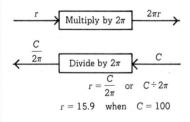

$$r = \frac{C}{2\pi} \quad \text{or} \quad C \div 2\pi$$

$r = 15.9$ when $C = 100$

4)

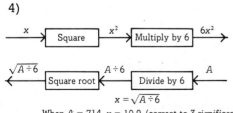

$$x = \sqrt{A \div 6}$$

When $A = 714$, $x = 10.9$ (correct to 3 significant figures)

5)

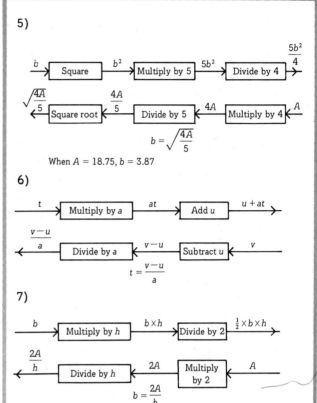

$$b = \sqrt{\frac{4A}{5}}$$

When $A = 18.75$, $b = 3.87$

6)

$$t \rightarrow \boxed{\text{Multiply by } a} \xrightarrow{at} \boxed{\text{Add } u} \xrightarrow{u+at}$$

$$\xleftarrow{\frac{v-u}{a}} \boxed{\text{Divide by } a} \xleftarrow{v-u} \boxed{\text{Subtract } u} \xleftarrow{v}$$

$$t = \frac{v-u}{a}$$

7)

$$b \rightarrow \boxed{\text{Multiply by } h} \xrightarrow{b \times h} \boxed{\text{Divide by 2}} \xrightarrow{\frac{1}{2} \times b \times h}$$

$$\xleftarrow{\frac{2A}{h}} \boxed{\text{Divide by } h} \xleftarrow{2A} \boxed{\text{Multiply by 2}} \xleftarrow{A}$$

$$b = \frac{2A}{h}$$

When $h = 10.4$ and $A = 200$, $b = 38.5$

CHAPTER 23

Exercise 1

1) (a) 13, 15   (b) 14, 16
   (c) 30, 35   (d) 36, 49
   (e) 13, 17
2) 36
3) 6 rectangular patterns:
   2×12, 3×8, 4×6, 6×4,
   8×3, 12×2
4) 4, 9, 16, 25, 36, 49, 64
   5, 7, 9, 11, 13, 15, 17
   9, 16, 25, 36, 49, 64, 81
   (a) squares
   (b) odd numbers
   (c) squares (except for 1)
5) (a) 16, 19, 22   (b) 36, 45, 54
   (c) 30, 38, 47   (d) 21, 34, 55
   (e) 3, $1\frac{1}{2}$, $\frac{3}{4}$
6) (a) 3, 6, 10, 15, 21, 28, 36, 45;
    triangular numbers
7) No

8) (a) true   (b) false
   (c) true   (d) true
   (e) false
9) 1, 5, 10, 10, 5, 1;
   1, 6, 15, 20, 15, 6, 1
   (a) 4   (b) 8   (c) 16
   (d) 32; 64
10) 104, 128

Exercise 2

1) (a)

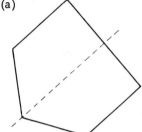

(b)

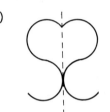

2) 8
3) (a) N   (b) E, T
   (c) X
4)

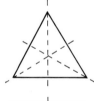

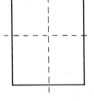

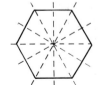

5)

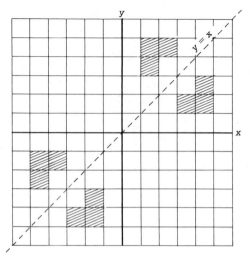

6)

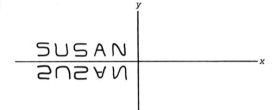

7) (a) 4   (b) 2
   (c) none   (d) 1
   (e) none   (f) 2

8) (a) $(3, -2)$ (b) $(-3, 2)$
   (c) $(-1, 2)$ (d) $(3, 0)$

9)

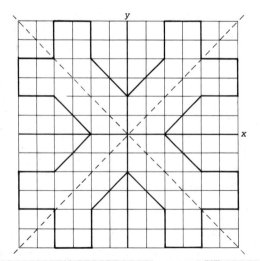

294

10)

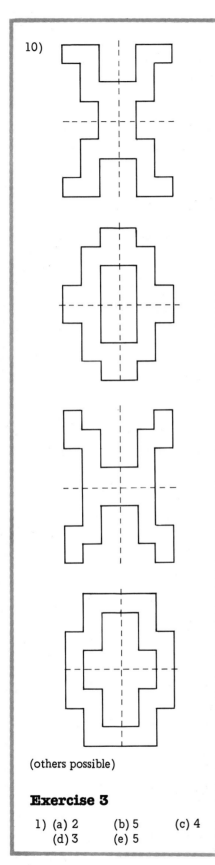

(others possible)

Exercise 3

1) (a) 2   (b) 5   (c) 4
   (d) 3   (e) 5

2)

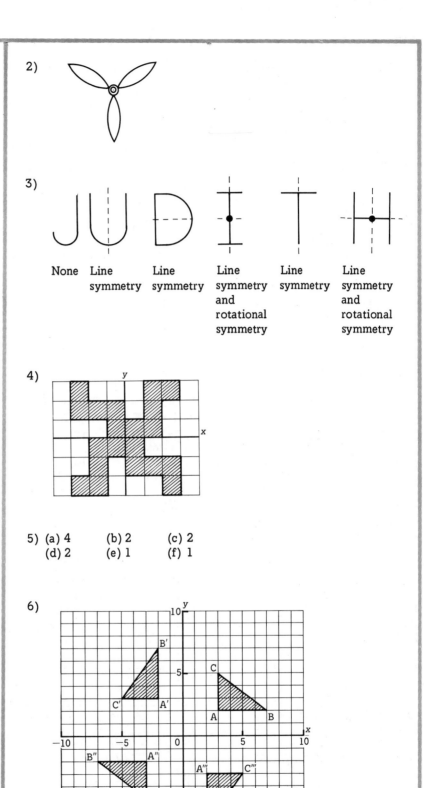

3)

JUDITH

None   Line symmetry   Line symmetry   Line symmetry and rotational symmetry   Line symmetry   Line symmetry and rotational symmetry

4)

5) (a) 4   (b) 2   (c) 2
   (d) 2   (e) 1   (f) 1

6)

7) (2, 1)

8)

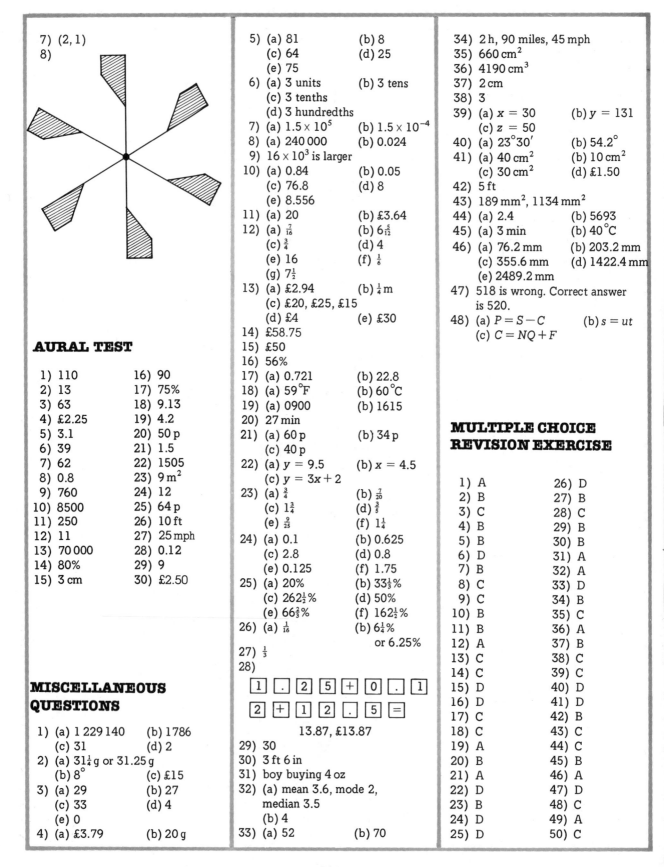

AURAL TEST

1) 110	16) 90
2) 13	17) 75%
3) 63	18) 9.13
4) £2.25	19) 4.2
5) 3.1	20) 50 p
6) 39	21) 1.5
7) 62	22) 1505
8) 0.8	23) 9 m^2
9) 760	24) 12
10) 8500	25) 64 p
11) 250	26) 10 ft
12) 11	27) 25 mph
13) 70 000	28) 0.12
14) 80%	29) 9
15) 3 cm	30) £2.50

MISCELLANEOUS QUESTIONS

1) (a) 1 229 140   (b) 1786
   (c) 31   (d) 2
2) (a) $31\frac{1}{4}$ g or 31.25 g
   (b) 8°   (c) £15
3) (a) 29   (b) 27
   (c) 33   (d) 4
   (e) 0
4) (a) £3.79   (b) 20 g

5) (a) 81   (b) 8
   (c) 64   (d) 25
   (e) 75
6) (a) 3 units   (b) 3 tens
   (c) 3 tenths
   (d) 3 hundredths
7) (a) 1.5×10^5 (b) 1.5×10^{-4}
8) (a) 240 000   (b) 0.024
9) 16×10^3 is larger
10) (a) 0.84   (b) 0.05
    (c) 76.8   (d) 8
    (e) 8.556
11) (a) 20   (b) £3.64
12) (a) $\frac{7}{16}$ (b) $6\frac{5}{12}$
    (c) $\frac{3}{4}$ (d) 4
    (e) 16   (f) $\frac{1}{6}$
    (g) $7\frac{1}{2}$
13) (a) £2.94   (b) $\frac{1}{4}$ m
    (c) £20, £25, £15
    (d) £4   (e) £30
14) £58.75
15) £50
16) 56%
17) (a) 0.721   (b) 22.8
18) (a) 59°F   (b) 60°C
19) (a) 0900   (b) 1615
20) 27 min
21) (a) 60 p   (b) 34 p
    (c) 40 p
22) (a) $y = 9.5$ (b) $x = 4.5$
    (c) $y = 3x + 2$
23) (a) $\frac{3}{4}$ (b) $\frac{7}{20}$
    (c) $1\frac{3}{4}$ (d) $\frac{3}{5}$
    (e) $\frac{9}{25}$ (f) $1\frac{1}{4}$
24) (a) 0.1   (b) 0.625
    (c) 2.8   (d) 0.8
    (e) 0.125   (f) 1.75
25) (a) 20%   (b) $33\frac{1}{3}$%
    (c) $262\frac{1}{2}$%   (d) 50%
    (e) $66\frac{2}{3}$%   (f) $162\frac{1}{2}$%
26) (a) $\frac{1}{16}$ (b) $6\frac{1}{4}$%
                          or 6.25%
27) $\frac{1}{3}$
28)

$$\boxed{1}\,\boxed{.}\,\boxed{2}\,\boxed{5}\,\boxed{+}\,\boxed{0}\,\boxed{.}\,\boxed{1}$$
$$\boxed{2}\,\boxed{+}\,\boxed{1}\,\boxed{2}\,\boxed{.}\,\boxed{5}\,\boxed{=}$$

13.87, £13.87
29) 30
30) 3 ft 6 in
31) boy buying 4 oz
32) (a) mean 3.6, mode 2,
    median 3.5
    (b) 4
33) (a) 52   (b) 70

34) 2 h, 90 miles, 45 mph
35) 660 cm^2
36) 4190 cm^3
37) 2 cm
38) 3
39) (a) $x = 30$ (b) $y = 131$
    (c) $z = 50$
40) (a) 23°30′   (b) 54.2°
41) (a) 40 cm^2 (b) 10 cm^2
    (c) 30 cm^2 (d) £1.50
42) 5 ft
43) 189 mm^2, 1134 mm^2
44) (a) 2.4   (b) 5693
45) (a) 3 min   (b) 40°C
46) (a) 76.2 mm   (b) 203.2 mm
    (c) 355.6 mm   (d) 1422.4 mm
    (e) 2489.2 mm
47) 518 is wrong. Correct answer
    is 520.
48) (a) $P = S - C$ (b) $s = ut$
    (c) $C = NQ + F$

MULTIPLE CHOICE REVISION EXERCISE

1) A	26) D
2) B	27) B
3) C	28) C
4) B	29) B
5) B	30) B
6) D	31) A
7) B	32) A
8) C	33) D
9) C	34) B
10) B	35) C
11) B	36) A
12) A	37) B
13) C	38) C
14) C	39) C
15) D	40) D
16) D	41) D
17) C	42) B
18) C	43) C
19) A	44) C
20) B	45) B
21) A	46) A
22) D	47) D
23) B	48) C
24) D	49) A
25) D	50) C

Index